MULTIVARIATE
STATISTICAL INFERENCE

MULTIVARIATE STATISTICAL INFERENCE

Proceedings of the Research Seminar at Dalhousie University, Halifax, March 23-25, 1972

Edited by

D. G. KABE
St. Mary's University

and

R. P. GUPTA
Dalhousie University

1973

NORTH-HOLLAND PUBLISHING COMPANY - AMSTERDAM • LONDON
AMERICAN ELSEVIER PUBLISHING COMPANY, INC. - NEW YORK

ISBN North-Holland 0 7204 2086 5
ISBN American Elsevier 0 444 105328

PUBLISHERS:

NORTH-HOLLAND PUBLISHING COMPANY – AMSTERDAM
NORTH-HOLLAND PUBLISHING COMAPNY, LTD.–LONDON

SOLE DISTRIBUTORS FOR THE U.S.A. AND CANADA:

AMERICAN ELSEVIER PUBLISHING COMPANY, INC.
52 VANDERBILT AVENUE
NEW YORK, N.Y. 10017

PRINTED IN THE NETHERLANDS

PREFACE

This book contains the proceedings of the research seminar on Multivariate Statistical Analysis held at Dalhousie University, Halifax, Nova Scotia, from March 23-25, 1972. The purpose of the seminar was to stimulate research and disseminate up-to-date knowledge in the field of multivariate analysis. Prominent theoretical and applied statisticians from United States and Canada presented invited and contributed papers on a wide spectrum of topics. Emphasis was placed on applicable methodology rather than theoretical developments.

The invited lectures were delivered by

1.	A. P. Dempster	Harvard University
2.	D. A. S. Fraser	University of Toronto
3.	I. Guttman	University of Montreal
4.	A. G. Laurent	Wayne State University
5.	I. Olkin	Stanford University
6.	S. R. Searle	Cornell University

Contributed papers were also an important part of the seminar. Some of these are included in the proceedings.

The editors wish to acknowledge responsibility for any mistakes, the reader may find in the manuscripts, since the authors did not have the opportunity to proofread their contributions.

D. G. Kabe

R. P. Gupta

ACKNOWLEDGEMENT

While we assume complete responsibility for all errors in the organization of the seminar and the editing of these proceedings, we wish to express our gratitude to several persons for their help and encouragement. Dr. Henry D. Hicks, President of Dalhousie University and Dr. D. Owen Carrigan, President of St. Mary's University were kind enough to make the opening remarks. Dr. A. J. Tingley, our colleague and chairman of the mathematics department, Dalhousie University took great personal interest in all phases of organizing the seminar and made it a great success. He encouraged and guided us in every activity. We take this opportunity to record our appreciation and gratitude to him. We also wish to acknowledge the help and advice of our colleagues, expecially Drs. C. A. Field, M. J. L. Kirby, W. R. S. Sutherland and S. Swaminathan.

Special thanks are due to the contributors to the present proceedings and to North-Holland Publishing Company for their excellent cooperation.

We also wish to thank Miss Paula Flemming for excellent typing of the manuscript.

CONTENTS

DIFFERENTIAL REPRESENTATION OF A BIVARIATE STOCHASTIC PROCESS

by P. Buckholtz and M. T. Wasan

In this paper we assign a set of conditions in terms of differentials which lead to the existence of a partial differential equation. When it is solved under suitable boundary conditions one can obtain the density function of a bivariate stochastic process. We also discuss illustrative examples of such processes.

Introduction

The problems of multivariate stochastic processes are arduous ones because they involve consideration of vector measures and also they embrace notational difficulties. For stochastic processes with one dimensional continuous time parameter, two extreme methods which are quite different in character have been considered by Ito and McKean and by I. Siegal respectively in their books.

Ito and McKean in their book "Diffusion Processes and Their Sample Paths," make a detailed investigation of sample paths of certain stochastic processes. I. Siegal, on the other hand, looks upon the sample paths as intermediates to getting information and has developed an Integration Algebra to bypass the sample paths.

Intermediate methods are the P. Levy construction of Brownian Motion and S. Bernstein's method. Bernstein's method uses differentials involving random variables to obtain a partial differential equation for a class of density functions. This is a very direct method. Kolmogorov translated the conditions in Bernstein's method into expressions involving conditional probabilities and from these derived the famous forward and backward diffusion equations.

In this paper we give differential representation of a pair of correlated Brownian motion processes as well as first passage time processes. We also indicate how the method is easily extended to multivariate cases.

Theorem 1A. Let $\{X(t), Y(s), t>0, s>0\}$ be a bivariate stochastic process with density function $f(x,y,t,s)$. Let

$$\begin{aligned} \Delta X &= A_1 \Delta t + B_1 \varepsilon \sqrt{\Delta t} \\ \Delta Y &= A_2 \Delta s + B_2 \eta \sqrt{\Delta s} \end{aligned} \qquad (1)$$

where A_1, A_2, B_1 and B_2 are functions of x,y,t and s and where $\frac{\partial A_1}{\partial x}$, $\frac{\partial A_2}{\partial y}$, $\frac{\partial^2 B_1}{\partial x^2}$, $\frac{\partial^2 B_2}{\partial y^2}$, $\frac{\partial^2 B_1}{\partial x \partial y}$, $\frac{\partial^2 B_2}{\partial x \partial y}$ exist and

$$\begin{aligned} &E(\varepsilon) = E(\eta) = 0 \\ &E(\varepsilon^2) = E(\eta^2) = 1 \\ &E(\varepsilon\eta) = \rho(x,y,t,s) \\ &E(|\varepsilon|^3),\ E(|\varepsilon^2\eta|),\ E(|\varepsilon\eta^2|),\ E(|\eta^3|) \end{aligned} \tag{2}$$

are all of order $o(\frac{1}{\sqrt{\Delta t}})$. We assume Δs and Δt are of the same order.

If $\lim_{\Delta t \to 0} \frac{\Delta s}{\Delta t} = \alpha$, α constant, $\alpha > 0$, then a density function of the process $f(x,y,t,s)$ satisfies the differential equation

$$\begin{aligned} &\frac{\partial f}{\partial t} + \frac{\partial (A_1 f)}{\partial x} + \alpha \frac{\partial f}{\partial s} + \frac{\partial (\alpha A_2 f)}{\partial y} \\ &= \frac{1}{2} \frac{\partial^2 (B_1^2 f)}{\partial x^2} + \frac{\partial^2 (\sqrt{\alpha}\rho B_1 B_2 f)}{\partial x \partial y} + \frac{1}{2} \frac{\partial (\alpha B_2^2 f)}{\partial y^2} \end{aligned} \tag{3}$$

Proof. Let $Q(x,y)$ be an arbitrary function of x and y with all partial derivatives up to third order continuous. Further let $Q(x,y)$ and all partial derivatives be zero on and outside the rectangle $\{(x,y) \mid a<x<b,\ c<y<d\}$.

Define

$$M(t,s) = E(Q(X(t),Y(s))) = \int_c^d \int_a^b Q(x,y) f(x,y,t,s)\,dx\,dy$$

We now find the directional derivative of $M(t,s)$ in the direction $\frac{ds}{dt} = \alpha$. Let $V = (1,\alpha)$, then

$$D_V M(t,s) = \lim_{\Delta t \to 0} E\left(\frac{Q(X(t+\Delta t),Y(s+\alpha\Delta t)) - Q(X(t),Y(s))}{\Delta t}\right) \tag{4}$$

$$D_V M(t,s) = \frac{\partial M}{\partial t} + \alpha \frac{\partial M}{\partial s} \tag{5}$$

also. We now expand Q in two variables:

$$\begin{aligned} &Q(X(t+\Delta t),Y(s+\Delta s)) \\ &= Q(X(t),Y(s)) + \frac{\partial Q}{\partial x}\Delta X + \frac{\partial Q}{\partial y}\Delta Y + \frac{1}{2}\frac{\partial^2 Q}{\partial x^2}(\Delta X)^2 \\ &+ \frac{\partial^2 Q}{\partial x \partial y}\Delta X \Delta Y + \frac{1}{2}\frac{\partial^2 Q}{\partial y^2}(\Delta Y)^2 + \frac{1}{6}\frac{\partial^3 Q}{\partial x^2}(\Delta X)^3 \\ &+ \frac{1}{2}\frac{\partial^3 Q}{\partial x^2 \partial y}(\Delta x)^2(\Delta Y) + \frac{1}{2}\frac{\partial^3 Q}{\partial x \partial y^2}\Delta X(\Delta Y)^2 \end{aligned}$$

$$+ \frac{1}{6} \frac{\partial^3 Q}{\partial y^3} (\Delta Y)^3 + o(\Delta X) \qquad (6)$$

where

$$\Delta X = A_1 \Delta t + B_1 \varepsilon \sqrt{\Delta t}$$

$$\Delta Y = \alpha A_2 \Delta t + B_2 \eta \sqrt{\alpha \Delta t}$$

$$(\Delta X)^2 = A_1^2 (\Delta t)^2 + 2A_1 B_1 \varepsilon (\Delta t)^{\frac{3}{2}} + B_1^2 \varepsilon^2 \Delta t$$

$$(\Delta Y)^2 = \alpha A_2^2 (\Delta t)^2 + 2A_2 B_2 \eta \alpha^{\frac{3}{2}} (\Delta t)^{\frac{3}{2}} + B_2^2 \eta^2 \alpha \Delta t$$

$$(\Delta X)(\Delta Y) = \alpha A_1 A_2 (\Delta t)^2 + (\alpha A_2 B_1 \varepsilon + \sqrt{\alpha} A_1 B_2 \eta)(\Delta t)^{\frac{3}{2}} + \sqrt{\alpha} B_1 B_2 \varepsilon \eta \Delta t$$

$$(\Delta X)^3 = A_1^3 (\Delta t)^3 + 3A_1 B_1 \varepsilon (\Delta t)^{\frac{5}{2}} + 3A_1 B_1^2 \varepsilon^2 (\Delta t)^2 + B_1^3 \varepsilon^3 (\Delta t)^{\frac{3}{2}}$$

$$(\Delta X)^2 (\Delta Y) = \alpha A_1^2 A_2 (\Delta t)^3 + (2\alpha A_1 A_2 B_1 \varepsilon + \sqrt{\alpha} A_1^2 B_2 \eta)(\Delta t)^{\frac{5}{2}}$$
$$+ (\alpha A_1 B_1^2 \varepsilon^2 + 2\sqrt{\alpha} A_1 B_1 B_2 \eta)(\Delta t)^2 + \sqrt{\alpha} B_1^2 B_2 \varepsilon^2 \eta (\Delta t)^{\frac{3}{2}}$$

$$\Delta X (\Delta Y)^2 = \alpha^2 A_1 A_2^2 (\Delta t)^3 + (\alpha^2 B_1 A_2^2 \varepsilon + 2\alpha^{\frac{3}{2}} A_1 B_2 A_2 \eta)(\Delta t)^{\frac{5}{2}}$$
$$+ (2\alpha^{\frac{3}{2}} A_2 B_1 B_2 \eta + \alpha A_1 B_2^2 \eta^2)(\Delta t)^2 + \alpha B_1 B_2^2 \varepsilon \eta^2 (\Delta t)^{\frac{3}{2}}$$

$$(\Delta Y)^3 = \alpha^3 A_2^3 (\Delta t)^3 + 3\alpha^{\frac{5}{2}} A_2^2 B_2 \eta (\Delta t)^{\frac{5}{2}} + 3\alpha^2 A_2 B_2^2 \eta^2 (\Delta t)^2$$
$$+ B_2^3 \eta^3 \alpha^{\frac{3}{2}} (\Delta t)^{\frac{3}{2}}$$

Substituting these in (6) and collecting like powers of Δt we get:

$$\Big[Q(X(t+\Delta t), Y(s+\alpha\Delta t)) - Q(X(t), Y(s))\Big] = \left(\frac{\partial Q}{\partial x} B_1 \varepsilon + \frac{\partial Q}{\partial y} B_2 \eta \sqrt{\alpha}\right) \sqrt{\Delta t}$$
$$+ \left(\frac{\partial Q}{\partial x} A_1 + \frac{\partial Q}{\partial y} \alpha A_2 + \frac{1}{2} \frac{\partial^2 Q}{\partial x^2} B_1^2 \varepsilon^2 + \frac{\partial^2 Q}{\partial x \partial y} B_1 B_2 \varepsilon \eta \sqrt{\alpha} + \frac{1}{2} \frac{\partial^2 Q}{\partial y^2} B_2^2 \alpha \eta^2\right) \Delta t$$
$$+ \left(\frac{\partial^2 Q}{\partial t^2} A_1 B_1 \varepsilon + \frac{\partial^2 Q}{\partial x \partial y} (\alpha B_1 A_2 \varepsilon + \sqrt{\alpha} A_1 B_2 \eta) + \frac{\partial^2 Q}{\partial y^2} A_2 B_2 \alpha^{\frac{3}{2}} \eta\right.$$
$$+ \frac{1}{6} \frac{\partial^3 Q}{\partial x^3} B_1^3 \varepsilon^3 + \frac{1}{2} \frac{\partial^3 Q}{\partial x^2 \partial y} \sqrt{\alpha} B_1^2 B_2 \varepsilon^2 \eta + \frac{1}{2} \frac{\partial^2 Q}{\partial x \partial y^2} \alpha B_1 B_2^2 \varepsilon \eta^2$$
$$\left. + \frac{1}{6} \frac{\partial^2 Q}{\partial y^3} B_2^3 \eta^3 \alpha^{\frac{3}{2}}\right) (\Delta t)^{\frac{3}{2}} + o(\Delta t)$$

Since conditions of interchanging differentiation and integration are satisfied when we substitute the above into (4) and use conditions (2) we get

$$D_V M(t,s) = \lim_{\Delta t \to 0} E\left[\frac{Q(X(t+\Delta t),\ Y(s+\alpha\Delta t)) - Q(X(t),Y(s))}{\Delta t}\right]$$

$$= E\left[\frac{\partial Q}{\partial x} A_1 + \frac{\partial Q}{\partial y} \alpha A_2 + \frac{1}{2}\frac{\partial^2 Q}{\partial x^2} B_2^2 + \frac{\partial^2 Q}{\partial x \partial y}\sqrt{\alpha}\rho B_1 B_2 + \frac{1}{2}\frac{\partial^2 Q}{\partial y^2}\alpha B_2^2 \eta^2\right]$$

$$= \int_c^d\int_a^b \left[\frac{\partial Q(x,y)}{\partial x} A_1(x,y,t,s) + \ldots +\right.$$

$$= \left.\frac{1}{2}\frac{\partial^2 Q(x,y)}{\partial y^2}\alpha B_2^2(x,y,t,s)\right] f(x,y,t,s)\,dxdy \tag{7}$$

Integrating by parts

$$\int_c^d\int_a^b \frac{\partial Q}{\partial x} A_1 f dxdy = \int_c^d \left[Q(x,y)A_1 f\Big|_a^b\right] dy - \int_c^d\int_a^b Q\frac{\partial(A_1 f)}{\partial x}\,dx\,dy$$

$$= -\int_c^d\int_a^b Q\frac{\partial(A_1 f)}{\partial x}\,dx\,dy \tag{8}$$

using the fact Q is zero on the boundary of the rectangle. Similarly

$$\int_c^d\int_a^b \frac{\partial^2 Q}{\partial x^2} B fdxdy = \int_c^d \left[\frac{\partial Q}{\partial x} B_1^2 f\Big|_a^b\right] dy - \int_c^d \left[Q\frac{\partial(B_1^2 f)}{\partial x}\Big|_a^b\right] dy$$

$$+ \int_c^d\int_a^b Q\frac{\partial^2(B_1^2 f)}{\partial x^2}\,dxdy \tag{9}$$

$$= \int_c^d\int_a^b Q\frac{\partial^2(B_1^2 f)}{\partial x^2}\,dxdy$$

etc.

Using (7), (8), etc. in (7) we get

$$D_V M(t,s) = \int_c^d\int_a^b Q\left[-\frac{\partial(A_1 f)}{\partial x} - \frac{\partial(\alpha A_2 f)}{\partial y} + \frac{1}{2}\frac{\partial^2(B_1^2 f)}{\partial x^2}\right.$$

$$\left.+ \frac{\partial^2(\sqrt{\alpha}\rho B_1 B_2 f)}{\partial x \partial y} + \frac{1}{2}\frac{\partial(\alpha B_2^2 f)}{\partial y^2}\right] dxdy \tag{10}$$

From (5) we have

$$D_V M(t,s) = \frac{\partial M}{\partial t} + \alpha\frac{\partial M}{\partial s}$$

$$= \frac{\partial}{\partial t}\int_c^d\int_a^b Q(x,y)f(x,y,t,s)\,dxdy$$

$$+ \alpha\frac{\partial}{\partial y}\int_c^d\int_a^b Q(x,y)f(x,y,t,s)\,dxdy$$

$$= \int_c^d \int_a^b Q(x,y) \left[\frac{\partial f}{\partial t} + \alpha \frac{\partial f}{\partial s} \right] dxdy \qquad (11)$$

Equating (10) and (11) and using the arbitrariness of Q we get the result (3)

Remark 1. In this equation if $\alpha = 1$ it means the components of the time parameter increase at the same rate. If, also, $t = s$ it means the components of the time parameter are the same. This being the case, the partial differential equation becomes

$$f_3 + \frac{\partial (A_1 f)}{\partial x} + f_4 + \frac{\partial (A_2 f)}{\partial y}$$

$$= \frac{1}{2} \frac{\partial^2 (B_1^2 f)}{\partial x^2} + \frac{\partial^2 (\rho B_1 B_2 f)}{\partial x \partial y} + \frac{1}{2} \frac{\partial^2 B_2^2 f)}{\partial y^2}$$

Now

$$\frac{df}{dt} (x,y,t,t) = f_3 (x,y,t,t) + f_4 (x,y,t,t)$$

and so letting $F(x,y,t) = f(x,y,t,t)$, we have

$$\frac{\partial F}{\partial t} + \frac{\partial (A_1 F)}{\partial x} + \frac{\partial (A_2 F)}{\partial y}$$

$$= \frac{1}{2} \frac{\partial^2 (B_1^2 F)}{\partial x^2} + \frac{\partial^2 (\rho B_1 B_2 F)}{\partial x \partial y} + \frac{1}{2} \frac{\partial^2 (B_2 F)}{\partial y^2} \qquad (12)$$

which is the differential equation for a bivariate stochastic process with one dimensional time parameter (see example 2A).

Example 1A. In equation (3) Theorem 1A., letting $\rho = 0$, $A_1 = A_2 = 0$, $B_1 = B_2 = 1$, the equation becomes

$$\frac{\partial f}{\partial t} + \alpha \frac{\partial f}{\partial s} = \frac{1}{2} \frac{\partial^2 f}{\partial x^2} + \frac{\alpha}{2} \frac{\partial^2 f}{\partial y^2} \qquad (13)$$

Boundary Conditions:

$$f(x,y,t,\infty) = f(x,y,\infty,s) = f(x,\pm\infty,t,s) = f(\pm\infty,y,t,s) = 0$$

$$\int_{-\infty}^{\infty} \int_{-\infty}^{\infty} f(x,y,t,s) dx\, dy = 1$$

For α constant and greater than zero but otherwise arbitrary (13) written in the form

$$\left[\frac{\partial f}{\partial t} - \frac{1}{2} \frac{\partial^2 f}{\partial x^2} \right] + \alpha \left[\frac{\partial f}{\partial s} - \frac{1}{2} \frac{\partial^2 f}{\partial y^2} \right] = 0$$

may be broken into a system of two uncoupled (independent) partial differential equations

$$\frac{\partial f}{\partial t} = \frac{1}{2}\frac{\partial^2 f}{\partial x^2} \quad \text{and} \quad \frac{\partial f}{\partial s} = \frac{1}{2}\frac{\partial^2 f}{\partial y^2}$$

The solution of these equations, satisfying the boundary conditions on f, is

$$f(x,y,t,s) = \left[\frac{1}{\sqrt{2\pi t}}\, e^{-\frac{1}{2}\frac{x^2}{t}}\right]\left[\frac{1}{\sqrt{2\pi s}}\, e^{-\frac{1}{2}\frac{y^2}{s}}\right]$$

$$= \frac{1}{2\pi\sqrt{st}} \exp\left(-\frac{1}{2}\left(\frac{x^2}{t} + \frac{y^2}{s}\right)\right) \qquad s,t > 0$$

<u>Example 2A.</u> In equation (12) Theorem 1A, let $A_1 = A_2 = 0$, $B_1 = B_2 = 1$, $\rho_{12} = \rho$ a constant $|\rho| \le 1$. It becomes

$$\frac{\partial f}{\partial t} = \frac{1}{2}\frac{\partial^2 f}{\partial x^2} + \rho\frac{\partial^2 f}{\partial x \partial y} + \frac{1}{2}\frac{\partial^2 f}{\partial y^2}$$

Boundary conditions:

$$f(x,y,\infty) = 0\,, \quad f(\pm\infty,y,t) = 0$$

$$f(x,\pm\infty,t) = 0\,, \quad \int_{-\infty}^{\infty}\int_{-\infty}^{\infty} f(x,y,t)\,dxdy = 1$$

<u>Solution</u>. Refer [4]

$$f(x,y,t) = \frac{1}{\pi\sqrt{1-\rho^2}\,t} \exp\left[-\frac{1}{2(1-\rho^2)t}(x^2-2\rho xy+y^2)\right], \quad t > 0$$

$$-\infty < x,\ y < \infty$$

<u>Theorem 1B.</u> Let $\{X(t), Y(s), t>0, s>0\}$ be a stochastic bivariate process with density function $f(x,y,t,s)$, and which is strictly increasing with probability 1.

Let

$$T_x = \inf\{t \mid t>0 \cap X(t)>x\}$$
$$S_y = \inf\{s \mid s>0 \cap Y(s)>y\} \tag{1}$$

be Borel measurable

$$\Delta T_x = T_{x+\Delta x} - T_x = A_1 \Delta X + B_1 \varepsilon\sqrt{\Delta x}$$
$$\Delta S_y = S_{y+\Delta y} - S_y = A_2 \Delta Y + B_2 \eta\sqrt{\Delta y} \tag{2}$$

where Δx and $\Delta y > 0$ and A_1, A_2, B_1 and B_2 are functions of x,y,t,s where $\frac{\partial A_1}{\partial t}$, $\frac{\partial A_2}{\partial s}$, $\frac{\partial^2 B_1}{\partial t^2}$, $\frac{\partial^2 B_1}{\partial t \partial s}$, $\frac{\partial^2 B_2}{\partial s^2}$, $\frac{\partial^2 B_2}{\partial t \partial s}$ exist and ε and η are random variables such that

$$\begin{aligned} &E(\varepsilon) = E(\eta) = 0 \\ &E(\varepsilon^2) = E(\eta^2) = 1 \\ &E(\varepsilon\eta) = \rho(x,y,t,s) \\ &E(|\varepsilon|^3),\ E(|\varepsilon^2\eta|),\ E(|\varepsilon\eta^2|),\ E(|\eta^3|) \end{aligned} \tag{3}$$

are all of order $o(\frac{1}{\sqrt{\Delta x}})$.

If $\lim_{\Delta x \to 0} \frac{\Delta y}{\Delta x} = \alpha(x,y,s,t)$, $\alpha > 0$, then a density function of the process $f(x,y,t,s)$ satisfies the differential equation

$$\frac{\partial f}{\partial x} + \frac{\partial(A_1 f)}{\partial t} + \alpha \frac{\partial f}{\partial y} + \frac{\partial(\alpha A_2 f)}{\partial s} \tag{4}$$

$$= \frac{1}{2}\frac{\partial^2(B_1^2 f)}{\partial t^2} + \frac{\partial^2(\sqrt{\alpha}\rho\beta B_1 B_2 f)}{\partial s \partial t} + \frac{1}{2}\frac{\partial^2(\alpha B_2^2 f)}{\partial s^2}$$

The proof of Theorem 1B is the same as that of Theorem 1A. The only difference is that the expected value is now taken of an arbitrary function of the random variables T_x and S_y . Thus we have

$$M(x,y) = E(Q(T_x, S_y))$$

as the function we take the directional derivative of M in the x,y plane.

We note here that Theorem 1B differs from 1A in that in 1B, α may be a function of x,y,t and s and that the process in 1B is strictly increasing with probability one. The variables in equation (4) Theorem 1B are exactly the reverse of those in equation (3) Theorem 1A. This is the same as the relationship of the differential equations associated with Brownian Motion with positive drift and the first passage time process of such Brownian Motion. These processes are, in fact, special cases of the processes in Theorems 1A and 1B respectively.

<u>Example 1B</u>. Let α constant, $\rho = 0$, $A_1 = A_2 = B_1 = B_2 = 1$. Equation (4) Theorem 1B becomes

$$\frac{\partial f}{\partial x} + \alpha \frac{\partial f}{\partial y} + \frac{\partial f}{\partial t} + \alpha \frac{\partial f}{\partial s} = \frac{1}{2} \frac{\partial^2 f}{\partial t^2} + \frac{\alpha}{2} \frac{\partial^2 f}{\partial s^2}$$

Boundary Conditions:

$$f(x,y,t,\infty) = 0, \quad f(x,y,\infty,s) = 0, \quad f(x,\infty,t,s) = 0,$$

$$f(\infty,y,t,s) = 0, \ f(x,y,0,0) = \delta(x)\delta(y), \ \int_0^\infty \int_0^\infty f(x,y,t,s)\,dxdy = 1$$

For α arbitrary in the same manner as in Example 1A, we get a system of two uncoupled equations and solve. (Refer [12].)

Solution

$$f(x,y,t,s) = \begin{cases} \dfrac{st}{2\pi(x^3y^3)^{\frac{1}{2}}} \exp\left[-\dfrac{(x-t)^2}{2x} - \dfrac{(y-s)^2}{2y}\right], & x,y,t,s > 0 \\ 0 & \text{elsewhere} \end{cases}$$

which is a bivariate inverse Gaussian with independent variables. This may not be the only solution.

Example 2B. Let $\alpha = \frac{1}{2}$, $\rho = \frac{1}{\sqrt{2}}$, $A_1 = A_2 = B_1 = B_2 = 1$.

Equation (4) Theorem 1B becomes

$$\frac{\partial f}{\partial x} + \frac{1}{2}\frac{\partial f}{\partial y} + \frac{\partial f}{\partial t} + \frac{1}{2}\frac{\partial f}{\partial s} = \frac{1}{4}\frac{\partial^2 f}{\partial s^2} + \frac{1}{2}\frac{\partial^2 f}{\partial s \partial t} + \frac{1}{2}\frac{\partial^2 f}{\partial t^2}$$

Boundary conditions:

$$x > y > 0, \quad t > s > 0, \quad f(x,y,\infty,s) = 0$$

$$f(\infty,y,t,s) = 0, \quad \int_0^\infty \int_0^x f(x,y,t,s)\,dydx = 1$$

The solution is Example 1A transformed by a translation

$$f(x,y,t,s) = \begin{cases} \dfrac{s(t-s)}{2\pi((x-y)^3y^3)^{\frac{1}{2}}} \exp\left[-\dfrac{(x-y-(t-s))^2}{2(x-y)} - \dfrac{(y-s)^2}{2y}\right] \\ \qquad\qquad x > y > 0, \quad t > s > 0 \\ 0 \qquad\qquad \text{elsewhere} \end{cases}$$

See [6, 11]. This is a special case of a bivariate inverse Gaussian Process.

Theorem 2. Let $(X(t), Y(t), Z(t))$, $t > 0$, be a trivariate stochastic process with density function $f(x,y,z,t)$. Let

$$\begin{aligned} \Delta X(t) &= A_1 \Delta t + B_1 \varepsilon\sqrt{\Delta t} \\ \Delta Y(t) &= A_2 \Delta t + B_2 \eta\sqrt{\Delta t} \\ \Delta Z(t) &= A_3 \Delta t + B_3 \nu\sqrt{\Delta t} \end{aligned} \qquad (1)$$

where $\Delta t > 0$ and $A_1, A_2 \ldots, B_3$ are functions of x,y,z and t and $\frac{\partial A_1}{\partial x}, \frac{\partial A_2}{\partial y}, \frac{\partial A_3}{\partial z}, \frac{\partial^2 B_1}{\partial x^2}, \frac{\partial^2 B_2}{\partial y^2}, \frac{\partial^2 B_3}{\partial z^2}, \frac{\partial^2 B_1 B_2}{\partial x \partial y}, \frac{\partial^2 B_1 B_3}{\partial x \partial z}, \frac{\partial^2 B_2 B_3}{\partial y \partial z}$ exist and where ε, η and ν are random variables such that

$$\begin{aligned} &E(\varepsilon) = E(\eta) = E(\nu) = 0 \\ &E(\varepsilon^2) = E(\eta^2) = E(\nu^2) = 1 \qquad (2) \\ &E(\varepsilon\eta) = \rho_{12}(x,y,z,t),\ E(\varepsilon\nu) = \rho_{13}(x,y,z,t),\ E(\eta\nu) = \rho_{23}(x,y,z,t) \end{aligned}$$

Finally we assume $f(x,y,z,t)$ is differentiable to order two and satisfies the Lindeberg condition:

$$\iiint_{|\Delta x|>\delta_1,\ |\Delta y|>\delta_2,\ |\Delta z|>\delta_3} |\Delta x|^h |\Delta y|^k |\Delta z|^\ell f(x,y,z,t+\Delta t)\,dxdydz = o(\Delta t) \qquad \Delta t \downarrow 0$$

for $h,k,\ell = 0,1,2$; $h+k+\ell = 2$ and where $\delta_1, \delta_2, \delta_3$ are arbitrarily small positive constants.

Then $f(x,y,z,t)$ satisfies the differential equation

$$\begin{aligned} &\frac{\partial f}{\partial t} + \frac{\partial (A_1 f)}{\partial x} + \frac{\partial (A_2 f)}{\partial y} + \frac{\partial (A_3 f)}{\partial z} \\ &= \frac{1}{2}\frac{\partial^2 (B_1^2 f)}{\partial x^2} + \frac{1}{2}\frac{\partial^2 (B_2^2 f)}{\partial y^2} + \frac{1}{2}\frac{\partial^2 (B_3^2 f)}{\partial z^2} + \frac{\partial^2 (B_1 B_2 \rho_{12} f)}{\partial x \partial y} \qquad (3) \\ &\quad + \frac{\partial^2 (B_1 B_3 \rho_{13} f)}{\partial x \partial z} + \frac{\partial^2 (B_2 B_3 \rho_{23} f)}{\partial y \partial z} \end{aligned}$$

Proof. Let $Q(x,y,z)$ be an arbitrary function which is continuous and has continuous derivatives to the third order in the region $\{(x,y,z) \mid a \le x \le b,\ c \le y \le d,\ e \le z \le g\}$ and takes the value zero outside this region. Let all the derivatives to the third order be zero on the boundary of the region. Consider the function.

$$\begin{aligned} M(t) &= E[Q(X(t),\ Y(t),\ Z(t))] \\ &= \int_{-\infty}^{\infty}\int_{-\infty}^{\infty}\int_{-\infty}^{\infty} Q(x,y,z)\ f(x,y,z,t)\,dx\ dy\ dz \end{aligned}$$

$$\frac{dM(t)}{dt} = \lim_{\Delta t \to 0} E\left[\frac{Q(X(t+\Delta t),Y(t+\Delta t),Z(t+\Delta t)) - Q(X(t),Y(t),Z(t))}{\Delta t}\right] \qquad (4)$$

also

$$\frac{dM(t)}{dt} = \frac{\partial}{\partial t}\int_{-\infty}^{\infty}\int_{-\infty}^{\infty}\int_{-\infty}^{\infty} Q(x,y,z)\ f(x,y,z,t)dxdydz$$

$$= \int_e^g\int_c^d\int_a^b Q(x,y,z)\ \frac{\partial}{\partial t}\ f(x,y,z,t)dxdydz \tag{5}$$

Expanding $Q(x+\Delta x,\ y+\Delta y,\ z+\Delta z)$ in a neighbourhood of x,y,z we obtain

$Q(x+\Delta x,\ y+\Delta y,\ z+\Delta z) - Q(x,y,z)$

$$= \frac{\partial Q}{\partial x}\Delta x + \frac{\partial Q}{\partial y}\Delta y + \frac{\partial Q}{\partial z}\Delta z + \frac{1}{2}\ \frac{\partial^2 Q}{\partial x^2}\Delta x^2 + \frac{1}{2}\ \frac{\partial^2 Q}{\partial y^2}\Delta y^2 + \frac{1}{2}\ \frac{\partial^2 Q}{\partial z^2}\Delta z^2$$

$$+ \frac{\partial^2 Q}{\partial x\partial y}\Delta x\Delta y + \frac{\partial^2 Q}{\partial x\partial z}\Delta x\Delta z + \frac{\partial^2 Q}{\partial y\partial z}\Delta y\Delta z + \frac{1}{6}\ \frac{\partial^2 Q}{\partial x^3}\Delta x^3 + \ldots$$

$$+ \frac{1}{2}\ \frac{\partial^3 Q}{\partial y\partial z^2}\Delta y\Delta z^2 + o(\Delta^3) \tag{6}$$

where $\Delta = \max(\Delta x,\ \Delta y,\ \Delta z)$ and where all the partial derivatives are evaluated at (x,y,z). Remembering that $X(t+\Delta t) - X(t) = \Delta X$, etc. We substitute expansion (6) into (4) and then using the Lindeberg condition which eliminates terms of third order Δ (see [7]) we get

$$\frac{dM(t)}{dt} = \lim_{\Delta t\to 0}\ E\left[\frac{\frac{\partial Q}{\partial x}\Delta x + \frac{\partial Q}{\partial y}\Delta y + \frac{\partial Q}{\partial z}\Delta z + \frac{1}{2}\ \frac{\partial^2 Q}{\partial x^2}\Delta x^2 + \ldots + \frac{\partial^2 Q}{\partial y\partial z^2}\Delta y\Delta z^2}{\Delta t}\right] \tag{7}$$

We deduce from expressions (1) for Δx, ΔY and ΔZ and conditions (2) $E(\varepsilon) = 0$ etc. that equation (7) becomes

$$\frac{dM(t)}{dt} = \lim_{\Delta t\to 0}\ E\left[\frac{\partial Q}{\partial x}A_1 + \frac{\partial Q}{\partial y}A_2 + \frac{\partial Q}{\partial z}A_3 + \frac{1}{2}\ \frac{\partial^2}{\partial x^2}B_1^2 + \frac{1}{2}\ \frac{\partial^2}{\partial y^2}B_2^2\right.$$

$$+ \frac{1}{2}\ \frac{\partial^2 Q}{\partial z^2}B_3^2 + \frac{\partial^2 Q}{\partial x\partial y}B_1B_2\rho_{12} + \frac{\partial^2 Q}{\partial x\partial z}B_1B_3\rho_{13} + \frac{\partial^2 Q}{\partial y\partial z}B_2B_3\rho_{23}$$

$$\left. + o(\Delta t)\right] \tag{8}$$

$$= \int_e^g\int_c^d\int_a^b\left[\frac{\partial Q}{\partial x}A_1 + \frac{\partial Q}{\partial A_2} + \ldots + \frac{\partial^2 Q}{\partial y\partial z}B_2B_3\rho_{23}\right]f(x,y,z,t)dxdydz$$

Integrating by parts and using the boundary conditions on Q and its derivatives, (8) becomes

$$\frac{dM(t)}{dt} = \int_{-\infty}^{\infty}\int_{-\infty}^{\infty}\int_{-\infty}^{\infty} Q(x,y,z)\left[-\frac{\partial(A_1 f)}{\partial x} - \frac{\partial(A_2 f)}{\partial y} - \frac{\partial(A_3 f)}{\partial z} + \frac{1}{2}\frac{\partial^2(B_1^2 f)}{\partial x^2} + \frac{1}{2}\frac{\partial^2(B_2^2 f)}{\partial y^2} + \frac{1}{2}\frac{\partial^2(B_3 f)}{\partial z^2} + \frac{\partial^2(B_1 B_2 \rho_{12} f)}{\partial x \partial y} + \frac{\partial^2(B_1 B_3 \rho_{13} f)}{\partial x \partial z} + \frac{\partial^2(B_2 B_3 \rho_{23} f)}{\partial y \partial z}\right] dxdydz$$

Equating this with the second part of (5) and using the arbitrariness of Q we get result (3).

Example 1. In equation (3) Theorem 2 letting $A_1 = A_2 = A_3 = 0$, $B_1 = B_2 = B_3 = 1$ and ρ_{12}, ρ_{13}, ρ_{23} constants such that $|\rho_{ij}| \leq 1$ where $i = 1$ and $j = 3$ the equation becomes

$$\frac{\partial f}{\partial t} = \frac{1}{2}\frac{\partial^2 f}{\partial x^2} + \frac{1}{2}\frac{\partial^2 f}{\partial y^2} + \frac{1}{2}\frac{\partial^2 f}{\partial z^2} + \rho_{12}\frac{\partial^2}{\partial x \partial y} + \rho_{13}\frac{\partial^2 f}{\partial x \partial z} + \rho_{23}\frac{\partial^2 f}{\partial y \partial z}$$

Boundary conditions:

$$f(x,y,z,\infty) = f(x,y,\pm\infty,t) = f(x,\pm\infty,z,t) = f(\pm\infty,y,z,t) = 0$$

$$\int_{-\infty}^{\infty}\int_{-\infty}^{\infty}\int_{-\infty}^{\infty} f(x,y,z,t)\,dxdydz = 1$$

Solution

$$f(x,y,z,t) = \frac{\sqrt{|A|}}{(2\pi t)^{3/2}} \exp -\frac{1}{2t}(X^T A X) \qquad t>0$$

where

$$X = \begin{pmatrix} x \\ y \\ z \end{pmatrix} \text{ and } A = \begin{pmatrix} 1 & \rho_{12} & \rho_{13} \\ \rho_{12} & 1 & \rho_{23} \\ \rho_{13} & \rho_{23} & 1 \end{pmatrix} .$$

REFERENCES

[1] Bartlett, M. S. (1966). An Introduction to Stochastic Processes. Cambridge University Press.

[2] Basu, A. K. and Wasan, M. T. On the First Passage Time Processes of Brownian Motion. Queen's Mathematical Preprints No. 1971-38, Kingston, Ontario.

[3] Buckholtz, P. and Wasan, M. T. Sufficient Conditions for a Cauchy Process. Queen's Mathematical Preprints No. 1972-2, Kingston, Ontario.

[4] Cox, D. R. and Miller, H. D. (1965). The Theory of Stochastic Processes. John Wiley & Sons, Inc., New York.

[5] Feller, W. (1966). An Introduction to Probability Theory and its Applications, Vol. II. John Wiley & Sons, Inc., New York.

[6] Gupta, R. P. and Wasan, M. T. Bivariate Time Distribution of Brownian Motion, Trabajos de Estadistica Y de Investigacion Operativa. Vol. XX, 1969, pp. 117-128.

[7] Ito, K. and McKean, H. (1965). Diffusion Processes and their Sample Paths. Springer-Verlag, Berlin.

[8] Levy, P. Processus Stochastiques et Mouvement Brownien. Gauthier-Villars, Paris, 1965.

[9] Petrovskii, I. (1967). Partial Differential Equations. Iliffe Books Ltd., London.

[10] Siegal, I. and Kunze, R. (1968). Integrals and Operators. McGraw-Hill, New York.

[11] Wasan, M. T. On an Inverse Gaussian Process, Skandinavisk Aktuarietidskift (1968), pp. 69-96.

[12] Wasan, M. T. Sufficient Conditions for a First Passage Time Process to be that of Brownian Motion, Journal of Applied Probability, Vol. 6, No. 1, pp. 218-223 (1969).

P. Buckholtz
Department of Mathematics
Queen's University
Kingston, Ontario
Canada

M. T. Wasan
Department of Mathematics
Queen's University
Kingston, Ontario
Canada

ON THE MULTIVARIATE GENERALIZATION OF THE FAMILY OF DISCRETE LAGRANGE DISTRIBUTIONS

P. C. Consul and L. R. Shenton

Good's generalization to several independent variables of Lagrange expansion of an inverse multivariate function has been modified to give a new form which seems to be a generalization of Poincare's form. The result is used to define a multivariate generalization of discrete Lagrange probability distributions. The mean vector as well as the variance-covariance matrix of the multivariate distributions have been determined. A multivariate generalization of Borel-Tanner distribution is also obtained as a particular case of the given family.

1. Introduction.

The multivariate generalizations of important discrete distributions, their applications and some of their properties have been discussed by a number of scientists which include Bhat and Kulkarni [1], Camp [2], Chatfield and others [3], Chang [4], Krishnamoorthy [9], Neyman [10], Olkin and Sobel [12], Patil and Bildikar [13], Steyn [16], Teicher [18] and Wishart [19]. A compact and systematic account of these generalizations of discrete distributions has been given by Johnson and Kotz [8]. The importance and vast scope for the use of discrete distributions and their generalizations is abundantley clear by the works cited above.

Recently Consul and Shenton [5] have written a basic paper defining a new class of univariate descrete distributions under the title of Lagrange distributions (on account of their relationship with the Lagrange's expansion of an inverse function). The two authors have further studied some interesting properties of the Lagrange distributions in [6, 14] and have opened almost a new field for research. The univariate families of Lagrange distributions include a large number of important discrete distributions.

In [15] Shenton and Consul used Poincare's generalization to two independent variables of Lagrange's expansion of an inverse function to give a bivariate form of discrete Lagrange probability distributions (LD) which include a bivariate form of Borel-Tanner distribution and studied some of their properties. A particular family of bivariate LD is precisely the distribution of the number of customers served in a busy period when there are two types of customers, with different arrival rates, and the queue, initiated by m customers of one type and n customers of another type, is served by a single server.

Good [7] has considered the generalizations to an arbitrary number of independent variables of Lagrange's expansion and has given the expansion in a number of forms. We shall use one of these forms to derive another form (which is more suitable for our purpose) and use it to define a multivariate generalization of discrete LD. We shall also indicate some of its probable uses, derive its mean vector and variance-covariance matrix and obtain the multivariate generalization of Borel-Tanner distribution as a particular case of our generalization.

2. Notation.

For the sake of conciseness and clarity we shall use the following symbols:

The letters $\underset{\sim}{t}$, $\underset{\sim}{T}$, $\underset{\sim}{u}$, $\underset{\sim}{\beta}$, $\underset{\sim}{0}$, $\underset{\sim}{m}$ represent column vectors so that

$$\underset{\sim}{t}' = (t_1, t_2, \ldots, t_n) \qquad \underset{\sim}{u}' = (u_1, u_2, \ldots, u_n)$$

$$\underset{\sim}{T}' = (T_1, T_2, \ldots, T_n) \qquad \underset{\sim}{\beta}' = (\beta_1, \beta_2, \ldots, \beta_n)$$

$$\underset{\sim}{0}' = (0, 0, \ldots, 0) \qquad \underset{\sim}{m}' = (m_1, m_2, \ldots, m_n)$$

and $$\underset{\sim}{T}^{-1} = (T_1^{-1}, T_2^{-1}, \ldots, T_n^{-1})' \qquad \underset{\sim}{\beta}^{-1} = (\beta_1^{-1}, \beta_2^{-1}, \ldots, \beta_n^{-1})'$$

Also,

$$g^{(i)}(\underset{\sim}{t}') = g^{(i)}, \qquad \left[g^{(i)}\right]^{m_\nu} = g_{m_\nu}^{(i)}$$

$$\frac{\partial}{\partial t_k} g^{(i)} = g_k^{(i)}, \quad m_1\left[g^{(i)}\right]^{m_1 - 1} \frac{\partial g^{(i)}}{\partial t_k} = \frac{\partial}{\partial t_k}\left[g^{(i)}\right]^{m_1} = g_{m_1 k}^{(i)}$$

and $D_i^r = \partial^r/\partial t_i^r$, $D_i^\nu = \partial^\nu/\partial t_i^\nu$ are differential operators which will be considered to be on the left of all the functions and will operate on all of them.

3. Multivariate Generalization of Lagrange Expansion.

Good [7] has proved that if $t_i = u_i g^{(i)}(\underset{\sim}{t}')$, where the multivariate functions $g^{(i)}(\underset{\sim}{t}')$ are analytic in the neighbourhood of the origin, such that $g^{(i)}(\underset{\sim}{0}') \neq 0 (i=1,2,\ldots,)$ and if $f(\underset{\sim}{t}')$ is another multivariate function, analytic in the neighbourhood of the origin, then

$$(1)\qquad f(\underset{\sim}{t}'(\underset{\sim}{u}'))=\Sigma\frac{u_1^{m_1}\cdots u_n^{m_n}}{m_1!\cdots m_n!}\left|D_1^{m_1}\cdots D_n^{m_n}\left\{f(t')\,g_{m_1}^{(i)}\cdots g_{m_n}^{(n)}\left\|\delta_i^{\nu}-u_i g_{\nu}^{(i)}\right\|\right\}\right|_{\underset{\sim}{t}=\underset{\sim}{0}}$$

where the summation is taken over all non-negative integers $m_1, m_2, \ldots, m_n$. (This convention of multiple summations will be taken for granted henceforth).

While quoting Poincare's result for $n = 2$, Good has also laid stress upon the fact that the factor multiplying $u_1^{m_1}\cdots u_n^{m_n}$ in (1) is not a proper 'coefficient' since it involves $u_1, u_2, \ldots, u_n$ that Poincare had picked out the coefficient from the four relevant terms and that Poincare's result does not suggest the result (1) proved by Good. We shall now modify Good's general result (1) so that the factor multiplying $u_1^{m_1}\cdots u_n^{m_n}$ becomes a true coefficient.

The determinant in (1), when written in full, is

$$(2)\qquad \begin{vmatrix} 1-u_1g_1^{(1)} & -u_1g_2^{(1)} & -u_1g_2^{(1)} & \cdots & -u_1g_n^{(1)} \\ -u_2g_1^{(2)} & 1-u_2g_2^{(2)} & -u_2g_3^{(2)} & \cdots & -u_2g_n^{(2)} \\ \vdots & \vdots & \vdots & & \vdots \\ -u_ng_1^{(n)} & -u_ng_2^{(n)} & \cdots & \cdots & 1-u_ng_n^{(n)} \end{vmatrix}$$

If $\Delta_1, \Delta_2, \ldots, \Delta_n$ are the cofactors of the elements in the first row of the determinant, the expression after the differential operators in (1) can be written as

$$(3)\qquad g_{m_2}^{(2)}g_{m_3}^{(3)}\cdots g_{m_n}^{(n)}\left[\left\{\left(g_{m_1}^{(1)}-u_1g_{m_1}^{(1)}g_1^{(1)}\right)\Delta_1+u_1g_{m_1}^{(1)}\sum_{\nu=2}^{n}g_{\nu}^{(1)}\Delta_{\nu}\right\}f(t')\right]$$

Now, by rearranging the terms and collecting the coefficient of $(u_1)^{m_1}/m_1!$, the expression (1) with (3) becomes

$$(4)\qquad f(\underset{\sim}{t}') = \Sigma\frac{u_1^{m_1}\cdots u_n^{m_n}}{m_1!\cdots m_n!}\left[D_1^{m_1-1}D_2^{m_2}\cdots D_n^{m_n}g_{m_2}^{(2)}\cdots g_{m_n}^{(n)}\left\{\left(D_1g_{m_1}^{(1)}-g_{m_1I}^{(1)}\right)\Delta_1+\sum_{\nu=2}^{n}g_{m_1\nu}^{(1)}\Delta_{\nu}\right\}f(\underset{\sim}{t}')\right]_{\underset{\sim}{t}=\underset{\sim}{0}}$$

wherein the last expression can be reconverted into a determinant. By using this simple algebra repeatedly each row of the determinant can be amended to provide a new form of Good's multivariate lagrange type expansion as

$$(5) \qquad f(\underset{\sim}{t}'(\underset{\sim}{u}')) = \frac{u_1^{m_1}\cdots u_n^{m_n}}{m_1!\cdots m_n!}\left[D_1^{m_1-1}\cdots D_n^{m_n-1}\left|\left|D_\nu g_{m_\nu}^{(\nu)} I - G\right|\right| f(\underset{\sim}{t}')\right]_{\underset{\sim}{t}=\underset{\sim}{0}}$$

where G is the $n\times n$ matrix $\left(g_{m_i j}^{(i)}\right)$.

The above result is a generalization of Poincare's result as the factor multiplying $u_1^{m_1}\cdots u_n^{m}$ in (5) is a true coefficient. When $n = 2$, the operational determinant in (5) becomes

$$\begin{vmatrix} D_1\left(g^{(1)}\right)^{m_1} - \dfrac{\partial\left(g^{(1)}\right)^{m_1}}{\partial t_1} & - \dfrac{\partial\left(g^{(1)}\right)^{m_1}}{\partial t_2} \\ - \dfrac{\partial\left(g^{(2)}\right)^{m_2}}{\partial t_1} & D_2\left(g^{(2)}\right)^{m_2} - \dfrac{\partial\left(g^{(2)}\right)^{m_2}}{\partial t_2} \end{vmatrix} f(t_1,t_2)$$

On evaluation of the determinant and simplification by the use of the operators D_1 and D_2 one gets

$$\left(g^{(1)}\right)^{m_1}\frac{\partial\left(g^{(2)}\right)^{m_2}}{\partial t_1}\frac{\partial f}{\partial t_2} + \left(g^{(2)}\right)^{m_2}\frac{\partial\left(g^{(1)}\right)^{m_1}}{\partial t_2}\frac{\partial f}{\partial t_1} + \left(g^{(1)}\right)^{m_1}\left(g^{(2)}\right)^{m_2}\frac{\partial^2 f}{\partial t_1\partial t_2}$$

which is exactly same as obtained by Poincare.

4. Multivariate Generalization of Discrete Lagrange Probability Distributions and Its Applications.

If $g^{(i)}(\underset{\sim}{t}')$, $i = 1,2,\ldots,n$ are n multivariate probability generating functions defined on non-negative integral vectors such that $g^{(i)}(t_1,t_2,\ldots,t_{\nu-1},0,t_{\nu+1},\ldots,t_n) \neq 0$, for all ν and i , then the transformations

$$(6) \qquad t_i = u_i\cdot g^{(i)}(\underset{\sim}{t}') \ , \ i = 1,2,\ldots,n$$

give $u_i = 0$ for $t_i = 0$ and $\underset{\sim}{u}' = (1,1,\ldots,1)$ for $\underset{\sim}{t}'=(1,1,\ldots,1)$. Since each u_i is a multivariate function of $(t_1,\ldots,t_n)$ having non-zero derivatives at $t_\nu = 0$ one can reasonably assume that there exists a multivariate power series expansion for each t_i in terms of $(u_1, u_2, \ldots, u_n)$ and is given by (5) when $f(\underset{\sim}{t}') = t_1 t_2 \ldots t_n$.

If $f(\underset{\sim}{t}')$ is another multivariate p.g.f., defined on non-negative integral vectors, then the multivariate power series (5) in u's must be another p.g.f. defined on non-negative integral vectors. Thus, a new multivariate discrete probability distribution is given by

$$(7)\quad \begin{cases} \Pr[\underset{\sim}{x}'=\underset{\sim}{0}'] = f(\underset{\sim}{0}') \\ \Pr[\underset{\sim}{x}'=\underset{\sim}{m}'] = \dfrac{1}{m_1!\ldots m_n!}\left| D_1^{m_1-1} \ldots D_n^{m_n-1} \left\| D_\nu g_{m_\nu}^{(\nu)} I - G \right\| f(\underset{\sim}{t}') \right|_{\underset{\sim}{t}=\underset{\sim}{0}} \end{cases}$$

for $(m_1, \ldots, m_n) \neq (0,0,\ldots 0)$, where G represents the $n \times n$ matrix

$$\left\{g_{m_i j}^{(i)}\right\} \quad \text{and} \quad g_{m_\nu}^{(\nu)} = \left\{g^{(\nu)}\right\}^{m_\nu}, \quad g_{m_i j}^{(i)} = \frac{\partial}{\partial t_j}\left\{g^{(i)}\right\}^{m_i}.$$

The above is the multivariate generalization of the univariate discrete Lagrange probability distribution defined by Consul and Shenton [6]. As $g^{(i)}(\underset{\sim}{t}')$ and $f(\underset{\sim}{t}')$ are replaced by particular sets of multivariate p.g.f.'s, defined on non-negative integral vectors, one can get many families of discrete multivariate probability distributions. We shall denote the multivariate discrete distribution (7) by $L(\underset{\sim}{g}'; f; \underset{\sim}{m}')$, its mean vector by $L_{(1)} = (L_1, L_2, \ldots, L_n)$.

The mean vectors of the discrete probability distributions given by $g^{(i)}(\underset{\sim}{t}')$ and $f(\underset{\sim}{t}')$ can be represented by the symbols $G_{(1)}^{(i)} = \left\{G_1^{(i)}, G_2^{(i)}, \ldots, G_n^{(i)}\right\}$ and $F_{(1)} = (F_1, F_2, \ldots, F_n)$. If some of the n multivariate probability distributions do not have the same number of variates, we can treat the particular means to be zero and the corresponding t's in p.g.f.'s can be replaced by unity.

APPLICATIONS:

(i) A busy airport is served everyday by a large number of airplanes of different types which arrive at different rates on account of the various flights (passenger, freight, charter, training, etc.). Each type of flight service has its own multivariate probability distribution for the arrival of different types of airplanes. Each airplane needs many kinds of services at the same airport. If there is only one set of crew (which looks after the service needs of the different airplanes) it can be treated as a single server giving multivariate service and the different types of airplanes may be supposed to form a queue for the services.

Let the average input vector of the i-th type flight service per unit time element be $\left\{\lambda_1^{(i)}, \lambda_2^{(i)}, \ldots, \lambda_n^{(i)}\right\}$ and the average service vector be $(\mu_1, \mu_2, \ldots, \mu_n)$, then the ratio vector $\left\{\lambda_1^{(i)}/\mu_1, \ldots, \lambda_n^{(i)}/\mu_n\right\}$ will denote the average rates of change per unit time element. Let the mean vector of the multivariate distribution given by $g^{(i)}(\underset{\sim}{t}')$ be $G_{(1)}^{(i)} = \left\{\lambda_1^{(i)}/\mu_1, \ldots, \lambda_n^{(i)}/\mu_n\right\}$ and let $f(\underset{\sim}{t}') = t_1^{\nu_1} t_2^{\nu_2} \ldots t_n^{\nu_n}$.

If there is a queue with r_1 airplanes of type $1, r_2$ airplanes of type $2, \ldots, r_n$ airplanes of type n when the service crew starts its services at the airport, then the probability distribution of the number of airplanes of different types served in a busy period will be given by

$$(8) \quad \Pr[\underset{\sim}{x}'=\underset{\sim}{m}'] = \frac{1}{m_1! \ldots m_n!}\left[D_1^{m_1-1} \ldots D_n^{m_n-1} \left\| D_\nu t_\nu^{r_\nu} g_m^{(\nu)} I - t_\nu^{r_\nu} G \right\| \right]_{\underset{\sim}{t}=\underset{\sim}{0}}$$

for $m_i \geq r_i$, $i = 1, 2, \ldots, n$.

(ii) If each atomic fission generates n different types of reactions and has its own multivariate probability distribution of generating such reactions then the probability that the whole process started by n such different fissions will contain m_1 reactions of type $1, m_2$ reactions of type $2, \ldots, m_n$ reactions of type n will probably be given by a multivariate discrete distribution of the type (7) with suitable choice of $g^{(i)}(\underset{\sim}{t}')$ and $f(\underset{\sim}{t}')$.

(iii) When the number of hits by each one of the sources, like X-Rays, cosmic rays, radio-active particles, shocks by static electricity generated by modern clothes and carpets etc., exceed certain levels $k_1, k_2, \ldots, k_n$, then their adverse effects in the form of different types of diseases may be visible. If $g^{(i)}(\underset{\sim}{t}')$ represents the p.g.f. of the multivariate probability distribution of the number of persons who contacted such visible diseases on account of the i-th source, then the probability distribution of the number of such attacks by different diseases in the whole process will be given by (7) with the particular values of $g^{(i)}(\underset{\sim}{t}')$.

(iv) If each of the sources like (i) bad sanitation (ii) food deficiency (iii) air pollution (iv) water pollution (v) consumption of food preservants has its own multivariate distribution of generating different types of epidemics (when each source pollution exceeds some specific limits), the multivariate probability distribution of the numbers of epidemics of each kind may also be of form (7).

5. Mean Vector and Variance-Covariance Matrix of Multivariate Lagrange Distribution.

Since $F_{(1)}$ and $G^{(i)}_{(1)}$ denote the mean row vectors of the multivariate distributions given by p.g.f.'s $f(\underset{\sim}{t}')$ and $g^{(i)}(\underset{\sim}{t}')$, let $\left\{G^{(i)}_{(1)}\right\}$ be a $n\times n$ matrix of mean values given by

(9)
$$\left\{G^{(i)}_{(1)}\right\} = \begin{bmatrix} G^{(i)}_1 & \cdots & G^{(i)}_n \\ \vdots & \cdots & \vdots \\ G^{(n)}_1 & & G^{(n)}_n \end{bmatrix}$$

Also, let the variance-covariance matrix of the multivariate distribution, given by $g^{(i)}(\underset{\sim}{t}')$, be represented by

(10)
$$G^{(i)}_{(2)} = \begin{bmatrix} G^{(i)}_{11} & \cdots & G^{(i)}_{1n} \\ \vdots & \cdots & \vdots \\ G^{(i)}_{n1} & & G^{(i)}_{nn} \end{bmatrix}$$

and, similarly, the variance-covariance matrix of the generalization of L.D. and of the distribution given by p.g.f. $f(\underset{\sim}{t}')$ be denoted by $L_{(2)}$ and $F_{(2)}$ respectively. Also, let $\left\{G^{(i)}_{(2)}\right\}$ represent a matrix whose element in ι-th row and m-th column is the column vector $\left\{G^{(1)}_{\iota m}, G^{(2)}_{\iota m}, \ldots, G^{(n)}_{\iota m}\right\}'$ for $\iota, m = 1,2,3,\ldots,n$.

Now, by replacing each t_i and u_i by e^{T_i} and e^{β_i} in the n transformations (6), taking the logarithms and by expanding, we get the relation in vectors,

$$\underset{\sim}{T} = \underset{\sim}{\beta} + \left\{G^{(i)}_{(1)}\right\}\underset{\sim}{T} + \frac{1}{2}\underset{\sim}{T}'\left\{G^{(i)}_{(2)}\right\}\underset{\sim}{T} + \cdots$$

which gives the value of vector as

(11) $$\underset{\sim}{\beta} = \left\{\underset{\sim}{I} - \left(G_{(1)}^{(i)}\right) - \frac{1}{2}\underset{\sim}{T}'\left(G_{(2)}^{(i)}\right) - \cdots\right\}\underset{\sim}{T}$$

Similarly, by replacing each t_i and u_i by e^{Ti} and $e^{\beta i}$ respectively in the relation

$$f(\underset{\sim}{t}') = \Sigma\, u_i^{m_1} \ldots u_n^{m_n} \cdot L(\underset{\sim}{g}'; f; \underset{\sim}{m}') = \iota(\underset{\sim}{u}') ,$$

taking logarithms on both sides and expanding in power series of T_i and β_i, we obtain

(12) $$F_{(1)}\underset{\sim}{T} + \frac{1}{2}\underset{\sim}{T}'F_{(2)}\underset{\sim}{T} + \ldots = L_{(1)}\underset{\sim}{\beta} + \frac{1}{2}\underset{\sim}{\beta}'L_{(2)}\underset{\sim}{\beta} + \ldots$$

By eliminating the column vector $\underset{\sim}{\beta}$ between (11) and (12),

(13) $$F_{(1)}\underset{\sim}{T} + \frac{1}{2}\underset{\sim}{T}'F_{(2)}\underset{\sim}{T} + \ldots = L_{(1)}\left\{I - \left(G_{(1)}^{(i)}\right) - \frac{1}{2}\underset{\sim}{T}'\left(G_{(2)}^{(i)}\right) - \ldots\right\}\underset{\sim}{T} +$$
$$+ \frac{1}{2}\underset{\sim}{T}'\left\{\underset{\sim}{I} - \left(G_{(1)}^{(i)}\right) - \frac{1}{2}\underset{\sim}{T}'\left(G_{(2)}^{(i)}\right) - \ldots\right\}' L_{(2)}\left\{\underset{\sim}{I} - \left(G_{(1)}^{(i)}\right) - \frac{1}{2}\underset{\sim}{T}'\left(G_{(2)}^{(i)}\right) - \ldots\right\}\underset{\sim}{T}$$

Since both sides of (13) must be identical, a simple term by term comparison of the terms yields the following two relations:

(14) $$F_{(1)} = L_{(1)}\left(\underset{\sim}{I} - \left(G_{(1)}^{(i)}\right)\right)$$

and

(15) $$F_{(2)} = -\left(\sum_{i=1}^{n} L_i G^{(i)}\right)_{(2)} + \left(I - \left(G_{(1)}^{(i)}\right)\right)' L_{(2)}\left(I - \left(G_{(1)}^{(i)}\right)\right)$$

where $\left(\sum_{i=1}^{n} L_i G^{(i)}\right)_2$ represents a $n \times n$ matrix with elements like $L_1 G_{jk}^{(1)} + \ldots + L_n G_{jk}^{(n)}$.

Hence the mean-vector of the multivariate generalization of L.D. is given by

(16) $$L_{(1)} = F_{(1)}\left(I - \left(G_{(1)}^{(i)}\right)\right)^{-1}$$

and the corresponding variance-covariance matrix, when $\left(I - \left(G_{(1)}^{(i)}\right)\right)$ is a non-singular $n \times n$ matrix, by

(17) $$L_{(2)} = \left(I - \left(G_{(1)}^{(i)}\right)\right)'^{-1}\left(F - \sum_{i=1}^{n} L_i G^{(i)}\right)_{(2)}\left(I - \left(G_{(1)}^{(i)}\right)\right)^{-1}$$

where the elements of the middle matrix are of the form

(18) $$F_{jk} - L_1 G_{jk}^{(1)} - L_2 G_{jk}^{(2)} - \ldots - L_n G_{jk}^{(n)} \ , \quad j,k = 1,2,\ldots,n \ .$$

6. Multivariate Generalization of Borel-Tanner Distribution.

Let the different types of customers (airplanes) generated by each one of the several sources be multivariate Poissionian and such that the average ratio vector, denoting the rate of change per unit time-element as indicated in example (i) of Section 4, for the i-th source is $(a_{i_1}, a_{i_2}, \ldots, a_{in})$. Thus, the Poissionian p.g.f.'s become

(19) $$g^{(i)}(\underset{\sim}{t}') = \exp\left[\sum_j a_{ij}(t_j - 1)\right], \quad i = 1,2,\ldots,n \ .$$

If the queue starts with r_1 customers of type $1,\ldots,r_n$ customers of type n, then the probability that the total number of customers $m_1, m_2, \ldots, m_n$ of type $1, \ldots,$ type n respectively served in a busy period can be obtianed from the multivariate distribution (8) in the form

$$\Pr[\underset{\sim}{x}'=\underset{\sim}{m}'] = \frac{\exp\left[-\sum_i m_i\left(\sum_j a_{ij}\right)\right]}{m_1!\ldots m_n!}\left[\prod_{i=1}^{n}\left\{\sum_k (m_k a_{ki}) + D_i\right\}^{m_i - 1}\right] \times$$
$$\left|\left|\left(\sum_k m_k a_{k\nu} + D_\nu\right) t_\nu^{r_\nu} \quad I - \left(m_\nu a_{\nu i} t_\nu^{r_\nu}\right)\right|\right|_{\underset{\sim}{t}=\underset{\sim}{0}}$$

for $m_i \geq r_i \ i = 1,2,\ldots,n$ which can be reduced, by simple algebra to the form

(20) $$\Pr[\underset{\sim}{x}'=\underset{\sim}{m}'] = \left[\prod_{i=1}^{n}\left\{\frac{e^{-m_i\left(\sum_j a_{ij}\right)}}{(m_i - r_i)!}\left(\sum_j m_j a_{ji}\right)^{m_i - r_i}\right\}\right] \cdot \left|\left| I - \frac{a_{\nu j}(m_\nu - r_\nu)}{\sum_i (m_i a_{i\nu})}\right|\right|$$

for $m_i \geq r_i \ , \ i = 1,2,\ldots,n$.

The above distribution is the multivariate generalization of Borel-Tanner distribution. Its mean vector $L_{(1)}$ is given by (16) as

(21) $$L_{(1)} = \underset{\sim}{m}'\left(I - \left(a_{ij}\right)\right)^{-1}$$

and the variance-covariance vector $L_{(2)}$, by (17) and (18), as

$$L_{(2)} = \left(I - \left(a_{i_\nu}\right)\right)'^{-1} \left(\left(-\sum_k L_k a_{k\nu}\right) I\right)\left(I - \left(a_{\nu i}\right)\right)^{-1} .$$

References

[1] Bhat, B. R. and Kulkarni, N. V. (1966). On efficient multinomial estimation, Jour. Roy. Stat. Soc., Sr. B, 28, 45.

[2] Camp, B. H. (1929). The multinomial solid and the chi-test, Trans. Amer. Math. Soc., 31, 133.

[3] Chatfield, C., Ehrenberg, A. S. C. and Goodhardt, G. J. (1966). Progress on a simplified model of stationary purchasing behaviour, Jour. Roy. Stat. Soc., Sr. A, 129, 317.

[4] Cheng Ping (1964). Minimax estimates of parameters of distributions belonging to the exponential family, Acta Mathematica Sinica, 5, 277.

[5] Consul, P. C. and Shenton, L. R. (1971). Use of Lagrange expansion for generating discrete generalized probability distributions, SIAM Jour. Appld. Math. to appear in 23 (2).

[6] Consul, P. C. and Shenton, L. R. (1971). A note on some properties of Lagrange distributions.

[7] Good, I. J. (1959). Generalizations to several variables of Lagrange's expansion with applications to stochastic processes, Trans. Camb. Phil. Soc., 56, 367.

[8] Johnson, N. L. and Kotz, S. (1969). Discrete distributions, Houghton Mifflin Company, Boston.

[9] Krishnamoorthy, A. S. (1951). Multivariate binomial and Poisson distributions, Sankhyā, 11, 117.

[10] Neyman J. and Bates, G. C. (1952). Contributions to the theory of accident proneness, University California, Publications in Statistics 1, 215.

[11] Neyman, J. (1963). Certain chance mechanisms involving discrete distributions (Inaugural Address), Proceedings of the International Symposium on Discrete Distributions, Montreal, 4.

[12] Olkin, I. and Sobel, M. (1965). Integral expressions for tail probabilities of the multinomial and negative multinomial distributions, Biometrika, 52, 167.

[13] Patil, G. P. and Bildikar, S. (1967). Multivariate logarithmic series distribution as a probability model in population and community ecology and some of its statistical properties, Jour. Amer. Stat. Association, 62, 655.

[14] Shenton, L. R. and Consul, P. C. (1971). On cumulants and moments of Lagrange distributions, Inst. Statist. Math. (Tokyo)(submitted).

[15] Shenton, L. R. and Consul, P. C. (1971). On bivariate Lagrange and Borel-Tanner distributions and their use in queueing theory, Sankhyā, Sr. A., (submitted).

[16] Steyn, H. S. (1951, '55, '56, '57, '58, '63). Proceedings Koninklijke Nederlandse Akademie van Wetenschappen, Sr. A, 54 (23-30), 58 (588-595), 59 (190-197), 60 (119-127), 61 (129-138), 66 (85-96).

[17] Steyn, H. S. (1959). On X^2-tests for contingency tables of negative multinomial types, Statistica Neerlandica, 13, 433.

[18] Teicher, H. (1954). On the multivariate Poisson distribution, Skandinavisk Aktuarietidskrift, 37, 1.

[19] Wishart, J. (1949). Cumulants of multivariate multinomial distributions, Biometrika, 36, 47.

P. C. Consul
Department of Mathematics and Statistics
University of Calgary
Calgary, Alberta
Canada

L. R. Shenton
Department of Statistics
University of Georgia
Athens, Georgia
U.S.A.

ALTERNATIVES TO LEAST SQUARES IN MULTIPLE REGRESSION

A. P. Dempster*

A broad and important class of alternatives to least squares estimation of regression coefficients may be characterized loosely as procedures which pull back or shrink least squares estimates towards a chosen origin. Specific choices of an alternative differ in the degree of pull-back and in the pattern of pull-back. Two directions are described in some detail, namely the STEIN and RIDGE methods where the pattern does not depend on the observed least squares estimates, and the variable selection and REGF methods where the pattern does so depend. A preliminary report is made on a simulation study which demonstrates the large improvements over least squares which can be expected in reasonably life-like situations.

1. Introduction.

A common procedure in applied regression analysis is to select a subset of the available predictor variables and to estimate regression coefficients on the subset by least squares. Such a procedure can be viewed as estimating regression coefficients on all the available predictor variables where the estimated regression coefficients are assigned zero values for the unselected subset. In short, a subset of the regression coefficients is pulled back all the way to zero, while the rest are estimated by least squares and are therefore regarded as not pulled back at all. This sharp cut-off pattern of pull-back is usually data-generated in practice, for example, through the familiar rules of forward or backward selection with rough significance test judgments on when to stop selecting or rejecting variables.

A different approach to pull-back is to modify the standard least squares formula $\underset{\sim}{b} = (\underset{\sim}{X}^T\underset{\sim}{X})^{-1}(\underset{\sim}{X}^T\underset{\sim}{Y})$ into $\underset{\sim}{b}^* = (\underset{\sim}{X}^T\underset{\sim}{X}+\underset{\sim}{Q})^{-1}\underset{\sim}{X}^T\underset{\sim}{Y}$ where $\underset{\sim}{Q}$ is a positive definite symmetric matrix. In practice, $\underset{\sim}{Q}$ is chosen *a priori* up to an unknown scale factor which is determined from the data. Two *a priori* choices have recently received prominent attention in the statistical literature. The first takes

---

*This work was facilitated by National Science Foundation Grant GP-31003X

$\underset{\sim}{Q}$ proportional to $\underset{\sim}{X}^T\underset{\sim}{X}$ so that $\underset{\sim}{b}^* = f\underset{\sim}{b}$ for some f on $0 < f < 1$. Estimators of this type were used by James and Stein (1960) to prove the inadmissibility of least squares estimates. They are advocated as practical tools by Stein and his colleagues, such as Efron and Morris (1972), at least in the context of estimating several means. For short they will be called here STEIN estimators. The second choice takes $\underset{\sim}{Q} = k\underset{\sim}{I}$ for some $k > 0$. These are the RIDGE estimators of Hoerl and Kennard (1970) which are further discussed below. In both the STEIN and RIDGE approaches, the <u>a priori</u> determination of $\underset{\sim}{Q}$ up to a scalar multiple can be naturally conceived as determining the pattern or shape of pull-back, while the degree of pull-back is data-dependent. By contrast, the variable selection approach uses the data to determine a pattern while the degree is fixed <u>a priori</u>.

The alternatives to least squares studied in this paper are intended to make sense under the standard linear model with normally distributed constant variance deviations. Another class of alternatives, designed to be effective when the distribution of deviations may be long-tailed, is no doubt equally important, but is conceptually quite different. Statisticians will in the long run wish to develop methods which are simultaneously effective against both kinds of weakness in least squares, but in the current state of the art the two difficulties are best studied separately. The objective of this paper is to make plain how serious the first problem alone can be, and to describe some simple remedies.

Within the standard normal linear model framework there are two major theoretical justifications of least squares, namely, the frequentist property of minimum variance among unbiased estimators, and the Bayesian property of minimum posterior mean squared error against a uniform prior pseudo-distribution of true regression coefficients. There are two corresponding positions on what is wrong with least squares. The frequentist statistician computes

$$(\text{mean squared error}) = (\text{bias})^2 + (\text{variance})$$

and recognizes that unbiased least squares estimators may show poor mean squared error when compared to biased estimators, since an upward revision in squared bias may be greatly outweighed by a larger reduction in variance, at least over <u>a priori</u> plausible regions of parameter space. The Bayesian view is simpler, namely that the uniform prior which justifies least squares is wrong and should be replaced with a realistic prior distribution.

The originators of STEIN estimators and of RIDGE estimators obviously prefer frequentist justifications, but they are of course aware of the corresponding Bayesian interpretations. Others, for example Lindley and Smith (1972), insist that the frequency properties of procedures are irrelevant to the analysis of particular data sets and that only Bayesian justifications are valid. I believe that the weight of the argument in the dispute lies strongly with the Bayesians, and that the Bayesian view deserves and will gradually attain greater prominence in practical work, not as a piece of dogma but as a natural framework for arriving at sensible inferences. In practice as well as in principle, I find the frequentist approach unworkable because the dependence of loss functions on many unknown parameters means that clear judgments on good procedures are unavailable unless some prior information is weighted into the judgment. The choice of an origin towards which to shrink is tantamount to selecting a prior "middle", and a pattern and degree of shrinking likewise must either conform to prior assumptions or be inadmissible. The direct approach is to put assumptions openly on the table in the form of the only well-developed formal mechanism for expressing prior knowledge of unknowns. As a dividend, one gains the suggestive imagery of subjective probability, including the tool of <u>posterior</u> expected loss which, unlike frequentist expected loss, is guaranteed relevant to a particular data set under analysis. These remarks are intended to explain the Bayesian slant of the subsequent discussion.

Variable selection methods may appear attractive for a variety of reasons, for example, because measurement costs may dictate the use of a predictor based on very few variables. Here, I assume that measurement costs may be ignored, and that the justification for setting an estimated regression coefficient to zero is that zero is judged to be closer to the true value than is some competing estimate. A serious difficulty with variable selection methods is that, unlike the STEIN or RIDGE methods, they often lack credibility as approximately Bayesian techniques. It will rarely be possible to believe that a particular set of estimated zeros corresponds to actual zeros in true regression coefficients, so that pull-back procedures which pull certain estimates nearly to zero in a flexible way offer greater hope for credibility than do sharp cut-off procedures. Such a modification of variable selection is proposed below under the name REGF.

Many relevant properties of specific variants of STEIN, RIDGE or REGF estimation are difficult to derive analytically, but can be assessed at least roughly by simulation techniques. After defining and displaying simple properties of these estimators in Section 2, I shall present in Section 3 a preliminary report on a substantial simulation study of these and other competing methods.

2. Some Details.

2.1 Notation. Consider an $n \times 1$ response vector $\underset{\sim}{Y}$ of observations assumed generated by the linear model

$$\underset{\sim}{Y} = \underset{\sim}{X}\,\underset{\sim}{\beta} + \underset{\sim}{e} \tag{2.1}$$

where $\underset{\sim}{X}$ is an $n \times p$ rank p matrix of n observations on p fixed variables, $\underset{\sim}{e}$ is an $n \times 1$ vector of independent $N(0,\sigma^2)$ random variables, and $\underset{\sim}{\beta}$ and σ^2 are unknown parameters of the model. In practice, the model (2.1) usually appears with a constant term inserted. For purposes of estimating of $\underset{\sim}{\beta}$, however, the more general model can be reduced with little information loss to the constant term free form (2.1) with $\underset{\sim}{Y}$ and $\underset{\sim}{X}$ replaced by deviations from sample means. In the interest of uncluttered notation, I shall assume the model in the form (2.1).

Sufficient statistics for the model (2.1) are given by the least squares estimator

$$\underset{\sim}{b} = (\underset{\sim}{X}^T\underset{\sim}{X})^{-1}(\underset{\sim}{X}^T\underset{\sim}{Y}) \tag{2.2}$$

and the residual mean square

$$s^2 = \frac{1}{n-p}(\underset{\sim}{Y}-\underset{\sim}{X}\underset{\sim}{b})^T(\underset{\sim}{Y}-\underset{\sim}{X}\underset{\sim}{b}) \ , \tag{2.3}$$

which are the most widely accepted point estimators for $\underset{\sim}{\beta}$ and σ^2, respectively. For present purposes, therefore, the model can be specified in terms of $\underset{\sim}{b}$, s^2 and

$$\underset{\sim}{P} = \underset{\sim}{X}^T\underset{\sim}{X} \ , \tag{2.4}$$

where $\underset{\sim}{b}$ and $(n-p)s^2$ are independently distributed as $N(\underset{\sim}{\beta},\sigma^2\underset{\sim}{P}^{-1})$ and $\sigma^2\tilde{\chi}^2_{n-p}$, while $\underset{\sim}{P}$ is fixed.

In order to clarify RIDGE estimators in particular, it is convenient to reexpress the above model in terms of principal components of the fixed variables, i.e., to express (2.1) in the form

$$\underset{\sim}{Y} = \underset{\sim}{X}^* \underset{\sim}{\alpha} + \underset{\sim}{e} \tag{2.5}$$

with

$$\underset{\sim}{X}^* = \underset{\sim}{X} \underset{\sim}{C}^T$$

and

$$\underset{\sim}{\alpha} = \underset{\sim}{C} \underset{\sim}{\beta} , \tag{2.6}$$

where $\underset{\sim}{C}$ is the matrix of eigenvectors of $\underset{\sim}{P}$ satisfying

$$\underset{\sim}{C}\underset{\sim}{P}\underset{\sim}{C}^T = \mathrm{diag}(\lambda_1,\lambda_2,\ldots,\lambda_p)$$

$$\text{and} \quad \underset{\sim}{C}\underset{\sim}{C}^T = \underset{\sim}{I} . \tag{2.7}$$

The least squares estimator $\underset{\sim}{b}$ of $\underset{\sim}{\beta}$ can likewise be reexpressed using the least squares estimator $\underset{\sim}{a}$ of $\underset{\sim}{\alpha}$ where

$$\underset{\sim}{a} = \underset{\sim}{C}\underset{\sim}{b} . \tag{2.8}$$

The sampling distributions of the components

$$a_1, a_2, \ldots, a_p \tag{2.9}$$

of $\underset{\sim}{a}$ are independent normals with means

$$\alpha_1, \alpha_2, \ldots, \alpha_p \tag{2.10}$$

and variances

$$\frac{\sigma^2}{\lambda_1}, \frac{\sigma^2}{\lambda_2}, \ldots, \frac{\sigma^2}{\lambda_p} . \tag{2.11}$$

Two quadratic loss functions will be used in the sequel, namely

$$\text{SEB} = (\hat{\underset{\sim}{\beta}}-\underset{\sim}{\beta})^T(\hat{\underset{\sim}{\beta}}-\underset{\sim}{\beta})/\sigma^2 \tag{2.12}$$

and

$$\text{SPE} = (\hat{\underset{\sim}{\beta}}-\underset{\sim}{\beta})^T\underset{\sim}{P}(\hat{\underset{\sim}{\beta}}-\underset{\sim}{\beta})/\sigma^2 . \tag{2.13}$$

In the first of these, SEB is a mnemonic for "sum of errors of betas", the concept being that the goal is estimation of the individual components $\beta_1,\beta_2,\ldots,\beta_p$ of $\underset{\sim}{\beta}$ which are assumed scaled in such a way that squared errors of a given size in each component are equally painful. SPE is a mnemonic for "sum of prediction errors", the reason being that repeated use of $\underset{\sim}{X}\hat{\underset{\sim}{\beta}}$ in place of $\underset{\sim}{X}\underset{\sim}{\beta}$ to predict $\underset{\sim}{Y}$ will result in a mean squared prediction error per prediction of $\sigma^2 + \sigma^2\text{SPE}$, as may be easily proved for any fixed $\hat{\underset{\sim}{\beta}}$.

In terms of principal components

$$SEB = \sum_{i=1}^{p} (\hat{\alpha}_i - \alpha_i)^2/\sigma^2 \tag{2.14}$$

while

$$SPE = \sum_{i=1}^{p} \lambda_i (\hat{\alpha}_i - \alpha_i)^2/\sigma^2 \quad . \tag{2.15}$$

2.2. Preselected pull-back patterns. It was remarked in Section 1 that the STEIN and RIDGE approaches to pull-back were tied to two choices of an inner-product $\underset{\sim}{Q}$ over the space of p independent variables, namely the choices $\underset{\sim}{Q}$ proportional to $\underset{\sim}{P}$ and to $\underset{\sim}{I}$, respectively. The first choice is mathematically natural in the context of the model (2.1) because the model is invariant under all one-one linear transformations of p-space into itself, and the class of $\underset{\sim}{Q}$ proportional to $\underset{\sim}{P}$ is exactly those invariant under the full linear group. On the other hand, this choice is dependent on the particular data set $\underset{\sim}{X}$. The second choice asserts that the particular coordinate system in p-space used in the representation (2.1) is orthonormal up to a single scale factor. This choice may be made a priori or may be made dependent on the observed $\underset{\sim}{X}$. For example, the rows of $\underset{\sim}{X}$ may be rescaled, and the element of $\underset{\sim}{\beta}$ inversely rescaled to correspond, so that $\underset{\sim}{P} = \underset{\sim}{X}^T\underset{\sim}{X}$ has identical diagonal elements or equivalently so that $\underset{\sim}{P}$ is proportional to the observed correlation matrix of an original set of p independent variables.

The "reason" for any particular choice of $\underset{\sim}{Q}$ can be formulated in frequentist or Bayesian terms. In the preferred Bayesian terms of this paper, the STEIN choice corresponds to

$$\underset{\sim}{\beta} \sim N(\underset{\sim}{0}, \omega^2 \underset{\sim}{P}^{-1}) \tag{2.16}$$

a priori while the RIDGE choice corresponds to

$$\underset{\sim}{\beta} \sim N(\underset{\sim}{0}, \tau^2 \underset{\sim}{I}) \tag{2.17}$$

a priori. In terms of $\underset{\sim}{\alpha}$ coordinates the prior distributions are

$$\alpha_i \text{ independently } N(0, \omega^2 \lambda_i^{-1}) \tag{2.18}$$

and

$$\alpha_i \text{ independently } N(0, \tau^2) \ , \tag{2.19}$$

respectively, for $i = 1,2,\ldots,p$. Assuming n large enough that σ^2 can be regarded as approximately known, the posterior distributions corresponding to (2.18) and (2.19) are approximately

$$\alpha_i \text{ independently } N(fa_i, f\sigma^2\lambda_i^{-1}) \tag{2.20}$$

and

$$\alpha_i \text{ independently } N(f_i a_i, f_i \sigma^2 \lambda_i^{-1}) \quad , \tag{2.21}$$

respectively, where

$$f = \frac{1}{1+(\frac{\sigma}{\omega})^2} \tag{2.22}$$

and

$$f_i = \frac{\lambda_i}{\lambda_i + (\frac{\sigma}{\tau})^2} \quad . \tag{2.23}$$

Hoerl and Kennard (1970) use the expression $\lambda_i/(\lambda_i+k)$ for the shrinking factor, and they advocate a judgmental frequency-based scheme for picking k. In a Bayesian framework the natural approach is to estimate the ratio $(\sigma/\tau)^2$ from $\underset{\sim}{a}$ and s^2. For example, the particular variant RIDGM used in Section 3 uses the fact that a_i are a priori independently $N(0, \tau^2 + \sigma^2\lambda_i^{-1})$ and chooses τ^2 given σ^2 and λ_i so that $\Sigma a_i^2/(\tau^2 + \sigma^2\lambda_i^{-1})$ equals its expected value p.

When the λ_i are roughly constant, and comparable methods are used to estimate the ratios $(\sigma/\omega)^2$ and $(\sigma/\tau)^2$, there is little practical difference between (2.20) and (2.21). But when the data are "collinear", meaning that approximate linear relations hold among the columns of $\underset{\sim}{X}$, then there are one or several very small λ_i. This circumstance implies one or several small f_i leading to substantial differences between the STEIN and RIDGE approaches.

In a final analysis, the correct choice among methods for a particular data set depends on the relative plausibility of the competing a priori assumptions. But it is equally important to examine the consequences of different choices in order to understand what is at stake. To illustrate, consider a hypothetical situation where $\lambda_1 = 20$ is the maximum λ_i, $\lambda_p = .5$ is the minimum λ_i, and $\bar{\lambda} = 6$ denotes a typical value of λ_i. Suppose that σ^2 is arbitrarily fixed at 1, and suppose that directly competing values of ω^2 and τ^2 are fixed in the ratio $\omega^2/\tau^2 = \bar{\lambda}$ so that the pull-back factors f and f_i are the same when $\lambda_i = \bar{\lambda}$, as will roughly be the case when ω^2 and τ^2 are estimated from the data. In a situation where $f = .9$, it follows that

$f_1 = .968$ while $f_p = .429$. In the case of α_1, the prior distributions (2.18) and (2.19) are $N(0,.45)$ and $N(0,1.5)$, which are noticeably different, but for a typical observed $a_1 = 1$ the posterior distributions (2.20) and (2.21) are $N(.9, .045)$ and $N(.968, .048)$, which differ little. In the case of α_p, the prior distributions (2.18) and (2.19) are $N(0, 18)$ and $N(0, 1.5)$ which imply prior distributions $N(0, 20)$ and $N(0, 3.5)$ for a_p. For a typical observed $a_p = 3$, the posterior distributions (2.20) and (2.21) are $N(2.70, 1.8)$ and $N(1.29, .86)$. Thus, the difference in posterior judgments for α_i corresponding to the smallest λ_i is quite substantial. The squared difference $(2.70 - 1.29)^2 = 1.99$ is the posterior squared bias of either Bayesian method relative to the other, showing that the differences between the methods in potential squared bias is about double the difference in assessed posterior variance. For comparison, the least squares estimator corresponding to $\omega^2 = \tau^2 = \infty$ leads to a posterior assessment of $N(3.00, 2.00)$ which shows that the STEIN estimator differs relatively little from least squares, by comparison with RIDGE, as indeed the proponents of RIDGE intend.

I believe the foregoing numbers are not atypical, but the reader who wishes can easily create similar analysis with his own numbers, taken perhaps from a set of data at his disposal. It should be noted that the data can be used to provide some check on prior assumptions. For example, if a_p in the above numerical example had turned out to be 10 instead of 3, then the implicit RIDGE prior $N(0, 3.5)$ for a_p would appear implausible. Failing this kind of help from data, the choice must be made on genuine *a priori* grounds. The attractiveness of RIDGE regression depends on a judgment about the relative plausibility of the competing prior distributions. In the regression context, one may well feel that (2.16) has a decidely peculiar shape, being very much elongated in directions corresponding to small λ_i.

The simulation study reported in Section 3 provides striking quantitative evidence of the utility of RIDGE regression. Readers interested in pursuing frequentist properties should consult Hoerl and Kennard (1970), or for Bayesian analysis see Lindley and Smith (1972).

2.3. *Data-selected pull-back patterns*. Suppose that I denotes a subset of the integers $\{1,2,\ldots,p\}$ and that the corresponding $p(I)$ columns of $\underset{\sim}{X}$ are denoted by $\underset{\sim}{X}(I)$ where $p(I)$ denotes the number of elements in I. The least squares analysis of $\underset{\sim}{Y}$ on $\underset{\sim}{X}(I)$ yields

$$\underset{\sim}{b}(I) = (\underset{\sim}{X}(I)^T\underset{\sim}{X}(I))^{-1}(\underset{\sim}{X}(I)^T\underset{\sim}{Y}) \tag{2.24}$$

and

$$s(I)^2 = \frac{1}{n-p(I)} (\underset{\sim}{Y}-\underset{\sim}{X}(I)\underset{\sim}{b}(I))^T(\underset{\sim}{Y}-\underset{\sim}{X}(I)\underset{\sim}{b}(I)) . \tag{2.25}$$

The traditional variable selection technique picks a subset I by one of a fairly rich class of selection methods, and then adopts the point estimator $\hat{\underset{\sim}{\beta}}_I$ defined by

$$\hat{\underset{\sim}{\beta}}_I(I) = \underset{\sim}{b}(I) \text{ , and}$$

$$\hat{\underset{\sim}{\beta}}_I(I^C) = \underset{\sim}{0} \tag{2.26}$$

where I^C denotes the complement of I.

The proposed REGF techniques take a step away from the sharp cut-off of (2.26) towards a more conceivably Bayesian procedure. Suppose that $\mathcal{I}_r$ for $r = 0,1,\ldots,p$ denotes the class of subsets of $\{1,2,\ldots,p\}$ containing r elements. A particular REGF technique requires a procedure for selecting r and a definition of a set of weights $w(I)$ for each $I \varepsilon \mathcal{I}_r$, from which the estimator

$$\hat{\underset{\sim}{\beta}} = \sum_{I\varepsilon\mathcal{I}_r} w(I)\hat{\underset{\sim}{\beta}}_I \tag{2.27}$$

may be computed. A Bayesian framework for (2.27) requires the a priori assumption for a specified r that there are exactly r nonzero components in $\underset{\sim}{\beta}$ corresponding to an unknown $I \varepsilon \mathcal{I}_r$. Then $w(I)$ can be interpreted as the posterior probability that I is the subset with nonzero components while $\hat{\underset{\sim}{\beta}}(I)$ is an approximate posterior mean estimator for these nonzero components. How to define $w(I)$, and how to choose r will now be discussed in order.

The formal approach to $w(I)$ is to specify a prior distribution $q(I)$ over $\mathcal{I}_r$ and a likelihood $\ell(I)$ given the data. Then one chooses $w(I)$ proportional to the product $q(I) \times \ell(I)$. In actual applications, a user may wish to specify unequal $q(I)$ for different I, but here I shall assume a priori symmetry among the components of $\underset{\sim}{\beta}$ so that $q(I)$ is the same for all $I \varepsilon \mathcal{I}_r$. The representation of $\ell(I)$ remains ambiguous because the precise likelihood specified by data and model is

$$\ell(I,\underset{\sim}{\beta}(I),\sigma^2) = \left(\frac{1}{2\pi\sigma^2}\right)^{\frac{n}{2}} e^{-\frac{1}{2\sigma^2}(Q_1+Q_2)} \tag{2.28}$$

where

$$Q_1 = (\underset{\sim}{b}(I)-\underset{\sim}{\beta}(I))^T\underset{\sim}{P}(I)(\underset{\sim}{b}(I)-\underset{\sim}{\beta}(I)) \tag{2.29}$$

and

$$\underset{\sim}{Q}_2 = (n-r)\ s(I)^{2} . \tag{2.30}$$

Various devices can be used to eliminate $\underset{\sim}{\beta}(I)$ and σ^2 from (2.28). A simple heuristic rule is to take

$$\ell(I) = \sup_{\underset{\sim}{\beta}(I),\sigma^2} \ell(I,\underset{\sim}{\beta}(I),\sigma^2) . \tag{2.31}$$

The right side of (2.31) is proportional to $s(I)^{-n}$, leading to the choice

$$w(I) = s(I)^{-n} \Big/ \sum_{J\varepsilon I_r} s(J)^{-n} \tag{2.32}$$

with weights summing to unity. The Bayesian approach to eliminating $\underset{\sim}{\beta}(I)$ and σ^2 is to weight (2.28) using a prior joint density of $\underset{\sim}{\beta}(I)$ and σ^2 given I, and then integrate out nuisance parameters. A common choice of pseudoprior density element is

$$K(I)d\underset{\sim}{\beta}(I)\times h^{a-1}dh \tag{2.33}$$

where $h = 1/\sigma^2$. Multiplying (2.28) and (2.33) and integrating yields $\ell(I)$ or $w(I)$ proportional to

$$K(I)[\det \underset{\sim}{P}(I)]^{\frac{1}{2}}\ s(I)^{\frac{r-a-n}{2}} . \tag{2.34}$$

The factor $K(I)$ in (2.33) allows for different spread factors in the prior distribution of $\underset{\sim}{\beta}(I)$ over different r-dimensional hyperplanes in the space of $\underset{\sim}{\beta}$. Two choices of $K(I)$ arise naturally:

$$K(I) = 1 \quad \text{or} \quad K(I) = [\det \underset{\sim}{P}(I)]^{-\frac{1}{2}} . \tag{2.35}$$

The first choice treats all subspaces alike, while the second choice weights according to the geometry determined by $\underset{\sim}{X}(I)$. More precisely, if the prior distribution of $\underset{\sim}{\beta}(I)$ is taken to be normal with covariance $k\ \underset{\sim}{P}(I)^{-1}$ proportional to the covariances of $\underset{\sim}{b}(I)$, then as $k\to\infty$ the prior density becomes

$[\det \underset{\sim}{P}(I)]^{-1/2} d\underset{\sim}{\beta}(I)$ as in the second choice. The first choice is more in the spirit of RIDGE while the second choice is in the spirit of STEIN, but the practical difference between the two methods will usually be less drastic than the difference between RIDGE and STEIN because collinearity will be less extreme within a subset I than in the whole set.

The specific REGF methods FREGF and 1FREGF reported on below use (2.32) which is (2.34) with $K(I) = [\det \underset{\sim}{P}(I)]^{-1/2}$ and $a = r$, while the FDRGF and 1FDRGF use $K(I) = 1$ and $a = r - 1$. The choices of a here are somewhat arbitrary but not important in the applications of Section 3 since r is commonly 1 or 2.

Various rules for selecting a subset I in variable selection methods have analogs for REGF techniques. For example, forward selection of variables chooses a sequence of subsets $I_1 \subset I_2 \subset I_3 \ldots$ of sizes 1,2,3,... such that the reduction in sum of squares at each step is maximized. Having determined I_{i+1} the question becomes: Does the reduction in residual sum of squares from I_i to I_{i+1} justify passing on from I_i to I_{i+1} as a candidate for the final selected subset? A commonly used rule is to compute

$$F = \frac{(n-i)s(I_i)^2 - (n-i-1)s(I_{i+1})^2}{s(I_{i+1})^2} \qquad (2.36)$$

which is regarded as sufficiently large to cause the advance to I_{i+1} if greater than the $1 - (.05)/(n-i+1)$ quantile of the F distribution on 1 and $n-i+1$ degrees of freedom. The method FSLA studied in Section 3 uses this scheme, while 1FSL uses the same criterion (2.36) but stops selecting unless $F \geq 1$.

In the case of REGF methods, the choice is among the classes of sets $\mathcal{I}_0, \mathcal{I}_1, \mathcal{I}_2, \ldots$ so that a rule is needed to decide whether to pass from $\mathcal{I}_i$ to $\mathcal{I}_{i+1}$ for $i=0,1,2,\ldots$. Apart from factors independent of i, the expression (2.31) for $\ell(I)$ may be written

$$[(n-i)s(I)^2]^{-n/2} \quad \text{for } I \varepsilon \mathcal{I}_i .$$

The expression

$$L(\mathcal{I}_i) = \sum_{I \varepsilon \mathcal{I}_i} w(I) \cdot n \cdot \log[(n-i)s(I)^2] \qquad (2.37)$$

is an approximate mean posterior of -2 log likelihood for the model $\mathcal{I}_i$. The difference

$$F_1 = L(I_i) - L(I_{i+1}) \tag{2.38}$$

can be roughly conceived as a change in 2 log likelihood and judged as a χ^2 on 1 degree of freedom. The methods FREGF and FDRGF use the .95 quantile of χ_1^2 as the stopping criterion, while 1FREGF and 1FDRGF use the value $F_1 = 1$ as the change point. Other kinds of selection rules which have been studied include rules which refuse to allow too great a difference between estimates based on I_r and those based on the complete least squares analysis I_p.

Several obvious practical difficulties stand in the way of REGF methods. One is that the task of computing all $\binom{p}{r}$ regressions for $I \varepsilon I_r$ may become impossibly great as r increases. Approximate methods for seeking out only the most important subsets will be needed in many applications. A more fundamental difficulty is that the sharp cut-off pattern of variable selection has been removed only one step because there remains a sharp cut-off in the choice of I_r. A fully Bayesian method might assign a prior distribution to r and weight further according to a posterior distribution over I_r. The refusal to go this far into Bayesian analysis reflected a judgment that dependence on a specific prior distribution over r might be regarded as too arbitrary by potential users.

3. A Simulation Study.

A substantial simulation study has been carried out by Wermuth (1972) to compare many alternatives to least squares. Further analysis of this study will appear in Dempster, Schatzoff and Wermuth (1973). Only a few preliminary results are reported here.

The approach in the study is to simulate a range of situations chosen such that basic parameter structures vary in ways known to be interesting and across ranges which may commonly arise in practice. The simulations are restricted to the normal model (2.1) with $p = 6$ and $n = 20$. Two series of data sets were constructed, Series 1 with 32 data sets in a 2^5 factorial structure, and Series 2 with 128 data sets in a quarter replicate of a 2^9 structure.

The 5 factors in Series 1 may be labelled EIG, ROT, COL, CEN and BET. EIG refers to an initial eigenvalue structure for the matrix $\underset{\sim}{P}$ at two levels, namely (32, 25, 16, 9, 4, 1) and (64, 16, 4, 1, .25, .0625). ROT refers to 2 random rotations of matrices with the above eigenvalues, followed by reductions to correlation matrices, thus producing 4 candidates for $\underset{\sim}{P}$. COL refers to the absence of special collinearity as in the 4 choices of $\underset{\sim}{P}$ versus the special introduction of correlation .99 between the first two independent variables, yielding 4 more choices of $\underset{\sim}{P}$. For each of the 8 choices of $\underset{\sim}{P}$, data were simulated 4 times, namely, with 2 levels of noncentrality CEN, specifically, 100 and 200, and 2 levels of regression coefficients BET, specifically, vectors of true regression coefficients (32, 25, 16, 9, 4, 1) and (64, 16, 4, 1, .25, .0625). The variance σ^2 was adjusted to yield the desired noncentrality. Note that the noncentrality parameters are high enough to virtually guarantee significance of the standard $F_{6,14}$ test of $\underset{\sim}{\beta} = \underset{\sim}{0}$. (The nonnull means of the F statistic are $(106/6)\times(14/12)$ and $(206/6)\times(14/12)$).

Series 2 differs from Series 1 in that ROT, COL, CEN and BET were used at 4 levels rather than 2, thus building the 2^5 to 2^9. Generally, the degree of collinearity is less extreme in Series 2, and otherwise the parameter ranges are broadened. The EIG vectors are (30, 30, 30, 20, 20, 20) and (64, 16, 4, 2, 1, .5). ROT has 4 random rotations. COL has a factor for the presence or absence of correlation .95 between X_1 and X_2, and another factor for the presence or absence of correlation .92 between X_3 and X_1-X_2. Then noncentrality parameters in CEN are 10, 50, 100, 500 and the regression coefficient vectors are (1,1,1,1,1,1), (32,16,8,8,8,8), (1,1,1,0,0,0) and (32,16,8,0,0,0).

No replication was attempted. That is, only one random drawing was taken from each of the 32 models in Series 1 and 128 models in Series 2. This means that no attempt is being made to assess frequency properties of estimation procedures for specific models, and even frequency properties averaged over the whole of Series 1 and Series 2 are estimated only with rough accuracy. Rather, the attempt has been to assess the actual errors in a plausible range of applications. The resulting error summaries should be of interest to both Bayesians and frequentists.

Tables 1 and 2 show overall summaries for 11 of the 57 methods included in the complete study. Most of the methods included in the 11 are described above in detail. OREG connotes least squares, STEINM is the version of STEIN analogous to the RIDGM version of RIDGE, and BSLA and IBSL are the backward selection analogs of FSLA and IFSL. Table 2 is included with Table 1 because the distributions of the criteria can be very long-tailed. Nevertheless, the two tables show very similar comparisons among methods.

The most striking aspect of Tables 1 and 2 is that drastic

differences among methods appear, especially on SEB. RIDGM turns in a uniformly strong performance, while STEINM is much closer to OREG. The best REGF methods compete with RIDGM on Series 1 but not on Series 2. REGF methods are generally a little better than forward and backward selection. The selection methods which use $F = 1$ as a criterion perform badly in a few instances in Series 1 on SEB, but otherwise are generally similar in overall performance to selection methods which use nominal .95 levels of F.

4. Concluding remarks.

The reasoning in Section 2.2 and the numerical results in Section 3 suggest that RIDGE methods in particular are likely to deserve widespread practical use. Further study of RIDGE and other methods are of course much to be desired. As the number of independent variables p increases, the basic RIDGE assumption of a normally distributed pattern of components in $\underset{\sim}{\beta}$ may tend to fail more observably, so that techniques which select out large components may come more into their own. We need to develop techniques which will combine the virtues of RIDGE with the ability to select large $\underset{\sim}{\beta}$ components. The resulting methods will be close in theory to techniques which correct for outliers as discussed by Dempster (1971). The special handling of large values in the context of STEIN methods for estimating several means has been studied by Efron amd Morris (1971, 1972) who suggest limits on the absolute value of the shift caused by pulling back large estimates. Developing and testing further methods of this sort in the regression context is a task for the future. Other situations with many parameters, such as the covariance situation opened up in Dempster (1972), deserve development combining selection methods with STEIN-RIDGE methods.

Acknowledgment.

I wish to thank my collaborators M. Schatzoff and N. Wermuth for their major contributions to the larger study for which this paper is a preliminary report. Computer time was supplied by IBM Cambridge Scientific Center under a joint study agreement with Harvard University. The research was also supported by the National Science Foundation under grant GP-31003X.

Method	Series 1		Series 2	
	SEB	SPE	SEB	SPE
OREG	542.88	5.70	78.37	6.26
STEINM	482.16	5.94	63.40	5.84
RIDGM	45.16	4.95	17.59	4.98
FSLA	93.59	8.26	68.42	16.37
BSLA	76.22	5.96	64.34	9.94
FREGF	49.32	4.15	42.89	7.46
FDRGF	52.20	4.14	54.80	7.48
1FSL	215.33	4.89	60.09	6.08
1BSL	461.50	5.16	71.31	6.23
1FREGF	379.35	4.75	65.87	5.92
1FDRGF	457.61	4.78	67.59	5.90

TABLE 1. Means for cirteria SEB and SPE over 32 runs in Series 1 and 128 runs in Series 2.

Method	Series 1		Series 2	
	SEB	SPE	SEB	SPE
OREG	178.83	5.57	25.94	5.05
STEINM	161.79	5.50	19.42	4.68
RIDGM	31.87	3.93	7.43	4.11
FSLA	62.15	5.28	31.11	8.23
BSLA	25.20	4.86	33.76	7.74
FREGF	18.88	2.77	18.59	4.89
FDRGF	19.77	2.78	17.65	4.79
1FSL	56.71	4.79	19.47	4.83
1BSL	54.93	5.43	25.30	4.90
1FREGF	32.59	4.66	19.09	4.82
1FDRGF	77.45	4.76	21.11	4.81

TABLE 2. Medians for criteria SEB and SPE over 32 runs in Series 1 and 128 runs in Series 2.

References

1. Dempster, A. P., Schatzoff, M. and Wermuth, N. (1973). A simulation study of alternatives to least squares. (In preparation.)

2. Dempster, A. P. (1971). Model searching and estimation in the logic of inference. Foundations of Statistical Inference; pp. 56-76, V. P. Godambe and D. A. Sprott (editors), Holt, Rinehart and Winston of Canada, Toronto.

3. Dempster, A. P. (1972). Covariance selection. Biometrics. 28, 157.

4. Efron, B. and Morris, C. (1971, 1972). Limiting the risk of Bayes and empirical Bayes estimators. Part I: the Bayes case. Part II: the empirical Bayes case. J. Am. Statist. Assoc. 66, 807, 130.

5. Efron, B. and Morris, C. (1972). Empirical Bayes on vector observations: An extension of Stein's method. Biometrika 59, 335.

6. Hoerl, Arthur E. and Kennard, Robert W. (1970). Ridge regression: biased estimation for nonorthogonal problems, and applications to nonorthogonal problems. Technometrics 12, 56, 69.

7. James, W. and Stein, C. (1960). Estimation with quadratic loss. Proc. Fourth Berkeley Symp. 1, 361.

8. Lindley, D. V. and Smith, A. F. M. (1972). Bayes estimates for the linear model. J. Roy. Statist. Soc. Series B 34, 1.

9. Wermuth, N. (1972). An empirical comparison of regression methods. Ph.D. thesis, Department of Statistics, Harvard University.

A. P. Dempster
Department of Statistics
Harvard University
Cambridge, Massachusetts
U.S.A.

THE ELUSIVE ANCILLARY

D. A. S. FRASER

SUMMARY

The concept of an ancillary statistic has been promoted by R.A. Fisher (e.g., 1925) as a basis for determining the accuracy of estimates and as a means for avoiding any loss of information. Basu (1959, 1964) has shown that a weaker definition of an ancillary leads to nonuniqueness. Barnard and Sprott (1971) have suggested a resolution of the nonuniqueness using invariance of the likelihood function. Cox (1971) suggests a more generally applicable resolution based on information. The two resolutions however may well be contradictory in larger contexts, and Cox's approach in fact provides a substantial argument against the validity of the ancillary concept for statistical inference. Ancillary statistic, Information, Accuracy, Conditional probability, Inference.

1. The Fisher Ancillary.

Fisher's continued promotion of the ancillary statistic dates from some of his earliest works on inference.

For example, Fisher (1925): "Since the original data cannot be replaced by a single statistic without loss of accuracy, it is of interest to see what can be done by calculating, in addition to our estimate, an ancillary statistic which shall be available in combination with our estimate in future calculations. If our two statistics specify the values of $\partial\ell/\partial\theta$ and $\partial^2\ell/\partial\theta^2$ for some central value of θ, such as $\hat{\theta}$, then the variance of $\partial\ell/\partial\theta$ over the sets of samples for which both statistics are constant, ... will ordinarily be of order n^{-1} at least. With the aid of such an ancillary statistic the loss of accuracy tends to zero for large samples."

And Fisher (1934): "... the sets of observations which provide the same estimate differ in their likelihood functions, and therefore in the nature and quantity of the information they supply (and) yet when samples alike in the information they convey exist for all values of the estimate and occur with the same frequency for corresponding values of the parameter (then) a second case (occurs that) need involve no loss of information".

And also: "the loss (of information) may be recovered by using as an ancillary statistic, in addition to the maximum likelihood estimate, the second and higher differential coefficients at the maximum. In general we can only hope to recover the total loss, by taking into account the entire course of the likelihood function."

These quotations give some indication of the complexity that surrounds Fisher's concept of an ancillary statistic. He depended more on the many examples throughout his writings and less on any formal delineation of the concept and its role. There are of course many people who urge earlier and sharper conceptualizations in statistics but the history of this and indeed certain other concepts shows the substantial disadvantages of what could be called premature conceptualization.

2. The Basu Ancillary.

Basu (1959) extracts one aspect of Fisher's characterization and defines an ancillary statistic as "one whose distribution is the same for all possible values of the unknown parameter." He investigates this form of the concept in some detail and finds that often there may be more than one maximal ancillary.

Basu (1964) then considers the role of an ancillary statistic in inference. He presents three examples that have more than one maximal ancillary, and notes that this produces difficulties for the recommended programmes of inference. He does suggest that the difficulties can be resolved by recognizing the difference between a real and a conceptual statistical experiment. But he does not elaborate on the difference and did "... not think it necessary to enter into a lengthy discussion on the reality or performability of a statistical experiment".

His examples do not clarify his intent -- in one case a coin determines a subexperiment, and in another case he hypothesizes a coin and then uses its nonexistence to discount the reality of the component experiments.

3. The Barnard-Sprott Resolution.

Barnard and Sprott (1971) argue "that if proper attention is paid to the likelihood function as the primary inference from an experiment and the role of an ancillary in describing its shape, (then) the difficulties raised by Basu disappear." They consider the family of possible likelihood functions and seek a location indicator T (as estimate) and a shape or scale indicator A (as ancillary).

They then obtain a unique ancillary for each of the three Basu examples.

Barnard and Sprott's procedure for choosing an ancillary is in accord with Fisher's recommendations (1925, 1934) concerning ancillaries. Fisher recommends a complex of statistics to determine "the entire course of the likelihood function". In the invariance case examined by Barnard and Sprott such a complex of statistics gives the orbit or shape ancillary.

In commenting on the Barnard and Sprott (1971) paper, Cox notes that it is "plausible that their particular choice of ancillaries partitions the overall distribution into components as distinct as possible in some sense. If this is indeed true, it provides a resolution of the non-uniqueness independent of group arguments."

4. The Cox Resolution.

Cox (1971) elaborates concerning his comments on the Barnard and Sprott paper, and recommends choosing "that ancillary statistic which separates the expected information into components that differ as widely as possible." Specifically the expected information is examined for each value of an ancillary and the variance of this expected information is used to measure the success of the ancillary in separating the expected information into the different components.

Let $\ell(\Theta)$ be the log-likelihood and

$$I(\Theta:a) = E\left(-\frac{\partial^2 \ell(\Theta)}{\partial\Theta^2} : a,\Theta\right)$$

be the mean information given the value a of an ancillary. Then Cox suggests using

$$\mathrm{Var}(I(\Theta:a):\Theta)$$

as the measure of an ancillary and recommends choosing that ancillary for which the variance measure is largest.

In his original 1925 paper on information Fisher suggests the use of the second derivative $\partial^2\ell/\partial\Theta^2$ as an ancillary statistic measuring information, and in part considers how the mean conditional variance

$$E\left(\mathrm{Var}\left(\frac{\partial^2 \ell(\Theta)}{\partial\Theta^2} : a,\Theta\right):\Theta\right)$$

measures the inadequacy of a grouping of sample points. The common formula for variance about regression gives

$$\mathrm{Var}\left(\frac{\partial^2 \ell}{\partial \Theta^2}\right) = \mathrm{E\ Var}\left(\frac{\partial^2 \ell}{\partial \Theta^2} : a\right) + \mathrm{Var\ E}\left(\frac{\partial^2 \ell}{\partial \Theta^2} : a\right) .$$

Thus maximizing the variance of the conditional information

$$\mathrm{Var}(I(\Theta:a):\Theta)$$

is equivalent to minimizing the mean conditional variance of the observed information $-\partial^2 \ell/\partial \Theta^2$. The Cox criteria then becomes one of assembling together points where observed information is as nearly the same as possible, which is the intent in Fisher's original paper.

The observed information $-\partial^2 \ell/\partial \Theta^2$ gives a measure of the curvature of the likelihood function. Thus in a more technical sense the Cox procedure tends to agree with the Barnard and Sprott procedure of grouping on the basis of similarly shaped likelihood functions.

5. Other Segregation Methods.

Cox notes that his procedure for segregating points on the basis of the variance of the mean information is "rather arbitrary", but nevertheless consistent with the general objective for ancillary statistics -- complete segregation of samples on the basis of the information they provide. Consider some alternative measures that can be used for selecting an ancillary.

For any value a of an ancillary the mean log-likelihood slope

$$\mathrm{E}\left(\frac{\partial \ell}{\partial \Theta} : a, \Theta\right) = 0$$

is equal to zero. Then in accord with the Barnard and Sprott recommendation to group on the basis of the likelihood function it seems reasonable to group together points that have approximately the same likelihood slope $|\partial \ell/\partial \Theta|$. The formula for variance about regression gives

$$\mathrm{Var}\left|\frac{\partial \ell}{\partial \Theta}\right| = I(\Theta) - \left(\mathrm{E}\left|\frac{\partial \ell}{\partial \Theta}\right|\right)^2 = \mathrm{E\ Var}\left(\left|\frac{\partial \ell}{\partial \Theta}\right| : a\right) + \mathrm{Var\ E}\left(\left|\frac{\partial \ell}{\partial \Theta}\right| : a\right) .$$

Thus the grouping together of points having approximately the same likelihood slope can be accomplished by maximizing the variance of the conditional mean slope $|\partial \ell/\partial \Theta|$; this is a simple variation on maximizing the variance of the conditional mean curvature $-\partial^2 \ell/\partial \Theta^2$.

Alternatively we could consider grouping more directly on the basis of information. In his original paper on information Fisher

(1925) presents a first-order formula for the information lost in using only the maximum likelihood from a conditional distribution:

$$\frac{\mathrm{Var}\left(\partial^2 \ell/\partial\Theta^2 : a, \Theta\right)}{E\left(-\partial^2 \ell/\partial\Theta^2 : a, \Theta\right)} .$$

The mean value

$$E\, \frac{\mathrm{Var}\left(\partial^2 \ell/\partial\Theta^2 : a, \Theta\right)}{E\left(-\partial^2 \ell/\partial\Theta^2 : a, \Theta\right)}$$

then provides a measure of how much an ancillary fails to be a full complement to the maximum likelihood estimate. This formula from Fisher then suggests a variation on the Cox procedure, a variation that seems more specific to the purposes of an ancillary. The choice of an ancillary is then based on minimizing the mean value of the conditional variance of the observed information *as a proportion of the conditional mean of the observed information.*

The invariance grouping of sample points, the maximum separation of the information, and the minimum loss of information in using the ancillary as sole ancillary may well be contradictory criteria in larger contexts than the simple multinomial examined by Basu, Barnard and Sprott, and Cox.

6. Grouping Sample Points.

The Basu ancillary is concerned with grouping sample points together so that the marginal distribution of the groups is independent of the parameter. The Barnard and Sprott and the Cox procedures are concerned in addition with grouping points together on the basis of common likelihood and information properties; in doing this these procedures are attempting to restore some of the ancillary properties envisaged by Fisher.

The argument used with these procedures is one of mathematically assembling points in various ways and then assessing the groups on the basis of some general criteria. It should be noted that this is the type of argument that is used in the Birnbaum analysis to deduce the likelihood principle from the sufficiency and conditionality principles. Thus the argument used with the present procedures for choosing an ancillary is an argument that reduces the problem to the likelihood function alone and thereby voids the problem of finding an ancillary. In effect the methods solve the problem by causing it to vanish.

7. A Case Against Ancillaries.

Consider two distinct populations A_1 , A_2 of experimental units. And suppose that a treatment produces a reaction with probability $(1+\theta)/2$ for a unit from the first population and produces a reaction with probability $(2+\theta)/4$ for a unit from the second population. Let x_1 be the number of positive reactions for a sample of n_1 from the first population and let x_2 be the number of positive reactions for a sample of n_2 from the second population. The statistical model then is a product of two binomial models with probabilities $(1+\theta)/2$ and $(2+\theta)/4$ determined by the single parameter θ .

Now suppose that the two populations are in fact mixed together and that their sizes are in the ratio one to two. Consider a random sample of n from the combined populations and suppose that n_1 belongs to the first population and n_2 belongs to the second population.

The principles of conditional probability prescribe that we condition on what we know concerning the population sampling -- specifically that we examine the model given n_1 and given n_2 . We thus obtain the product of two binomial models as described for the direct sampling of the individual populations.

Alternatively consider two distinct populations B_1 , B_2 of experimental units, and suppose that a treatment produces a reaction with probability $(1+\theta)/3$ for a unit from the first population and produces a reaction with probability $(2+\theta)/3$ for a unit from the second population. Let x_1 be the number of positive reactions for a sample of m_1 from the first population and let x_2 be the number of positive reactions for a sample of m_2 from the second population. The statistical model then is a product of two binomial models with probabilities $(1+\theta)/3$ and $(2+\theta)/3$ determined by the single parameter θ .

Now suppose that the two populations are in fact mixed together and that their sizes are in the ratio one-to-one. Consider a random sample of n from the combined population and suppose that m_1 belongs to the first population and m_2 belongs to the second population.

The principles of conditional probability again prescribe that we condition on what we know concerning the population sample -- specifically that we examine the model given m_1 and given m_2 . We thus obtain the product of two binomial models as described for the direct sampling of the individual populations.

Note however that the multinomial model for the sampling from the mixed populations A_1 and A_2 is the same as from the mixed populations B_1 and B_2 :

	B_1	B_2	
A_1	x_1	$n-x_1$	n_1
A_2	n_2-x_2	x_2	n_2
	m_1	m_2	

from

$(1+\Theta)/6$	$(1-\Theta)/6$	$\frac{1}{3}$
$(2-\Theta)/6$	$(2+\Theta)/6$	$\frac{2}{3}$
$\frac{1}{2}$	$\frac{1}{2}$	

If we know the investigation is within populations A_1 , A_2 then we use the first product binomial. If we know the investigation is within populations B_1 , B_2 then we use the second product binomial.

If however we do not know about a population separation A_1 to A_2 or B_1 to B_2 then the ancillary discussions of Barnard and Sprott, and Cox prescribe that we choose one or other of the two separations depending on similar likelihoods or on information separation -- and in fact prescribe that we choose the A_1 to A_2 separation because of its preferred characteristics.

Clearly we could be in a A_1 to A_2 situation or in a B_1 to B_2 situation and accordingly have the conditional models determined for us. But if we are in the position of not knowing which applies -- if either, then no valid principle of inference can argue that we act as if we are in that situation for which the inference procedures would be more attractive. Inference cannot be preference in an objective analysis.

Conditioning on the sample size does of course provide the common paradigm for the presentation of the ancillary concept. And the ancillary concept accordingly acquires much of the force that goes with the paradigm. Part of the difficulty lies in the mathematical approach that treats all probabilities in the same way. In fact, however, different probabilities can be associated with quite distinct physical operations and any valid analysis must acknowledge the physical distinction.

If we have a probability space then any realization leads to conditional probabilities. If however we have a class of probability measures on a measurable space, then a realization even if it has constant probability is not grounds for the conditional model. The indeterminacy of the ancillary is the cautioning counter example.

8. Addendum.

Consider a probability space (S,A,P) and a class G of continuous transformations Θ that map S into S and form a group. Suppose that an observed response X has been obtained from some unknown transformation applied to a realization E on the probability space. The observed X identifies the set GX as the set of possible values for the realization E. Probability principles accordingly prescribe the conditional model $\left(S,A,P_{GX}\right)$ given the event GX. This reduction has been viewed as an ancillary reduction by some commentators. In fact it is the basic reduction of probability theory. An ancillary would need a class of probability measures. The model however presents a class of random variables on a probability space.

References

1. Barnard, G.A. and Sprott, D. (1971). A note on Basu's examples of anomalous ancillary statistics. Waterloo Symposium on Foundations of Statistical Inference. Toronto: Holt, Rhinehart, and Winston.

2. Basu, D. (1959). The family of ancillary statistics, Sankhya A, 21, 247.

3. Basu, D. (1964). Recovery of ancillary information, Sankhya A, 26, 3.

4. Cox, D. R. (1971). The choice between alternative ancillary statistics. J. R. Statist. Soc. B, 33, 251.

5. Fisher, R.A. (1925). Theory of statistical estimation, Proc. Camb. Phil. Soc., 22, 700.

6. Fisher, R.A. (1934). Two new properties of mathematical likelihood, Proc. Royal Soc., A, 144, 285.

7. Fisher, R.A. (1956). Statistical methods and scientific inference. Edinburgh: Oliver and Boyd.

D. A. S. Fraser
Department of Mathematics
University of Toronto
Toronto, Ontario
Canada

MULTIPLE BIRTH DISCRIMINATION

Seymour Geisser*

0. Introduction.

A statistical model for multiple birth discrimination is presented. Part of the model was first presented for like sexed twins based on univariate normal assumptions by Richter and Geisser (1960). They also obtained methods for like sexed triplets and quadruplets. The model is now extended to multivariate normal assumptions for like sexed t-tuplets where t is arbitrary. In addition the procedure is improved by obtaining relative weights for the various cases by devising a simple but plausible model from which the weights are derived. A finer discrimination is also obtained in that the new procedure identifies particular individuals with particular eggs. A further significant feature of the model is that the whole discriminatory procedure for the t-tuplet case depends only on the parameters involved in the twin situation. In what follows we shall first deal with the twin case and then the arbitrary t-tuplet case.

1. The Twin Case.

Analogously as in Richter and Geisser (1961) we let $R_1, R_2, \ldots$ now be a sequence of r-dimensional independent random variables such that R_i is $N(\mu_i, \Sigma_W)$. Further let μ_i be an observation on a random variable M which is $N(\mu, \Sigma_B)$. Let x_1, x_2 be r-dimensional observations on a pair of like sexed twins from the same mother. Then x_1, x_2 are interpreted as observations on R_i, R_j respectively. If $i = j$, the pair are monozygotic (one egg) twins; if $i \neq j$ the pair are dizygotic (two egg) twins. Assume that Σ_W, the within-egg covariance matrix is constant for all mothers. Hence $x_2 - x_1$ is an observation from a $N(0, 2\Sigma_W)$ population if x_1 and x_2 are each independent observations on R_i (the same egg). Now suppose x_1, x_2 are observations on R_i and R_j (different eggs) respectively. Then $x_2 - x_1$ is $N(\mu_i - \mu_j, 2\Sigma_W)$ given $\mu_i - \mu_j$. Assuming μ_i and μ_j are independent, then $\mu_i - \mu_j$ is distributed ad $N(0, 2\Sigma_B)$ and further $x_2 - x_1$ is unconditionally distributed as $N(0, 2\Sigma_W + 2\Sigma_B)$.

Hence the posterior probability that a future twin pair

---

*This work was supported in part by an NIH research grant.

$z = x_2 - x_1$, based on the r characteristics is dizygotic is $\varphi_d(z)/(\varphi_d(z) + \gamma\varphi_m(z))$ where φ_d and φ_m represent the density of a $N(0,2\Sigma_W + 2\Sigma_B)$ and a $N(0,2\Sigma_W)$ variable respectively while γ is the relative frequency of monozygotic twins to dizygotic like sexed twins in the population from which the new twin pair has been drawn. We assume that Σ_W is positive definite which guarantees the existence of both densities φ_d and φ_m .

2. Arbitrary Number of Offspring.

Now suppose that a birth gives rise to t offspring, $x_1,\ldots,x_t$. Further let $z_i = x_t - x_i$, $i=1,\ldots,t-1$. Since $x_1,\ldots x_t$ are observations assumed normal and independent, the joint set $z_1,\ldots z_{t-1}$ is multivariate normal $r(t-1)$ dimensional, conditional on $\Delta' = (\Delta_1',\ldots,\Delta_{t-1}')$ where $\Delta_i = \mu_t - \mu_i$, $i=1,\ldots,t-1$. Clearly

$$(2.1)\qquad \begin{cases} E(z_i|\Delta) = \Delta_i \\ \mathrm{Cov}(z_i|\Delta) = 2\Sigma_W \\ \mathrm{Cov}(z_j,z_k|\Delta) = \Sigma_W \end{cases}$$

Further a simple computation demonstrates that Δ is multivariate normal such that

$$(2.2)\qquad \begin{cases} E(\Delta_i) = 0 \\ \mathrm{Cov}(\Delta_i) = \begin{cases} 0 & \text{if } R_i = R_t \\ 2\Sigma_B & \text{if } R_i \neq R_t \end{cases} \end{cases}$$

and

$$(2.3)\qquad \mathrm{Cov}(\Delta_j,\Delta_k) = \begin{cases} \Sigma_B & \text{if } R_t \neq R_j \neq R_k \\ 2\Sigma_B & \text{if } R_t \neq R_k = R_j \\ 0 & \text{otherwise.} \end{cases}$$

In order to calculate the unconditional distribution of $z_1,\ldots,z_{t-1}$ we need the following well-known result:

Lemma 1. If $X \sim N(Y,\Sigma)$ conditional on Y where $Y \sim N(\theta,\Lambda)$ then unconditionally $X \sim N(\theta,\Lambda + \Sigma)$.

Proof: The assumptions of the lemma imply

$$\varphi_{X|Y}(t) = e^{it'Y-\frac{1}{2}t'\Sigma t}$$

$$\varphi_Y(u) = e^{iu'\theta-\frac{1}{2}u'\Lambda u} .$$

Hence substitution of the above in

$$\varphi_{X,Y}(t,u) = E_Y[e^{iu'Y} E_{X|Y}(e^{it'X})]$$

yields after some trivial algebra

$$\varphi_{X,Y}(t,u) = e^{i(u+t)'\theta - \frac{1}{2}(t+u)'\Lambda(t+u) - \frac{1}{2}t'\Sigma t}$$

Hence setting $u = o$, we obtain

$$\varphi_X(t) = e^{it'\theta - \frac{1}{2}t'(\Sigma+\Lambda)t}$$

which verifies the statement of the lemma. We note that this proof holds even when Σ and Λ are both positive semi-definite.

Application of the lemma implies that unconditionally the joint set $Z' = (z'_1, \ldots z'_{t-1})$ is multivariate normal such that $E(z_i) = 0$ for all i and

$$(2.4) \qquad Cov(z_i) = \begin{cases} 2\Sigma_W & \text{if } R_t = R_i \\ 2\Sigma_W + 2\Sigma_B & \text{if } R_t \neq R_i \end{cases}$$

while

$$(2.5) \qquad Cov(z_j, z_k) = \begin{cases} \Sigma_W + \Sigma_B & \text{if } R_t \neq R_j \neq R_k \\ \Sigma_W + 2\Sigma_B & \text{if } R_t \neq R_j = R_k \\ \Sigma_W & \text{otherwise} \end{cases} .$$

The assumption that Σ_W is positive definite guarantees that $Cov(Z|\Delta)$ is positive definite which in turn implies that $Cov(Z)$ is also positive definite even when $Cov(\Delta)$ is only positive semi-definite. Hence the density of $z_1, \ldots, z_{t-1}$ exists and is evaluated from (2.4) and (2.5) for any given t and any particular egg configuration and depends only on Σ_W and Σ_B the set of parameters appearing in the twin case. Estimates of Σ_W and Σ_B then are obtainable from twin data, if any are unknown.

3. A Model for the Relative Weights.

Let Y be the number of eggs at a birth that yield only like sexed offspring and assume

$$(3.1) \qquad Pr(Y=y) = p^{y-1}(1-p) \qquad y=1,2,\ldots,$$

(it is of course tacitly assumed that there is at least one egg at birth) where p is the relative chance of an additional egg. Further let q be the probability that an egg divides and let D be the number of divisions. Then we assume

$$(3.2) \qquad Pr(D=d \mid Y=y) = \binom{y+d-1}{y-1} q^d (1-q)^y$$

for $d=0,1,2,\ldots$ and $y=1,2,\ldots$. This then represents the chance that amongst y eggs there will be d divisions allowance being made for multiple divisions by every egg. Another way of viewing this is that the eggs represent urns and the offspring are distributed amongst the urns with no urn being empty. Hence the joint chance that a birth yields exactly Y eggs and D divisions is

$$(3.3)\qquad \Pr[Y=y, D=d] = \binom{y+d-1}{y-1} q^{d}(1-q)^{y} p^{y-1}(1-p)$$

We shall now obtain the probability that a birth yields Y eggs given that there are a total of T (like sexed) offspring. It is clear that $T = Y+D$ so that for $y=1,\ldots,t$

$$(3.4)\qquad \Pr(Y=y, T=t) = \binom{t-1}{y-1} q^{t-y}(1-q)^{y} p^{y-1}(1-p) \quad .$$

Summing both sides over y yields

$$(3.5)\qquad \Pr(T=t) = (1-q)(1-p)[q+(1-q)p]^{t-1} \quad .$$

Hence dividing (3.4) by (3.5) we obtain

$$(3.6)\qquad \Pr(Y=y \mid T=t) = \binom{t-1}{y-1} \frac{q^{t-y}(1-q)^{y-1}p^{y-1}}{[q+(1-q)p]^{t-1}}$$

$$= \binom{t-1}{y-1} \frac{\gamma^{y-1}}{(1+\gamma)^{t-1}}$$

where $y=1,\ldots,t$ and

$$(3.7)\qquad \gamma = \frac{(1-q)p}{q} \quad .$$

Therefore the chance of Y eggs given T offspring depends on the single parameter γ. Consider now its interpretation: Suppose $T=2$ i.e., the twin case. Now for monozygotic (one egg) twins $Y=1$ and

$$(3.8)\qquad \Pr(Y=1 \mid T=2) = (1+\gamma)^{-1} \ ;$$

for like sexed dizygotic twins (2 eggs), $Y=2$ and

$$(3.9)\qquad \Pr(Y=2 \mid T=2) = \gamma/(1+\gamma)^{-1} \quad .$$

Hence the relative frequency of like sexed dizygotic to monozygotic twins is simply γ, the ratio of (3.9) to (3.8). This quantity too then is clearly estimable from only twin data.

Now from the relations (2.4) and (2.5) and (3.6) we can determine the posterior probability that the t offspring were derived from $1,\ldots,t$ eggs. In addition for any of the values 2 to $t-1$, we assume that a priori every configuration or assignment of individuals is equally likely. This then will permit us to obtain the actual posterior probability for particular individuals associated with particular eggs i.e., a complete probabilistic ascertainment of which individuals are identical twins and which are fraternal is then feasible. We then choose that case which has maximum posteriori probability. If Σ_W, Σ_B and γ are unknown, they can be estimated from twin data alone and we may choose the case that has maximum estimated posteriori probability. We shall illustrate this with triplets: Here the triplet x_1, x_2, x_3 is transformed to $x_3 - x_1 = z_1$ and $x_3 - x_2 = z_2$ and (z_1, z_2) is unconditionally multivariate normal with zero mean and covariance matrix given below for each case. Individuals within the parentheses are presumed to be from the same egg.

Case	Covariance Matrix of the Joint Distribution of (z_1,z_2)	Relative Frequency
1 egg(x_1,x_2,x_3)	$\left(\begin{array}{c\|c} 2\Sigma_W & \Sigma_W \\ \hline \Sigma_W & 2\Sigma_W \end{array}\right)$	1
2 egg$(x_3),(x_1,x_2)$	$\left(\begin{array}{c\|c} 2\Sigma_W + 2\Sigma_B & \Sigma_W + 2\Sigma_B \\ \hline \Sigma_W + 2\Sigma_B & 2\Sigma_W + 2\Sigma_B \end{array}\right)$	$\frac{2}{3}\gamma$
2 egg$(x_3,x_1),(x_2)$	$\left(\begin{array}{c\|c} 2\Sigma_W & \Sigma_W \\ \hline \Sigma_W & 2\Sigma_W + 2\Sigma_B \end{array}\right)$	$\frac{2}{3}\gamma$
2 egg$(x_3,x_2),(x_1)$	$\left(\begin{array}{c\|c} 2\Sigma_W + 2\Sigma_B & \Sigma_W \\ \hline \Sigma_W & 2\Sigma_W \end{array}\right)$	$\frac{2}{3}\gamma$
3 egg$(x_3),(x_1),(x_2)$	$\left(\begin{array}{c\|c} 2\Sigma_W + 2\Sigma_B & \Sigma_W + \Sigma_B \\ \hline \Sigma_W + \Sigma_B & 2\Sigma_W + 2\Sigma_B \end{array}\right)$	γ^2

Note for the two egg case the original relative frequency is 2γ but since there are three cases which exhaust the discriminatory possibilities we assume that they are all equally likely beforehand so that the relative frequency now becomes $\frac{2}{3}\gamma$ for each of these

2 egg "subcases."

The equally likely a priori configurations for 2 egg-triplets obviously presents no difficulty. However for higher order births the same number of eggs can lead to different partitions. For example in quadruplets there are two distinct partitions in the two egg case i.e. 3 from one egg and one from the other (3,1) or two from each egg (2,2). The partition (3,1) is twice as likely as (2,2) and this must be taken into account. In general then for a fixed t and y there are $\binom{t-1}{y-1}$ compositions $(C_1,\ldots,C_y)$ for integral $C_j \geq 1$, $\sum_j C_j = t$ where C_j represents the number of the individuals belonging to the jth egg, as it were. We then must calculate the frequency f_p of every distinct ordered partition $p = (i_1, i_2,\ldots,i_y)$, subsets of the compostions, where $i_1 \geq i_2 \geq \ldots \geq i_y \geq 1$; $\sum_p f_p = \binom{t-1}{y-1}$ and the summation is over the distinct ordered partitions. Further if we define a_{pk} = number of i_j's equal to k for $k=1,2,\ldots,t-y+1$, for a partition p then it is clear from elementary combinatorial considerations that

$$(3.10) \qquad f_p = y!/\prod_{k=1}^{t-y+1} a_{pk}!$$

where $\sum_{k=1}^{t-y+1} k\, a_{pk} = t$.

Hence the a priori chance of a particular partition p of a y egg and t offspring case is

$$(3.11) \qquad f_p \frac{\gamma^{y-1}}{(1+\gamma)^{t-1}} .$$

Now for each distinct partition there will be b_p equally likely exhaustive assignments, the "numbered" individuals assigned to the different eggs. Hence for a particular partition p , given y and t the b_p exhaustive assignments each have prior chance

$$(3.12) \qquad \frac{f_p \gamma^{y-1}}{b_p (1+\gamma)^{t-1}} .$$

Elementary combinatorial analysis yields

$$(3.13) \qquad b_p = \binom{t}{i_1}\binom{t-i_1}{i_2}\binom{t-i_1-i_2}{i_3}\cdots\binom{i_y}{i_y} \Big/ \prod_{k=1}^{t-y+1} a_{pk}!$$

Hence each assignment for a given partition composed of t offspring and y eggs has prior probability

$$(3.14) \qquad y!\ \gamma^{y-1}/(1+\gamma)^{t-1} \binom{t}{i_1}\binom{t-i_1}{i_2}\binom{t-i_1-i_2}{i_3}\cdots\binom{i_y}{i_y} .$$

Of course first the distinct partitions must be enumerated (tables for their enumeration are available, see Riordan (1958, p. 108) and then the assignments for each partition must also be enumerated.

In order to illustrate this point, we present the case of quadruplets arising from 2 eggs, giving the prior probabilities

Partition	Assignments	Prior Probability
(2,2)	$(x_1,x_2),(x_3,x_4)$	$\gamma/3(1+\gamma)^3$
	$(x_1,x_3),(x_2,x_4)$	$\gamma/3(1+\gamma)^3$
	$(x_1,x_4),(x_2,x_3)$	$\gamma/3(1+\gamma)^3$
(3,1)	$(x_2,x_3,x_4),(x_1)$	$\gamma/2(1+\gamma)^3$
	$(x_1,x_2,x_4),(x_2)$	$\gamma/2(1+\gamma)^3$
	$(x_1,x_2,x_4),(x_3)$	$\gamma/2(1+\gamma)^3$
	$(x_1,x_2,x_3),(x_4)$	$\gamma/2(1+\gamma)^3$

All higher order cases can be handled in precisely the same way. Hence all the tools are at hand for constructing the discriminatory apparatus in the t-tuplet case. This technique should prove useful when there are available multivariate physical measurements that are approximately normal while blood type data is either lacking or inconclusive.

References

1. Richter, D. L. and Geisser, S. (1960). "A statistical model for diagnosing zygosis by ridge count." Biometrics 16, 110.

2. Riordan, J. (1958). An Introduction to Combinatory Analysis, John Wiley and Sons.

Seymour Geisser
School of Statistics
University of Minnesota
Minneapolis, Minn., 55455, U.S.A.

ON A CLASS OF UNBIASED TESTS FOR THE EQUALITY OF K COVARIANCE MATRICES

N. Giri

1. Introduction and Summary: Let $N(\xi_i, \Sigma_i)$, $i=1,\ldots,k$ be independent p-variate normal populations with unknown means ξ_i and unknown positive definite covariance matrix Σ_i . Suppose

$$X_j^i = (X_{j1}^i, \ldots, X_{jp}^i)' \; ; \; j=1,\ldots,N_i \; ; \; i=1,\ldots,k$$

are N_i independent observations from $N(\xi_i, \Sigma_i)$.

$$\text{Let} \quad N_i \bar{X}^i = \sum_{j=1}^{N_i} X_j^i \; , \; S_i = \sum_{j=1}^{N_i} (X_j^i - \bar{X}^i)(\bar{X}_j^i - \bar{X}^i)'$$

and $N_i > p$ for all i so that S_i's are positive definite with probability one and S_i's are independently distributed Wishart random variables with $n_i = N_i - 1$ degrees of freedom and parameter Σ_i .

Let $\sum_{i=1}^{k} n_i = n$.

For any square matrix A we will denote the determinant of A , the trace of A , the transpose of A , the exponential trace of A by $|A|$, tr A , A^1 , etr(A) respectively.

We are interested here in the problem of testing

$$H_o : \Sigma_1 = \ldots = \Sigma_k \; . \tag{1.1}$$

The likelihood ratio test for this problem is, to reject H_o if

$$\prod_{i=1}^{k} | S_i \; (S_i + \ldots + S_k)^{-1}|^{N_i} \leq C_1 \tag{1.2}$$

and the modified likelihood ratio test for this problem is, to reject H_o if

$$\prod_{i=1}^{k} | S_i \; (S_i + \ldots + S_k)^{-1}|^{n_i} \leq C_2 \tag{1.3}$$

where C_1, C_2 are positive constants depending on the size of the tests. We will consider here a general class of critical regions given by

$$\omega(a_1,\ldots,a_k \; ; \; b) \; : \; \prod_{i=1}^{k} | S_i|^{a_i} | S_1 + \ldots + S_k|^{-b} \leq \lambda \tag{1.4}$$

where $b = \sum_1^k a_i = Cn$, C being a positive constant. Kiefer and Schwartz (1965) have shown that if $0 < a_i \leq n_i - p$ for all i and b (not necessarily equal to $\sum_1^k a_i$) $\leq n - p$, then the above critical region is admissible Bayes and similar for testing H_o against the alternatives that all Σ_i's are unequal. It is fully invariant if $b = \sum_1^k a_i$ (hence admissible).

For $k = 2$, Dasgupta and Giri (1971) has shown that the test with the rejection region $\prod_{i=1}^2 |S_i(S_1 + S_2)^{-1}|^{a_i} \leq \lambda$ is unbiased for testing $H_o : \Sigma_1 = \Sigma_2$ against the alternatives $|\Sigma_1| \geq |\Sigma_2|$ when $a_1 \leq n_1$, $a_1 + a_2 = n_1 + n_2$ (i.e. $C = 1$) and against the alternatives $|\Sigma_1| \leq |\Sigma_2|$ when $n_1 \leq a_1$, $a_1 + a_2 = n_1 + n_2$. From theorem 3.1 of Dasgupta and Giri (1971) it is easy to show that if $a_1 + a_2 = C(n_1 + n_2)$, $C > 0$ then the test with critical region $\prod_{i=1}^2 |S_i(S_1 + S_2)^{-1}|^{a_i} \leq \lambda$ is unbiased for testing H_o against the alternatives $|\Sigma_1| \leq |\Sigma_2|$ when $\frac{a_1}{c} \leq n_1$ and against the alternatives $|\Sigma_1| \leq |\Sigma_2|$ when $\frac{a_1}{c} \geq n_1$. For the general case $(k > 2)$ the author is unable to get a straightforward generalization of this result. However, we will be able to get the generalization if $\Sigma_1,\ldots,\Sigma_k$ are such that they can be diagonalized by the same orthogonal matrix. A condition which is necessary and sufficient for this to be true is that $\Sigma_i\Sigma_j = \Sigma_j\Sigma_i$ for all i,j. A special case of this, which arises in the analysis of variance component, is the alternatives $H_1^1: \Sigma_1 = \ell_2\Sigma_2 = .. = \ell_k\Sigma_k$ where ℓ_i's are unknown scalar constants. Federer (1951) has pointed out that this type of model is also meaningful in certain genetical problems.

2. Unbiasedness of $\omega(a_1,\ldots,a_k;b)$: Let

$$T = \prod_{i=1}^k |S_i(S_1 + .. + S_k)^{-1}|^{a_i}, \quad \sum_1^k a_i = cn . \tag{2.1}$$

We will assume throughout that $\Sigma_1,\ldots,\Sigma_k$ can be diagonalized by the same orthogonal matrix 0_1. Let

$$A_i = \Sigma_1\Sigma_i^{-1} = 0_1\Theta_i 0_1 , \quad i=2,\ldots,k . \tag{2.2}$$

where Θ_i is a diagonal matrix and

$$\omega = \{(S_1,\ldots,S_k) | \ S_i\text{'s are pd and } T \geq \lambda\} . \tag{2.3}$$

Write $P_H(\omega)$ for the probability of the region ω when H is true. Under the alternatives $H_1 : \Sigma_i$'s are unequal

$$P_{H_1}(\omega) = M \int_\omega \prod_{i=1}^{k} \left\{ |S_i|^{\frac{n_i-p-1}{2}} |\Sigma_i|^{\frac{-n_i}{2}} \right\} \tag{2.4}$$
$$\{etr(-1/2\Sigma_1^{-1}(S_1 + A_2S_2 + .. + A_kS_k)\} \prod_1^k dS_i$$

where

$$M = \prod_1^k C_{p,n_i}$$

$$C_{p,n_i}^{-1} = \pi^{\frac{p(p-1)}{4}} 2^{\frac{n_ip}{2}} \prod_{j=1}^{p} \Gamma(\frac{n_i-j-1}{2}) .$$

Let $T_i = 0_1'S_i0_1$, $i=1,\ldots,p$. Then from (2.4) we get

$$P_{H_1}(\omega) = M \int_\omega |T_1|^{\frac{n_1-p-1}{2}} \left\{ \prod_{i=2}^{k} |T_i|^{\frac{n_i-p-1}{2}} |\Theta_i|^{\frac{n_i}{2}} \right\} \tag{2.5}$$
$$etr(-1/2(T_1 + \Theta_2T_2 + \ldots + \Theta_kT_k)) \prod_{i=1}^{k} dT_i .$$

Since T_1 is positive definite we can write

$$T_1 = 0 \,\Phi\, 0' = 0\, \Phi^{1/2}\Phi^{1/2}0' = PP'$$

where 0 is a orthogonal matrix, Φ is a diagonal matrix with positive diagonal elements; $\Phi=\Phi^{1/2}\Phi^{1/2}$ and $P = 0\,\Phi^{1/2}$. Let

$$U_i = P'^{-1}T_iP^{-1},\; V_i = 0'U_i0' \;;\; i=2,\ldots,k . \tag{2.6}$$

Then from (2.5) (as $\Theta_i\Phi^{1/2} = \Phi^{1/2}\Theta_i$)

$$P_{H_1}(\omega) = \int_{\bar\omega} |PP'|^{\frac{n-p-1}{2}} \prod_{i=2}^{k} \left\{ |V_i|^{\frac{n_i-p-1}{2}} |\Theta_i|^{\frac{n_i}{2}} \right\} \tag{2.7}$$
$$etr\{-1/2P'(I + \Theta_2^{1/2}V_2\Theta_2^{1/2} + .. + \Theta_k^{1/2}V_k\Theta_k^{1/2})P\}dT_1 \prod_{i=2}^{k} dV_i$$

$$=b \int_{\bar\omega} \prod_{i=2}^{k} \left\{ |V_i|^{\frac{n_i-p-1}{2}} |\Theta_i|^{\frac{n_i}{2}} \right\} |I+\Theta_2^{1/2}V_2\Theta_2^{1/2}+..+\Theta_k^{1/2}V_k\Theta_k^{1/2}|^{-n/2} \prod_{i=2}^{k} dV_i$$

$$= b \int_{\omega^*} \prod_{i=2}^{k} \left\{ |W_i|^{\frac{n_i-p-1}{2}} \right\} |I + W_2 + .. + W_k|^{-n/2} \prod_{i=2}^{k} dW_i$$

where

$$b = M\, C_{n,p}^{-1} ;$$

$$\bar{\omega} = \left\{ (I,V_2,\ldots,V_k) \mid V_i\text{'s are pd, } \prod_{i=2}^{k} |V_i(I+V_2+\ldots+V_k)^{-1}|^{a_i} \geq \lambda \right\} ;$$

$$\omega^* = \left\{ (I,W_2,\ldots W_k) \mid W_i\text{'s are pd; } (I,\Theta_2^{-1/2}W_2\Theta_2^{-1/2},\ldots,\Theta_k^{-1/2}W_k\Theta_k^{-1/2}) \in \bar{\omega} \right\}$$

and $\Theta_i = \Theta_i^{1/2}\Theta_i^{1/2}$, $W_i = \Theta_i^{1/2}V_i\Theta_i^{1/2}$

If $\Theta_i = I$ for all i then $W_i = V_i$ for $i=2,\ldots,k$.

Now

$$P_{H_o}(\omega) - P_{H_1}(\omega) \tag{2.8}$$

$$= b\left[\int_{\bar{\omega}} - \int_{\omega^*}\right] \prod_{i=2}^{k} \left\{ |W_i|^{\frac{n_i-p-1}{2}} \right\} |I + W_2 + \ldots + W_k|^{\frac{-n}{2}} \prod_{i=2}^{k} dW_i$$

$$\geq \lambda b\left[\int_{\bar{\omega}} - \int_{\omega}\right] \prod_{2}^{k} \left\{ |W_i|^{\frac{(n_i-a_i-p-1)/2}{c}} dW_i \right\}$$

$$= \lambda b\left[1 - \prod_{2}^{k} |\Theta_i|^{\frac{(n_i-a_i)/2}{c}}\right] \int_{\bar{\omega}} \prod_{2}^{k} |V_i|^{\frac{(n_i-a_i-p-1)/2}{c}} dV_i$$

$$= \lambda b\left[1 - \prod_{2}^{k} |\Sigma_1\Sigma_i^{-1}|^{\frac{(n_i-a_i)/2}{c}}\right] \int_{\bar{\omega}} \prod_{2}^{k} |V_i|^{\frac{n_i-a_i-p-1)/2}{c}} dV_i \quad .$$

The desired result follows from the fact that

$$\int_{\bar{\omega}} \prod_{2}^{k} \left\{ |V_i|^{\frac{(n_i-a_i-p-1)/2}{c}} dV_i \right\}$$

$$\leq \frac{1}{\lambda} \int_{\omega} \prod_{2}^{k} \left\{ |V_i|^{(n_i-p-1)/2} \right\} |I + V_2 + \ldots + V_k|^{-n/2} \prod_{2}^{k} dV_i < \infty \; .$$

Hence we have the following theorem:

<u>Theorem 1</u>. If $\Sigma_1,\ldots,\Sigma_k$ can be diagonalized by the same orthogonal matrix, then $\omega(a_1,\ldots,a_k;\, cn)\,(c > 0)$ is unbiased for testing $H_o\colon \Sigma_1 = \ldots = \Sigma_k$ against alternatives $|\Sigma_1| \geq |\Sigma_i|$ when $0 < a_i \leq cn_i$ for all i and against the alternatives $|\Sigma_1| \leq |\Sigma_i|$ when $a_i \geq c\,n_i$.

From this theorem it follows trivially that for testing H_o

against H_1', $\omega(a_1,\ldots,a_k,cn)$ $(c > 0)$ is unbiased if $\ell_i \geq 1$ when $0 < a_i < c\, n_i$ and if $\ell_i \leq 1$ when $a_i < c\, n_i$.

REFERENCES

1. Dasgupta, S. and Giri, N. (1971), Properties of tests concerning covariance matrices of normal distributions, Technical report no 150, University of Minnesota, School of Statistics, (To be published).

2. Federer, W. T. (1951), Testing Proportionality of Covariance Matrices. Annals Math. Statist., 22, 102.

3. Kiefer, J. and Schwarta, R. (1965), Admissible Bayes character of F^2, R^2 - and other fully invariant tests for classical multivariate normal problems, Annals Math. Statist, 36, 747.

N. Giri
Department of Mathematics
University of Montreal
Montreal, Quebec
Canada

EXAMINATION OF PSEUDO-RESIDUALS OF OUTLIERS FOR DETECTING SPUROSITY IN THE GENERAL UNIVARIATE LINEAR MODEL

GENE H. GOLUB, IRWIN GUTTMAN and RUDOLF DUTTER

0. Introduction and Summary.

Most of the recent and not so recent literature that deals with the "care and treatment" of outliers is concerned with observations on a univariate random variable, say y, whose expectation is given by $E(y) = \mu$. This simple model is usually accompanied, but not always, with the assumption of normality of y.

However, the care and treatment of outliers in the designed experiment situation, or in general, in situations in which the general univariate linear model provides the necessary framework, seems rather badly neglected. (A recent notable exception is the paper by Andrews (1971), and the subject is touched upon in the great pioneering paper of Anscombe (1960)).

In this paper, we propose rules that guard against the possibility of a spurious observation (that is, an observation that is generated from a source other than that intended) when the vector of independent observations $\underset{\sim}{y} = (y_1,\ldots,y_n)'$ is, hopefully, generated from the intended source described by (using usual notation) $\underset{\sim}{y} = N(X\underset{\sim}{\theta},\sigma^2 I)$, and where n is moderate or large. The rules are analogous to the A, S and W type discussed in Guttman and Smith (1969, 1971) for the simple model $\underset{\sim}{y} = N(\underset{\sim}{1}\mu,\sigma^2 I)$. (See also Tiao and Guttman (1967) and Guttman (1971)). We derive expressions for the premiums of the rules, and also discuss the concept of protection when spurosity is due to a "shifted-mean". The rules utilize so-called adjusted residuals, which are such that they are independently distributed as $N(0,\sigma^2)$, if $\underset{\sim}{y} = N(X\underset{\sim}{\theta},\sigma^2 I)$. Some tables of constants and values of certain functions necessary to put the rules in practice are supplied. Also, we discuss the protections afforded by these rules.

1. Some Assumptions, Definitions and Notation.

Suppose a vector of observations $\underset{\sim}{y} = (y_1,\ldots,y_n)'$ is to be generated (hopefully) from $N(X\underset{\sim}{\theta},\sigma^2 I)$, with X a $(n \times k)$ matrix of arbitrary constants, $n > k$, and where $\underset{\sim}{\theta}$ is a $(k \times 1)$ vector of unknown parameters. In this, the first part of the paper, we assume that the rank of X is full, that is, $r(X) = k$, and that the common variance of the y_i is known, say equal to σ^2.

Let $(X'X)^{-1} = (C_{rs})$, and denote $(X'X)^{-1}X'$ by X^+. We note that X^+ is the Moore-Penrose Inverse of X when X is of full rank. Further, let $M = I - XX^+$. As is well known, the residuals $\underset{\sim}{z} = \underset{\sim}{y} - \hat{\underset{\sim}{y}} = \underset{\sim}{y} - X\hat{\underset{\sim}{\theta}}$, where $\hat{\underset{\sim}{y}} = X\hat{\underset{\sim}{\theta}}$ and $\hat{\underset{\sim}{\theta}} = X^+\underset{\sim}{y}$ is the least squares estimator of $\underset{\sim}{\theta}$, can be written as $\underset{\sim}{z} = M\underset{\sim}{y}$. If indeed $E(\underset{\sim}{y}) = X\underset{\sim}{\theta}$, then it is easy to see that $E(\underset{\sim}{z}) = \underset{\sim}{0}$; further, as the variance-covariance matrix of $\underset{\sim}{y}$ is $\sigma^2 I$, then the variance-covariance matrix of $\underset{\sim}{z}$, say $V(\underset{\sim}{z})$, is given by

$$V(\underset{\sim}{z}) = \sigma^2 MM' = \sigma^2 M , \tag{1.1}$$

since M is symmetric and idempotent. Because of the normality assumption, we then have that

$$\underset{\sim}{z} = N(\underset{\sim}{0}, \sigma^2 M) \tag{1.2}$$

In general, $M \neq I$, so that the z_i's are correlated, and in the various calculations needed to determine premium and protection of rules based on $\underset{\sim}{z}$, the correlation pattern makes the performance of the computations unduly difficult. Accordingly, in the rules given below, which we emphasize are proposed for use when n is moderate to large, we make use of "adjusted residuals" ω, which are defined by

$$\underset{\sim}{\omega} = \underset{\sim}{z} + \sigma\, XP\underset{\sim}{u} , \tag{1.3}$$

where the $(k \times 1)$ random vector $\underset{\sim}{u}$ is independent of $\underset{\sim}{y}$ and has the $N(\underset{\sim}{0}, I_k)$ distribution (the u_i's, $i=1,\ldots,k$, could be obtained by random selection from a table of Standard Normal deviates, such as given in the tables of the Rand Corporation (1955)), and where P is any $(k \times k)$ matrix that satisfies

$$PP' = (X'X)^{-1} . \tag{1.3a}$$

It is easily seen that if $\underset{\sim}{y} = N(X\underset{\sim}{\theta}, \sigma^2 I)$, then $\underset{\sim}{\omega}$ is such that

$$\underset{\sim}{\omega} = N(\underset{\sim}{0}, \sigma^2 I) , \tag{1.4}$$

that is, the ω_j, $j = 1,\ldots,n$ are n independent $N(0,1)$ variables. Now in view of the fact that $X'M = 0$, and since $X^+ = (X'X)^{-1}X'$, we have from (1.3) that

$$XX^+\underset{\sim}{\omega} = XX^+M\underset{\sim}{y} + \sigma\, X(X'X)^{-1}X'XP\underset{\sim}{u} = \sigma\, XP\underset{\sim}{u} \tag{1.5}$$

Subtracting (1.5) from (1.3) yields

$$(I - XX^{+})\underset{\sim}{\omega} = \underset{\sim}{z} \tag{1.6}$$

that is,

$$\underset{\sim}{z} = M\underset{\sim}{\omega} \tag{1.6a}$$

We have chosen to generate adjusted residuals using (1.3), but it should be pointed out that this is by no means unique. In the first place, many $(k \times k)P$ satisfy (1.3a). Also, we could have generated adjusted residuals, say $\underset{\sim}{\omega}^*$, using the formula

$$\underset{\sim}{\omega}^* = \underset{\sim}{z} + \sigma XX^{+}\underset{\sim}{u}^* \tag{1.7}$$

where here the random vector $\underset{\sim}{u}^*$ is $(n \times 1)$ and such that $\underset{\sim}{u}^* = N(\underset{\sim}{0}, I_n)$. It is easy to see that $\underset{\sim}{\omega}^*$ is distributed as in (1.4), that is, $\underset{\sim}{\omega}^* = N(\underset{\sim}{0}, \sigma^2 I_n)$, and also that $\underset{\sim}{z} = M\underset{\sim}{\omega}^*$ (see (1.6a)).

The "corrections" added to $\underset{\sim}{z}$ to calculate $\underset{\sim}{\omega}$ and $\underset{\sim}{\omega}^*$, namely $\sigma XP\underset{\sim}{u}$ and $\sigma XX^{+}\underset{\sim}{u}^*$, have the same distribution, that is, $N(0, \sigma^2 X(X'X)^{-1}X')$ and so can be expected to be of the same magnitude. However, since $n > k$, only k independent $N(0,1)$'s need be selected for $\underset{\sim}{\omega}$ in contrast to n $N(0,1)$'s for $\underset{\sim}{\omega}^*$. However, it is to be admitted that the calculation of P, needed for $\underset{\sim}{\omega}$, adds numerical work, but the existence of efficient programs does minimize that aspect.

Most rules designed for the "care and treatment" of outliers are based on an examination of the residuals $\underset{\sim}{z}$. In fact, what is feared, for example, may be that the presence of some inordinately large (in absolute value) residual is due to the corresponding observation, say y_j, having been generated not from $N(\underset{\sim}{x}_{j\cdot}'\underset{\sim}{\theta}, \sigma^2)$ as intended, but from a spurious source, such as $N(\underset{\sim}{x}_{j\cdot}'\underset{\sim}{\theta} + a\sigma, \sigma^2)$, or $N(\underset{\sim}{x}_{j\cdot}'\underset{\sim}{\theta}, (1+b)\sigma^2)$ etc. (Here, $\underset{\sim}{x}_{j\cdot}'$ is the jth row of the matrix X). We base our rules on the residuals $\underset{\sim}{\omega}$, but it should be pointed out that for many X-matrices, $\underset{\sim}{\omega}$ will lie very close to $\underset{\sim}{z}$ when n is moderate to large, because the elements XP turn out to be of order $1/\sqrt{n}$. This means that the ith component of the random vector $\sigma XP\underset{\sim}{u}$ used to "break" the correlation pattern of the residuals, is, approximately, $N(0, \sigma^2 k/n)$ $(i = 1, \ldots, n)$. Thus, if n is moderate or large in relation to k, the ith correction is of order $\sigma k/\sqrt{n}$, that is, we expect the correction for all i to be small so that $\underset{\sim}{z} \simeq \underset{\sim}{\omega}$. This is illustrated in the examples of the ensuing sections.

2. The proposed rules - definitions and discussion.

Suppose we intend to generate $\underset{\sim}{y}$ from $(NX\underset{\sim}{\theta}, \sigma^2 I)$, and that our main interest is to estimate $\underset{\sim}{\theta}$. Further, suppose we wish to guard against the possibility that one of the y_j will be generated from a spurious source. We will conduct the discussion of this section under the setting of Section 1, e.g., σ is assumed known, X is of full rank, etc.

We propose rules of the type discussed in Guttman and Smith (1969, 1971). Each of these rules examines the adjusted residuals of the observations and implies the use of the least squares estimator $\hat{\underset{\sim}{\theta}} = X^+\underset{\sim}{y}$ for $\underset{\sim}{\theta}$ if all the adjusted residuals are "small" in absolute value. If, however, the largest adjusted residual in absolute value is "inordinately large," the various rules imply the use of an alternative estimator, alternative to $\underset{\sim}{\theta}$. We will compare the performance of these rules through their protections (see Section 4).

Specifically, the definition of the rules considered in this paper are as follows:

(i) The Anscombe - rule (A-rule).

Use the estimator $\hat{\underset{\sim}{\theta}}_{\underset{\sim}{A}}$ for $\underset{\sim}{\theta}$, where

$$\hat{\underset{\sim}{\theta}}_{\underset{\sim}{A}} = \hat{\underset{\sim}{\theta}} + \underset{\sim}{A} \tag{2.1}$$

with the $(k \times 1)$ vector $\underset{\sim}{A}$ given by

$$\underset{\sim}{A} = \begin{cases} \underset{\sim}{0} & \text{if all } |\omega_j| \leq C\,\sigma, \text{ all } j \\ \\ -\underset{\sim}{d}_i \underset{\sim}{m}'_{i\cdot}\omega & \text{if } |\omega_i| > C\,\sigma, \text{ and } |\omega_i| > |\omega_j|, \text{ all } j \neq i, \end{cases} \tag{2.2}$$

and where

$$\underset{\sim}{d}_i = \frac{1}{m_{ii}}(X'X)^{-1}\underset{\sim}{x}_{i\cdot}$$

$$m_{ii} = (i,i)\text{-th element of } M \tag{2.2a}$$

$$\underset{\sim}{m}'_{i\cdot} = i\text{-th row of } M\,,$$

with $C > 0$ an arbitrary constant (see section 3 for discussion of one basis for the selection of C).

Discussion. What does the A-rule do? Effectively, it implies that if an observation has an adjusted residual which is inordinately large in absolute value, throw it out and take the least squares estimator based on the remianing $(n-1)$ observations, and using as "X-matrix," a matrix that is formed by dropping the i-th row of the original matrix. To see this, we proceed as follows.

We note first, from (1.6a), that $-d_i m'_{i\cdot} \omega = -d_i z_i$, and since $z = My$, we have that $-d_i m'_{i\cdot} \omega = -d_i m'_{i\cdot} y$. Further, it is shown in the Appendix that the $k \times n$ matrix $-d_i m'_{i\cdot}$ is such that

$$-d_i m'_{i\cdot} = \tilde{X}_i^+ - X^+ \tag{2.3}$$

where the $(n \times k)$ matrix $\tilde{X}_i$ is assumed to be of rank k and given by

$$\tilde{X}_i = X - e_i x'_{i\cdot} \tag{2.3a}$$

with e_i a $(n \times 1)$ column vector of zeroes, except for the i-th component whose value is 1. (The "+" operation is the same as for X^+, that is $\tilde{X}_i^+ = (\tilde{X}_i'\tilde{X}_i)^{-1}\tilde{X}_i'$, since we assume the rank of $\tilde{X}_i$ to be k).

Hence, if indeed $|\omega_i| > C\sigma$ with $|\omega_i| > |\omega_j|$, all $j \neq i$, then A has the value $(\tilde{X}_i^+ - X^+)y$. In this situation then, the A-rule calls for the use of the point estimate of θ given by

$$\hat{\theta}_A = \hat{\theta} + (\tilde{X}_i^+ - X^+)y \tag{2.3b}$$

or, simply

$$\hat{\theta}_A = \tilde{X}_i^+ y \tag{2.3c}$$

But what is $\tilde{X}_i^+$? We note from (2.3a) and the definition of $\tilde{X}_i^+$, that $\tilde{X}_i^+$ is a matrix which has as i^{th} column, k zeroes. Hence, y_i, the <u>suspect</u> observation (since $|\omega_i| > C\sigma$ etc.) does not get into the act. In fact, because of the construction of $\tilde{X}_i$ itself, (2.3c) is equivalent to removing $x'_{i\cdot}$ from X, removing y_i from y, and going through the least square procedure with the reduced $[(n-1) \times k]$ matrix, say X_{oi} as the "X-matrix," and the reduced observation vector y_{oi} as "y-vector," where

$$X'_{oi} = (x_{1\cdot}, \ldots, x_{i-1,\cdot}, x_{i+1}, \ldots, x_{n\cdot})$$

and (2.4)

$$y'_{oi} = (y_1, \ldots, y_{i-1}, y_{i+1}, \ldots, y_n)$$

The only open question at this point is how to choose C, and thereby answer the question "how large is large for an adjusted residual?" Throughout this paper we choose C as that value that gives the premium (see section 3 for definition and discussion) a predetermined value (such as .005, .01, .02,..., .05).

Now the A-rule effectively throws out a "suspect" observation. This is not always desired, that is, we might wish to modify a suspect observation and save some of the information contained in it. The following two rules do just that.

(ii) The Semi-Winsorization Rule (S-rule).

Use the estimator $\hat{\underset{\sim}{\theta}}_{\underset{\sim}{S}}$ for $\underset{\sim}{\theta}$, where

$$\hat{\underset{\sim}{\theta}}_{\underset{\sim}{S}} = \hat{\underset{\sim}{\theta}} + \underset{\sim}{S} \tag{2.5}$$

with the $(k \times 1)$ vector $\underset{\sim}{S}$ given by

$$\underset{\sim}{S} = \begin{cases} \underset{\sim}{0} & \text{if } |\omega_j| \le C\sigma, \text{ all } j \\ X^+\hat{\underset{\sim}{\delta}}_i & \text{if } \omega_i > C\sigma, -\omega_i < \omega_j < \omega_i, \text{ all } j \ne i \\ X^+\hat{\underset{\sim}{\gamma}}_i & \text{if } \omega_i < -C\sigma, \omega_i < \omega_j < -\omega_i, \text{ all } j \ne i, \end{cases} \tag{2.5a}$$

and where the $(n \times 1)$ vector $\hat{\underset{\sim}{\delta}}_i$ has all components zero except for $(C\sigma - \underset{\sim}{m}'_{i\cdot}\underset{\sim}{\omega})$ in the i-th position, while the $(n \times 1)$ vector $\hat{\underset{\sim}{\gamma}}_i$ is also a vector of zeroes, except for $-(C\sigma + \underset{\sim}{m}'_{i\cdot}\underset{\sim}{\omega})$ in the i-th position.

Discussion. The S-rule implies the use of the least squares estimator $\hat{\underset{\sim}{\theta}} = X^+\underset{\sim}{y}$ if all observations are "well behaved," that is, if their adjusted residuals are all less than or equal to $C\sigma$ in absolute value. If, however, the i-th observation has a large positive adjusted residual ω_i which makes it suspect, that is, $\omega_i > C\sigma$ and ω_i is the largest adjusted residual in absolute value so that $\omega_i > |\omega_j|$, all $j \ne i$, then the S-rule calls for the replacement of $(\underset{\sim}{x}'_{i\cdot}, y_i)$ by $(\underset{\sim}{x}'_{i\cdot}, \hat{y}_i + C\sigma)$, where $\hat{\underset{\sim}{y}} = XX^+\underset{\sim}{y}$, that is, $\hat{y}_i = \underset{\sim}{x}'_{i\cdot}X^+\underset{\sim}{y} = \underset{\sim}{x}'_{i\cdot}\hat{\underset{\sim}{\theta}}$, and hence the use of the point estimate $X^+\tilde{\underset{\sim}{y}}$ for $\underset{\sim}{\theta}$, instead of $X^+\underset{\sim}{y}$, where $\tilde{\underset{\sim}{y}}' = (y_1, \ldots, y_{i-1}, \hat{y}_i + C\sigma, y_{i+1}, \ldots, y_n)$. Geometrically, this may be described as follows: if the "observation point" $(\underset{\sim}{x}'_{i\cdot}, y_i)$ is suspect because $\omega_i > C\sigma$ and $\omega_i > |\omega_j|$, all $j \ne i$,

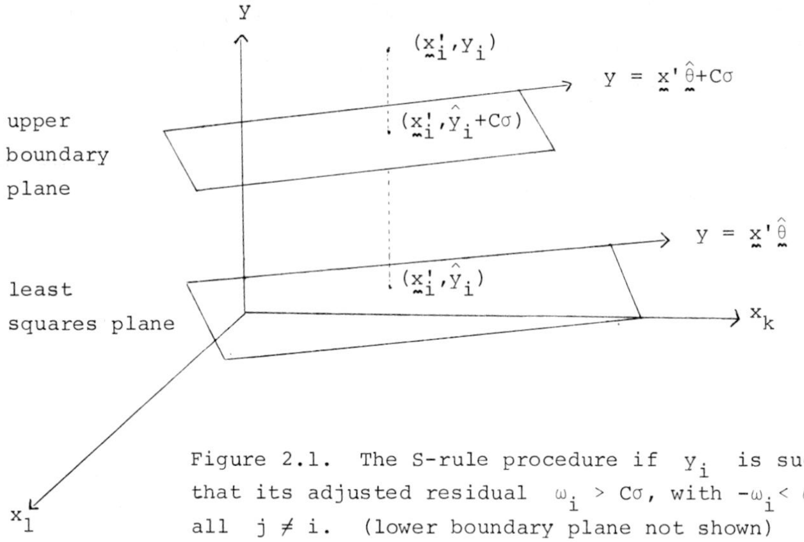

Figure 2.1. The S-rule procedure if y_i is such that its adjusted residual $\omega_i > C\sigma$, with $-\omega_i < \omega_j < \omega_i$, all $j \neq i$. (lower boundary plane not shown)

then "bring" this point to a point on the "boundary" plane $y = \underset{\sim}{x}'\hat{\underset{\sim}{\theta}} + C\sigma$, with coordinates $(\underset{\sim}{x}'_{i\cdot}, \hat{y}_i + C\sigma)$, and re-compute, that is, evaluate $X^+\tilde{\underset{\sim}{y}}$, where $\tilde{\underset{\sim}{y}}' = (y_1, \ldots, y_{i-1}, \hat{y}_i + C\sigma, y_{i+1}, \ldots, y_n)$ - see Figure 2.1.

Of course, if y_i is 'suspect' because $\omega_i < -C\sigma$ and $|\omega_i| > |\omega_j|$, all $j \neq i$, then we bring $(\underset{\sim}{x}'_{i\cdot}, y_i)$ to the plane $y = \underset{\sim}{x}'\hat{\underset{\sim}{\theta}} - C\sigma$, and re-compute, that is, evaluate $X^+\approx{\underset{\sim}{y}}$, where $\approx{\underset{\sim}{y}}' = (y_1, \ldots, y_{i-1}, \hat{y}_i - C\sigma, y_{i+1}, \ldots, y_n)$.

Now the above may be seen as follows. Dwelling on the case where y_i is suspect because $\omega_i > C\sigma$ with $\omega_i > |\omega_j|$, all $j \neq i$, and consulting (2.5) - (2.5a), we see that

$$\hat{\underset{\sim}{\theta}}_S = X^+(\underset{\sim}{y} + \hat{\underset{\sim}{\delta}}_i) \tag{2.6}$$

But $(\underset{\sim}{y} + \hat{\underset{\sim}{\delta}}_i)' = (y_1, \ldots, y_{i-1}, y_i + C\sigma - \underset{\sim}{m}'_i\underset{\sim}{\omega}, y_{i+1}, \ldots, y_n)$, from the definition of $\hat{\underset{\sim}{\delta}}_i$. Now, recall that $\underset{\sim}{z} = M\underset{\sim}{\omega}$, so that, in particular, $\underset{\sim}{m}'_{i\cdot}\underset{\sim}{\omega} = z_i$. Since $\underset{\sim}{z} = \underset{\sim}{y} - \hat{\underset{\sim}{y}}$, where $\hat{\underset{\sim}{y}} = XX^+\underset{\sim}{y} = X\hat{\underset{\sim}{\theta}}$, we know that $z_i = y_i - \hat{y}_i$, where $\hat{y}_i = \underset{\sim}{x}'_{i\cdot}\hat{\underset{\sim}{\theta}} = \underset{\sim}{x}'_{i\cdot}X^+\underset{\sim}{y}$. Hence, the i-th component of $(\underset{\sim}{y} + \hat{\underset{\sim}{\delta}}_i)$ is

$$y_i + C\sigma - \underset{\sim}{m}'_i\underset{\sim}{\omega} = y_i + C\sigma - z_i = y_i + C\sigma - (y_i - \hat{y}_i) \tag{2.7}$$

or

$$y_i + C\sigma - \underset{\sim}{m}'_i\underset{\sim}{\omega} = \hat{y}_i + C\sigma \tag{2.7a}$$

This means that (2.6) may be re-written as $\hat{\underset{\sim}{\theta}}_S = X^+\tilde{\underset{\sim}{y}}$ if $\omega_i > C\sigma$ and $\omega_i > |\omega_j|$, and similar reasoning leads to $\hat{\underset{\sim}{\theta}}_S = X^+\tilde{\underset{\sim}{y}}$ if $\omega_i < -C\sigma$ and $|\omega_i| > |\omega_j|$.

As with the A-rule, we have not yet discussed the choice of C - we will do so (in the next section) by choosing C so that the premium of the S-rule has a desired pre-selected value.

(iii) The Winsorization Rule (W-rule).

Use the estimator $\hat{\underset{\sim}{\theta}}_W$ for $\underset{\sim}{\theta}$, where

$$\hat{\underset{\sim}{\theta}}_W = \hat{\underset{\sim}{\theta}} + \underset{\sim}{W} \tag{2.8}$$

where the (k × l) vector $\underset{\sim}{W}$ is given by

$$\underset{\sim}{W} = \begin{cases} \underset{\sim}{0} & \text{if all } |\omega_i| \leq C\sigma \\ X^+\underset{\sim}{g}_{ij} & \text{if } \omega_i > C\sigma, \ -\omega_i < \omega_t < \omega_j < \omega_i \ \text{all } t \neq i,j, \\ & \text{or, if } \omega_i < -C\sigma, \ \omega_i < \omega_j < \omega_t < -\omega_i \ \text{all } t \neq i,j \end{cases} \tag{2.8a}$$

and where $\underset{\sim}{g}_{ij}$ is a (n × l) vector of zeroes, except for the i-th component which has value $(m'_{j\cdot} - m'_{i\cdot})\underset{\sim}{\omega}$.

Discussion: We have seen that the A-rule calls for a suspect observation to "be thrown out," while the S-rule calls for a suspect observation to be modified. In fact, if a modification is called for, the S-rule "brings" the suspect observation to one of the boundary planes $y = x'\hat{\underset{\sim}{\theta}} \pm C\sigma$, depending on whether the suspect observation has an adjusted residual which is greater than $C\sigma$ or less than $-C\sigma$, etc.

The W-rule proceeds in a similar manner as in the S-rule. From (2.8) - (2.8a), we first note that if all observations are well behaved, then the least squares estimator $X^+\underset{\sim}{y}$ is to be quoted as point estimate of $\underset{\sim}{\theta}$. Now if an adjusted residual is too large, for example, if $\omega_i > C\sigma$ with $-\omega_i < \omega_t < \omega_j < \omega_i$, all $t \neq i,j$, then the point estimate of $\underset{\sim}{\theta}$ that is to be quoted is

$$\hat{\underset{\sim}{\theta}}_W = \hat{\underset{\sim}{\theta}} + X^+\underset{\sim}{g}_{ij} = X^+(\underset{\sim}{y} + \underset{\sim}{g}_{ij}) , \tag{2.9}$$

and from the definition of $\underset{\sim}{g}_{ij}$, we have that this point estimate is

$$\hat{\underset{\sim}{\theta}}_W = X^+\underset{\sim}{y}_{ij} \tag{2.9a}$$

where

$$\underset{\sim}{y}_{ij} = (y_1,\ldots,y_{i-1},\ y_i + \underset{\sim}{m}_j'\underset{\sim}{\omega} - \underset{\sim}{m}_i'\underset{\sim}{\omega},\ y_{i+1},\ldots,y_n)'. \qquad (2.9b)$$

But the i-th component of $\underset{\sim}{y}_{ij}$ can be written as $y_i + z_j - z_i$ in view of the relation $\underset{\sim}{z} = M\underset{\sim}{\omega}$. Thus, the i-th component may be written as

$$y_i + z_j - z_i = y_i + z_j - (y_i - \hat{y}_i) = \hat{y}_i + z_j \qquad (2.10)$$

In summary then, the W-rule calls for the suspect observation y_i to be replaced by $\hat{y}_i + z_j$, <u>or</u>, the "suspect point" $(x_{i\cdot}', y_i)$ to be replaced by $(x_{i\cdot}', \hat{y}_i + z_j)$. This latter point may be arrived at in the following geometric manner.

Suppose we do find that y_i is suspect in that $\omega_i > C\sigma$ and $-\omega_i < \omega_t < \omega_j < \omega_i$, all $t \neq i,j$. We will then say that the point corresponding to ω_j , that is $(\underset{\sim}{x}_{j\cdot}', y_j)$ is the "neighbour" point of $(\underset{\sim}{x}_{i\cdot}', y_i)$. We proceed by constructing the plane $y = \underset{\sim}{x}'\hat{\underset{\sim}{\theta}}+D$ which passes through this "neighbour point" (see Figure 2.2), that is, the plane

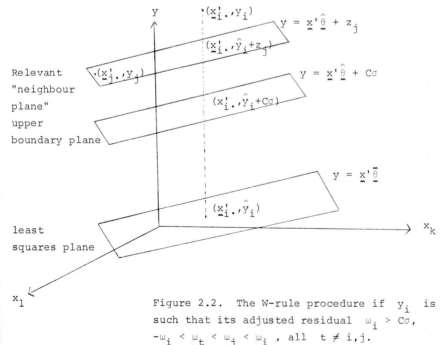

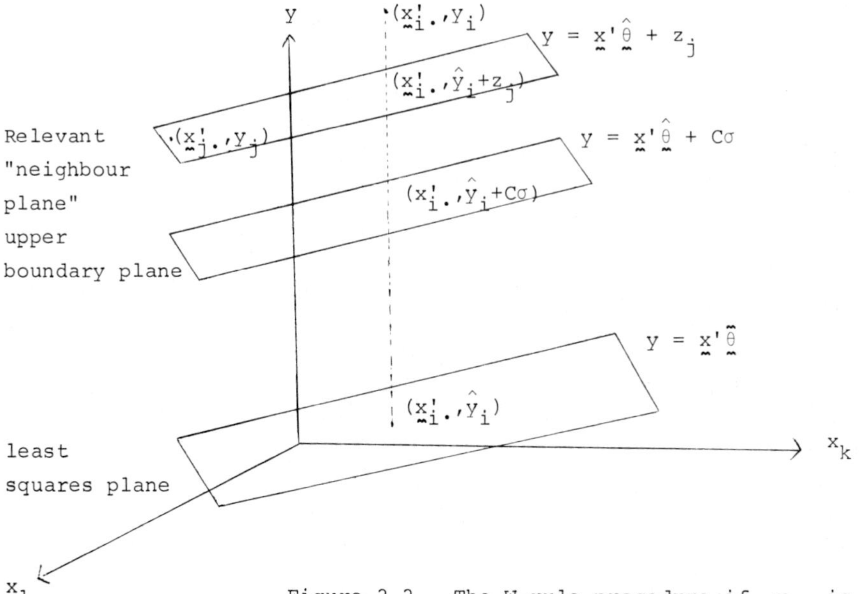

Figure 2.2. The W-rule procedure if y_i is such that its adjusted residual $\omega_i > C\sigma$, $-\omega_i < \omega_t < \omega_j < \omega_i$, all $t \neq i,j$.
(lower boundary plane not shown)

$$y = (\underset{\sim}{x}' - \underset{\sim}{x}'_{j\cdot})\hat{\underset{\sim}{\theta}} + y_j = \underset{\sim}{x}'\hat{\underset{\sim}{\theta}} + z_j \tag{2.11}$$

Now project the suspect point $(\underset{\sim}{x}'_{i\cdot}, y_i)$ along the direction of the y-axis onto this plane. It is easy to see that the foot of this projection on the plane (2.11) is $(\underset{\sim}{x}'_{i\cdot}, \hat{y}_i + z_j)$, for substituting $\underset{\sim}{x}' = \underset{\sim}{x}'_{i\cdot}$ in (2.11), we find $y = \underset{\sim}{x}'_{i\cdot}\hat{\underset{\sim}{\theta}} + z_j = \hat{y}_i + z_j$, as stated.

3. The Premiums of the rules - σ known, X full rank.

In this section, we develop the formulas for the Premiums of the rules described in Section 2. We employ the same assumptions made in Sections 1 and 2, namely that σ is known, X and $\tilde{X}_i$ are of full rank. Further, throughout this section, we assume that the "null situation" obtains, that is, there are no spurious observations, which means that $\underset{\sim}{y} = N(X\underset{\sim}{\theta}, \sigma^2 I)$ is the intended source.

Now from (2.2), (2.5a) and (2.8a), it is straightforward to show, utilizing the symmetry that abounds when the null case obtains, that

$$E(\underset{\sim}{A}) = E(\underset{\sim}{S}) = E(\underset{\sim}{W}) = \underset{\sim}{0} \tag{3.1}$$

This in turn implies that the estimators $\hat{\underset{\sim}{\theta}}_A$, $\hat{\underset{\sim}{\theta}}_S$ and $\hat{\underset{\sim}{\theta}}_W$ are unbiased for $\underset{\sim}{\theta}$. Now a convenient overall measure of dispersion of any estimator of $\underset{\sim}{\theta}$ is tr V, the trace of its variance-covariance matrix V. In fact, as is well known, the least squares estimator $\hat{\underset{\sim}{\theta}}$ is unbiased for $\underset{\sim}{\theta}$, and has variance-covariance matrix given by

$$V(\hat{\underset{\sim}{\theta}}) = (X'X)^{-1}\sigma^2 , \tag{3.2}$$

and we also know that if the null situation obtains, that

$$\text{tr } V(\hat{\underset{\sim}{\theta}}) = \sigma^2 \text{ tr}(X'X)^{-1} \tag{3.3}$$

is the best one can do. To determine the "price" one paid for the use of an alternative estimator to $\hat{\underset{\sim}{\theta}}$ for $\underset{\sim}{\theta}$, and guided by Anscombe (1960), we define the premium P_R of a given rule with its intended estimator $\hat{\underset{\sim}{\theta}}_R$ as

$$P_R = [\text{tr } V(\hat{\underset{\sim}{\theta}}_R) - \text{tr } V(\hat{\underset{\sim}{\theta}})] \,/\, \text{tr } V(\hat{\underset{\sim}{\theta}}) \tag{3.4}$$

given that the null case obtains. In words, the premium is the price one pays for using a rule R to guard against the possibility of a spurious observation (or observations) - the fire - when indeed all observations come from the intended source (that is, there is no fire, and $\hat{\underset{\sim}{\theta}}$ should then be used.) We now develop formulas for premium of the various rules of section 2.

(Before we do this, we wish to re-emphasize that when the observations are generated according to the null situation, the A, W and S rules provide unbiased estimators for $\underset{\sim}{\theta}$, for any n and k. These estimators are not as efficient as $\hat{\underset{\sim}{\theta}}$, but for those who use unbiasedness as a justification and are concerned with the possibility of occurrence of a spurious observation, then these rules may be used for any n and k.)

(i) Premium of the A-rule.

Without loss of generality, we put the known value of $\sigma^2=1$. From (2.2), we have that $\underset{\sim}{A} = \underset{\sim}{A}(\underset{\sim}{\omega})$, so that from (1.3), $\underset{\sim}{A} = \underset{\sim}{A}(\underset{\sim}{z},\underset{\sim}{u})$. But, as is well known, $\underset{\sim}{z}$ is independent of $\hat{\underset{\sim}{\theta}}$, and since $\underset{\sim}{u}$ is observed independently to the vector $\underset{\sim}{y}$, we have that $\underset{\sim}{A}$ is independent of $\hat{\underset{\sim}{\theta}}$. But $\hat{\underset{\sim}{\theta}}_{\underset{\sim}{A}} = \hat{\underset{\sim}{\theta}} + \underset{\sim}{A}$, which implies that the variance-covariance matrix of $\underset{\sim}{\theta}_{\underset{\sim}{A}}$ is

$$\begin{aligned} V(\hat{\underset{\sim}{\theta}}_{\underset{\sim}{A}}) &= V(\hat{\underset{\sim}{\theta}}) + V(\underset{\sim}{A}) \\ &= (X'X)^{-1} + E(\underset{\sim}{A}\underset{\sim}{A}') \end{aligned} \tag{3.5}$$

since $E(\underset{\sim}{A}) = \underset{\sim}{0}$. Substituting in (3.2), we have then that

$$P_{\underset{\sim}{A}} = \operatorname{tr} E(\underset{\sim}{A}\underset{\sim}{A}') \,/\, \operatorname{tr}(X'X)^{-1}, \tag{3.6}$$

and it remains to determine $E(\underset{\sim}{A}\underset{\sim}{A}')$.

Now we have that $\underset{\sim}{A} = \underset{\sim}{0}$ if $|\omega_i| < C$, all i, and $A = -\underset{\sim}{d}_i\underset{\sim}{m}_i'\underset{\sim}{\omega}$ if $\omega_i > C$, $-\omega_i < \omega_j < \omega_i$, or, $\omega_i < -C$, $\omega_i < \omega_j < \omega_i$, all $j \neq i$, and of course, $\underset{\sim}{\omega} = N(\underset{\sim}{0},I)$. Because of the symmetry, then, we may write

$$E(\underset{\sim}{A}\underset{\sim}{A}') = 2\sum_{i=1}^{n} \underset{\sim}{d}_i\underset{\sim}{m}_i' \cdot \mathcal{E}\, \underset{\sim}{m}_i \cdot \underset{\sim}{d}_i' \tag{3.7}$$

where the $(n \times n)$ matrix $\mathcal{E}$ has as r-s element $\mathcal{E}_{rs}$ given by

$$\mathcal{E}_{rs} = \int_{R_i^+} \omega_r\omega_s \left[\prod_{j=1}^{n} \phi(\omega_j)\right]\left[\prod_{j\neq i} d\omega_j\right] d\omega_i, \tag{3.8}$$

$$\text{with } R_i^+ = \left\{\omega \,\middle|\, \omega_i > C,\ -\omega_i < \omega_j < \omega_i,\ \text{all } j \neq i\right\}, \tag{3.8a}$$

$$\text{and} \qquad \phi(\omega) = (2\pi)^{-\frac{1}{2}} \exp\{-\omega^2/2\}.$$

Thus, $\mathrm{tr}\, E(\underset{\sim}{A}\underset{\sim}{A}')$ may be written as

$$\mathrm{tr}\, E(\underset{\sim}{A}\underset{\sim}{A}') = 2 \sum_{i=1}^{n} \left(\sum_{t=1}^{k} d_{it}^2 \right) e_i \tag{3.9}$$

where d_{it} is the t-th component of the $k \times 1$ vector $\underset{\sim}{d}_i$, and

$$e_i = \int_{R_i^+} (\underset{\sim}{m}_i' \cdot \underset{\sim}{\omega})^2 \left[\prod_{j=1}^{n} \phi(\omega_j) \right] \left[\prod_{j \neq i} d\omega_j \right] d\omega_i \ . \tag{3.9a}$$

Of course, we have that

$$(m_i' \cdot \omega)^2 = m_{ii}\omega_i^2 + \sum_{j \neq i} m_{ij}^2 \omega_j^2 + 2 \sum_{j \neq i} m_{ii} m_{ij} \omega_i \omega_j + 2 \sum_{\substack{j<t \\ t,j \neq i}} \sum m_{ij} m_{it} \omega_j \omega_t \ , \tag{3.10}$$

where m_{ij} is the j-th element of the $(1 \times n)$ vector $m_{i\cdot}'$, that is, the $(i - j)$-th element of the matrix M . The contribution to e_i of the cross product terms in (3.8) may easily be seen to be zero, since, they will at one stage of the multiple integration, involve an inner integral of the form

$$\int_{-\omega_i}^{\omega_i} \omega_j \phi(\omega_j) d\omega_j = 0 \tag{3.10a}$$

causing the entire multiple integral over R_i^+ to vanish. Hence (3.9a) may be written as

$$e_i = \int_{R_i^+} \left[m_{ii}^2 \omega_i^2 + \sum_{j \neq i} m_{ij}^2 \omega_j^2 \right] \left[\prod_{j=1}^{n} \phi(\omega_j) \right] \left[\prod_{j \neq i} d\omega_j \right] d\omega_i \tag{3.11}$$

Consulting the definition of R_i^+ given in (3.8a), it is again easy to verify that we may re-write (3.11) as

$$e_i = m_{ii}^2 \int_C^{\infty} \omega_i^2 b^{n-1}(\omega_i) \phi(\omega_i) d\omega_i + \sum_{j \neq i} m_{ij}^2 \int_C^{\infty} \int_{-\omega_i}^{\omega_i} \omega_j^2 \phi(\omega_j) \phi(\omega_i) \times b^{n-2}(\omega_i) d\omega_j d\omega_i \tag{3.12}$$

where

$$b(\mu) = \Phi(\mu) - \Phi(-\mu) = 2\Phi(\mu) - 1$$

$$\text{with} \qquad \Phi(\mu) = \int_{-\infty}^{\mu} \phi(t) dt \ . \tag{3.12a}$$

Now a simple integration by parts yields the fact that

$$\int_{-\omega}^{\omega} \mu^2 \phi(\mu)\, d\mu = b(\omega) - 2\omega\phi(\omega) \tag{3.13}$$

Hence

$$e_i = m_{ii}^2 \int_C^{\infty} \omega^2 b^{n-1}(\omega)\phi(\omega)\, d\omega + \sum_{j\neq i} m_{ij}^2 \left\{ \int_C^{\infty} \phi(\omega_i) b^{n-1}(\omega_i)\, d\omega_i - \right.$$
$$\left. -2 \int_C^{\infty} \omega_i \; \phi^2(\omega_i) b^{n-2}(\omega_i)\, d\omega_i \right\} \tag{3.14}$$

or

$$e_i = m_{ii}^2 \int_C^{\infty} \omega^2 b^{n-1}(\omega)\phi(\omega)\, d\omega + \sum_{j\neq i} m_{ij}^2 \left\{ \frac{1}{2n} [1 - b^n(C)] - \right.$$
$$\left. -2 \int_C^{\infty} \omega\phi^2(\omega) b^{n-2}(\omega)\, d\omega \right\} \tag{3.14a}$$

Introducing the functions

$$G(C) = 2 \int_C^{\infty} \omega^2 b^{n-1}(\omega)\phi(\omega)\, d\omega \; ,$$
$$H(C) = \frac{1}{n} [1 - b^n(C)] - 4 \int_C^{\infty} \omega\phi^2(\omega) b^{n-2}(\omega)\, d\omega \; , \tag{3.15}$$

we may write $2e_i$ as

$$2e_i = m_{ii}^2 \; G(C) + H(C) \sum_{j\neq i} m_{ij}^2 \tag{3.16}$$

Substituting (3.16) into (3.9), we have that

$$\text{tr } E(\underset{\sim}{A}\underset{\sim}{A}') = \sum_{i=1}^{n} \left(\sum_{t=1}^{k} d_{it}^2\right) [m_{ii}^2 G(C) + H(C) \sum_{j\neq i} m_{ij}^2] \tag{3.17}$$

Hence, the premium of the A-rule is given by (3.6) with (3.17) as the numerator, and of course, $\text{tr}(X'X)^{-1}$ as the denominator. (If $k=1$ and $E(\underset{\sim}{y}) = 1\underset{\sim}{\mu}$, the result here may be easily seen to reduce to that obtained by Tiao and Guttman (1967)).

This is an interesting result. Firstly, we note that $P_{\underset{\sim}{A}} = P_{\underset{\sim}{A}}(C)$, that is, the premium is a function of C. If we wish to "pay a premium" of a predetermined fixed amount, say p, then we may set $P_{\underset{\sim}{A}}(C) = \text{tr } E(\underset{\sim}{A}\underset{\sim}{A}')/\text{tr}(X'X)^{-1}$ equal to p, and solve for $C = C(p \vdots X)$, where, as noted, the constant C will depend on the design X and the value of the desired premium. If however, an experimenter has a "favorite" constant C and he wishes to employ the A-rule, then his premium will be given by substituting "his C" in (3.17) and dividing by $tr(X'X)^{-1}$.

In any case, once C is chosen, note that the i-th residual is judged large by taking into account characteristics of the "design of the experiment" that is, hopefully, generating the observations, characteristics that pertain to the i-th residual through the elements d_{it} of $\underset{\sim}{d}_i$ and the elements m_{ij} of $\underset{\sim}{m}_{i\cdot}'$, the i-th row of $M = I - X(X'X)^{-1}X'$.

As an illustration of how the A-rule might work in practice, consider the situation in which a 2^3 - factorial replicated design is used to generate 16 normal observations, with common known variance, put equal to 1. The model is

$$\underset{\sim}{y} = X\underset{\sim}{\theta} + \underset{\sim}{\epsilon} \tag{3.18}$$

where the (16×4) matrix X is such that

$$X' = \begin{bmatrix} 1 & 1 & 1 & 1 & 1 & 1 & 1 & 1 & 1 & 1 & 1 & 1 & 1 & 1 & 1 & 1 \\ -1 & -1 & 1 & 1 & -1 & -1 & 1 & 1 & -1 & -1 & 1 & 1 & -1 & -1 & 1 & 1 \\ -1 & -1 & -1 & -1 & 1 & 1 & 1 & 1 & -1 & -1 & -1 & -1 & 1 & 1 & 1 & 1 \\ -1 & -1 & -1 & -1 & -1 & -1 & -1 & -1 & 1 & 1 & 1 & 1 & 1 & 1 & 1 & 1 \end{bmatrix} \tag{3.18a}$$

and $\underset{\sim}{\epsilon} = N(\underset{\sim}{0}, I)$. Later on, for illustration purposes, we will generate $\underset{\sim}{y}$ by letting $\underset{\sim}{\theta}' = (.1,.2,.3,.4)$; now $X'X$ is easily seen to be such that

$$X'X = 16I$$

so that (3.19)

$$(X'X)^{-1} = \frac{1}{16} I \; . \tag{3.19a}$$

This in turn means that $P = \frac{1}{4} I$, so that

$$XP = \frac{1}{4} X \tag{3.20}$$

Note - the elements of XP are of order $1/\sqrt{n}$. Further, $X^+ = \frac{1}{16} X'$. For reasons of space, we do not display $M = I - XX^+ = I - \frac{1}{16} XX'$, but this matrix is readily computed.

Now suppose $\underset{\sim}{y}$ is generated as in (3.18), with, unknown to the experimenter, $\underset{\sim}{\theta}' = (.1,.2,.3,.4)$. We in fact have generated such a $\underset{\sim}{y}$ and it turns out that the $\underset{\sim}{\epsilon}$ vector is such that $\underset{\sim}{y}$ is as shown in Table 3.1.1. Determining M yields the residual vector $\underset{\sim}{z}$, also shown in Table 3.1.1. Finally, we generated

$\underset{\sim}{u} = N(0, I_4)$ and found the adjusted residuals $\underset{\sim}{\omega} = \underset{\sim}{z} + XP\underset{\sim}{u} = z + \frac{1}{4} X\underset{\sim}{u}$. We tabulate the $\underset{\sim}{\omega}$ vector in Table 3.1.1.

Table 3.1.1.

$\underset{\sim}{y}$	$\underset{\sim}{z}$	$\underset{\sim}{\omega}$
y_1 = -.480	z_1 = .307	ω_1 = .815
y_2 = -1.995	z_2 = -1.208	ω_2 = -.699
y_3 = .555	z_3 = .737	ω_3 = 1.134
y_4 = -1.521	z_4 = -1.339	ω_4 = -.941
y_5 = 2.491	z_5 = 2.260	ω_5 = 2.771
y_6 = -.290	z_6 = -.521	ω_6 = -.010
y_7 = .603	z_7 = -.233	ω_7 = .167
y_8 = .833	z_8 = -.003	ω_8 = .398
y_9 = .661	z_9 = 1.000	ω_9 = 1.170
y_{10} = -.253	z_{10} = .087	ω_{10} = .256
y_{11} = -.487	z_{11} = -.753	ω_{11} = -.694
y_{12} = 1.435	z_{12} = 1.169	ω_{12} = 1.228
y_{13} = .284	z_{13} = -.395	ω_{13} = -.222
y_{14} = -.852	z_{14} = -1.530	ω_{14} = -1.358
y_{15} = 2.700	z_{15} = 1.416	ω_{15} = 1.477
y_{16} = .291	z_{16} = -.993	ω_{16} = -.932

We note from Table 3.1.1. that for the above data, the fourteenth and fifth observations give rise to the minimum and maximum of the z_i's, and <u>also</u> to the minimum and maximum of the ω_i's. This is a typical result. In fact, we performed 100 experiments using the design (3.18a) with (text continues page 85).

Table 3.1.2

Values of C corresponding to various values of P_A for a 2^3 factorial design with r replicates $(k \cong r)$.

r	$n \diagdown P_{\underset{\sim}{A}}$	.005	.01	.02	.03	.04	.05	.10	max $P_A(C) = P_{\underset{\sim}{A}}(\tilde{0})$
2	16	3.5896	3.3753	3.1425	2.9955	2.8854	2.7949	2.4787	.3002
3	24	3.5867	3.3713	3.1366	2.9875	2.8747	2.7817	2.4451	.2244
4	32	3.5852	3.3689	3.1327	2.9817	2.8663	2.7707	2.4114	.1822
5	40	3.5842	3.3672	3.1295	2.9765	2.8586	2.7602	2.3729	.1547
6	48	3.5835	3.3658	3.1265	2.9716	2.8510	2.7495	2.3259	.1352
7	56	3.5828	3.3644	3.1237	2.9667	2.8438	2.7390	2.2634	.1205
8	64	3.5822	3.3632	3.1210	2.9618	2.8364	2.7279	2.1642	.1090
9	72	3.5817	3.3620	3.1182	2.9568	2.8287	2.7160		.0997
10	80	3.5812	3.3609	3.1155	2.9517	2.8206	2.7034		.0920
11	88	3.5807	3.3598	3.1127	2.9468	2.8122	2.6900		.0855
12	96	3.5802	3.3586	3.1099	2.9418	2.8033	2.6754		.0799
16	128	3.5784	3.3542	3.0985	2.9200	2.7635	2.6013		.0639

Table 3.1.3

Values of C corresponding to various values of $P_{\underset{\sim}{A}}$ for optimal regression designs in the interval $\{x \mid -1 \le x \le 1\}$ for fitting a polynomial of degree $d(d = k-1)$, (a) $d = 1$; $k = 2$. The design is: $n/2$ observations are to be made at each of $x = \pm 1$.

n \ $P_{\underset{\sim}{A}}$	.005	.01	.02	.03	.04	.05	.10	$\max P_{\underset{\sim}{A}}(C) = P_{\underset{\sim}{A}}(0)$
6	3.5932	3.3803	3.1495	3.0055	2.8977	2.8109	2.5154	.5549
8	3.5900	3.3762	3.1446	2.9990	2.8907	2.8023	2.5003	.4592
10	3.5883	3.3740	3.1417	2.9956	2.8865	2.7972	2.4907	.3972
12	3.5873	3.3726	3.1398	2.9931	2.8833	2.7934	2.4825	.3525
14	3.5865	3.3716	3.1383	2.9911	2.8806	2.7902	2.4748	.3183
16	3.5860	3.3708	3.1371	2.9894	2.8783	2.7873	2.4674	.2912
18	3.5855	3.3701	3.1361	2.9879	2.8762	2.7846	2.4600	.2689
20	3.5852	3.3696	3.1351	2.9864	2.8742	2.7820	2.4526	.2503
22	3.5849	3.3691	3.1342	2.9851	2.8723	2.7795	2.4453	.2344
24	3.5846	3.3686	3.1334	2.9838	2.8704	2.7770	2.4379	.2207
26	3.5844	3.3682	3.1326	2.9826	2.8686	2.7746	2.4302	.2087
28	3.5841	3.3678	3.1319	2.9814	2.8668	2.7722	2.4223	.1981
30	3.5839	3.3674	3.1312	2.9802	2.8650	2.7697	2.4141	.1886
32	3.5837	3.3671	3.1305	2.9790	2.9633	2.7673	2.4056	.1802
34	3.5836	3.3667	3.1298	2.9779	2.8615	2.7649	2.3968	.1726
36	3.5834	3.3664	3.1291	2.9767	2.8597	2.7624	2.3875	.1656
38	3.5832	3.3661	3.1284	2.9756	2.8579	2.7599	2.3779	.1593
40	3.5831	3.3658	3.1277	2.9744	2.8562	2.7574	2.3678	.1535
42	3.5829	3.3654	3.1271	2.9733	2.8544	2.7549	2.3571	.1481
44	3.5828	3.3651	3.1264	2.9721	2.8526	2.7523	2.3459	.1432
46	3.5827	3.3649	3.1257	2.9710	2.8508	2.7497	2.3337	.1386
48	3.5825	3.3646	3.1251	2.9698	2.8490	2.7473	2.3208	.1343
50	3.5824	3.3643	3.1244	2.9687	2.8473	2.7448	2.3071	.1304
52	3.5823	3.3640	3.1238	2.9675	2.8456	2.7422	2.2921	.1266
54	3.5821	3.3637	3.1231	2.9663	2.8439	2.7396	2.2755	.1231
56	3.5820	3.3634	3.1225	2.9652	2.8421	2.7370	2.2578	.1199
58	3.5819	3.3632	3.1218	2.9640	2.8403	2.7374	2.2374	.1168
60	3.5818	3.3629	3.1212	2.9628	2.8385	2.7316	2.2144	.1139
62	3.5817	3.3626	3.1205	2.9616	2.8367	2.7289	2.1879	.1111
64	3.5815	3.3623	3.1199	2.9604	2.8349	2.7260	2.1562	.1085
66	3.5814	3.3621	3.1192	2.9592	2.8330	2.7232	2.1153	.1060
68	3.5813	3.3618	3.1186	2.9680	2.8311	2.7203	2.0595	.1037
70	3.5812	3.3615	3.1179	2.9568	2.8292	2.7173	1.9608	.1014
72	3.5811	3.3613	3.1172	2.9556	2.8273	2.7143		.0993
74	3.5810	3.3610	3.1166	3.9544	2.8253	2.7113		.0972
76	3.5809	3.3607	3.1159	2.9531	2.8233	2.7082		.0953
78	3.5807	3.3605	3.1153	2.9515	2.8213	2.7050		.0934
80	3.5806	3.3602	3.1146	2.9506	2.8193	2.7018		.0917
82	3.5805	3.3599	3.1139	2.9494	2.8173	2.6985		.0899
84	3.5840	3.3597	3.1132	2.9482	2.8152	2.6952		.0883
86	3.5803	3.3594	3.1126	2.9470	2.8131	2.6919		.0867
88	3.5802	3.3591	3.1119	2.2458	2.8110	2.6885		.0852
90	3.5801	3.3589	3.1112	2.9446	2.8088	2.6849		.0838
92	3.5800	3.3586	3.1105	2.9434	2.8066	2.6814		.0824
94	3.5799	3.5853	3.1099	2.9421	2.8044	2.6777		.0810
96	3.5798	3.3581	3.1092	2.9409	2.8022	2.6740		.0797
98	3.5797	3.3578	3.1085	2.9396	2.7999	2.6702		.0785
100	3.5796	3.3575	3.1078	2.9383	2.7977	2.6663		.0772
128	3.5781	3.5358	3.0980	2.9192	2.7625	2.5998		.0638

Table 3.1.3. (b)

d = 2; k = 3

The design is: n/3 observations are to be made at each of $x = \pm 1, 0$

n \ $P_{\underset{\sim}{A}}$	.005	.01	.02	.03	.04	.05	.10	$\max P_{\underset{\sim}{A}}(C) = P_{\underset{\sim}{A}}(\tilde{0})$
9	3.5913	3.3800	3.1488	3.0041	2.8958	2.8081	2.5068	.4413
12	3.5898	3.3758	3.1435	2.9973	2.8881	2.7986	2.4899	.3605
15	3.5881	3.3735	3.1404	2.9933	2.8830	2.7926	2.4768	.3089
18	3.5870	3.3719	3.1382	2.9903	2.8791	2.7877	2.4647	.2721
21	3.5862	3.3708	3.1365	2.9878	2.8756	2.7833	2.4529	.2444
24	3.5856	3.3699	3.1349	2.9856	2.8725	2.7793	2.4413	.2224
27	3.5851	3.3691	3.1336	2.9835	2.8695	2.7754	2.4295	.2046
30	3.5847	3.3684	3.1323	2.9816	2.8666	2.7715	2.4170	.1898
33	3.5844	3.3678	3.1312	2.9797	2.8638	2.7677	2.4040	.1772
36	3.5840	3.3672	3.1300	2.9778	2.8610	2.7639	2.3902	.1664
39	3.5837	3.3666	3.1289	2.9760	2.8583	2.7600	2.3754	.1570
42	3.5835	3.3661	3.1279	2.9742	2.8555	2.7562	2.3596	.1487
45	3.5832	3.3656	3.1268	2.9725	2.8527	2.7522	2.3424	.1413
48	3.5830	3.3651	3.1258	2.9707	2.8500	2.7484	2.3233	.1348
51	3.5828	3.3647	3.1248	2.9689	2.8474	2.7445	2.3025	.1288
54	3.5825	3.3642	3.1238	2.9671	2.8447	2.7406	2.2782	.1235
57	3.5823	3.3638	3.1228	2.9653	2.8420	2.7366	2.2512	.1186
60	3.5821	3.3633	3.1217	2.9635	2.8393	2.7326	2.2176	.1141
63	3.5819	3.3629	3.1207	2.9617	2.8365	2.7284	2.1763	.1100
66	3.5817	3.3625	3.1197	2.9599	2.8337	2.7241	2.1203	.1062
69	3.5816	3.3621	3.1187	2.9580	2.8309	2.7197	2.0280	.1027
72	3.5814	3.3616	3.1177	2.9562	2.8280	2.7152		.0995
75	3.5812	3.3612	3.1167	2.9543	2.8250	2.7105		.0964
78	3.5810	3.3608	3.1157	2.9524	2.8220	2.7058		.0936
81	3.5808	3.3604	3.1147	2.9505	2.8189	2.7009		.0909
84	3.5807	3.3600	3.1137	2.9487	2.8158	2.6960		.0884
87	3.5805	3.3596	3.1126	2.9469	2.8126	2.6909		.0861
90	3.5803	3.3592	3.1116	2.9451	2.8094	2.6857		.0839
93	3.5802	3.3588	3.1106	2.9432	2.8061	2.6803		.0818
96	3.5800	3.3584	3.1096	2.9413	2.8028	2.6747		.0798
99	3.5798	3.3579	3.1085	2.9394	2.7994	2.6690		.0779

Table 3.1.3 (c)

d = 3; k = 4

The design is: n/4 observations are to be made at each of $x = \pm 1, \pm .4472$.

n \ $P_{\underset{\sim}{A}}$	.005	.01	.02	.03	.04	.05	.10	$\max P_{\underset{\sim}{A}}(C) = P_{\underset{\sim}{A}}(\tilde{0})$
12	3.5929	3.3796	3.1480	3.0027	2.8939	2.8053	2.4988	.3705
16	3.5896	3.3753	3.1425	2.9955	2.8854	2.7949	2.4787	.3002
20	3.5879	3.3729	3.1391	2.9910	2.8795	2.7878	2.4615	.2558
24	3.5867	3.3713	3.1366	2.9875	2.8747	2.7817	2.4451	.2244
28	3.5859	3.3700	3.1345	2.9845	2.8704	2.7761	2.4287	.2008
32	3.5852	3.3689	3.1327	2.9817	2.8663	2.7707	2.4114	.1822
36	3.5847	3.3680	3.1311	2.9791	2.8625	2.7655	2.3930	.1672

(Table 3.1.3 (c) continued on following page.)

(Table 3.1.3 (c) cont'd)

$n \backslash P_{\underset{\sim}{A}}$	.005	.01	.02	.03	.04	.05	.10	$\max P_{\underset{\sim}{A}}(C) = P_{\underset{\sim}{A}}(0)$
40	3.5842	3.3672	3.1295	2.9765	2.8586	2.7602	2.3729	.1547
44	3.5838	3.3665	3.1280	2.9740	2.8548	2.7549	2.3510	.1442
48	3.5835	3.3658	3.1265	2.9716	2.8510	2.7495	2.3259	.1352
52	3.5831	3.3651	3.1251	2.9691	2.8474	2.7443	2.2976	.1274
56	3.5828	3.3644	3.1237	2.9667	2.8438	2.7390	2.2634	.1205
60	3.5825	3.3638	3.1223	2.9642	2.8401	2.7335	2.2209	.1144
64	3.5822	3.3632	3.1210	2.9618	2.8364	2.7297	2.1642	.1090
68	3.5820	3.3626	3.1196	2.9593	2.8326	2.7220	2.0730	.1041
72	3.5817	3.3620	3.1182	2.9568	2.8287	2.7160		.0997
76	3.5814	3.3615	3.1168	2.9543	2.8247	2.7098		.0956
80	3.5812	3.3609	3.1155	2.9517	2.8206	2.7034		.0920
84	3.5809	3.3603	3.1141	2.9492	2.8164	2.6968		.0886
88	3.5807	3.3598	3.1127	2.9468	2.8122	2.6900		.0855
92	3.5805	3.3592	3.1113	2.9443	2.8078	2.6829		.0826
96	3.5802	3.3586	3.1099	2.9418	2.8033	2.6754		.0799
100	3.5800	3.3581	3.1085	2.9392	2.7988	2.6677		.0774
128	3.5784	3.3542	3.0985	2.9200	2.7635	2.6013		.0639

Further tables for (a) $d = 4$; $k = 5$; Design: $\frac{n}{5}$ observations to be made at each of $x = \pm 1, \pm .6547, 0$ (b) $d = 5$; $k = 6$; Design $\frac{n}{6}$ observations to be made at each $x = \pm 1, \pm .7651, \pm .2852$ and (c) $d = 6$; $k = 7$; Design $\frac{n}{7}$ observations to be made at each of $x = \pm 1, \pm .8302, \pm .4689, 0$ can be obtained from the authors. (This same remark applies equally to Tables 3.2.2 and 3.3.2, for $P_{\underset{\sim}{S}}$ and $P_{\underset{\sim}{W}}$.)

Table 3.1.4

Values of C for certain second order response surface designs, and for designs which are formed by replicating these r times (see pages 348, 350 and 371 of Cochran and Cox (1957) for the basic X matrices), corresponding to specified values of $P_{\underset{\sim}{A}}$.

Design (a): Central composite rotatable in 2-x variables (k = 6)

Replicates	$n \backslash P_{\underset{\sim}{A}}$	.005	.01	.02	.03	.04	.05	.10	$\max P_{\underset{\sim}{A}}(C) = P_{\underset{\sim}{A}}(0)$
r = 1	13	3.6167	3.3954	3.1682	3.0254	2.9185	2.8318	2.5339	.3938
2	26	3.5895	3.3746	3.1403	2.9915	2.8789	2.7859	2.4481	.2171
3	39	3.5861	3.3696	3.1326	2.9803	2.8633	2.7657	2.3835	.1597

Design (b): Central composite rotatable in 3-x variables (k = 10)

Replicates	$n \backslash P_{\underset{\sim}{A}}$	.005	.01	.02	.03	.04	.05	.10	$\max P_{\underset{\sim}{A}}(C) = P_{\underset{\sim}{A}}(0)$
r = 1	20	3.6169	3.4067	3.1809	3.0385	2.9321	2.8415	2.5438	.3174
2	40	3.5903	3.3749	3.1387	2.9874	2.8714	2.7745	2.3993	.1616
3	60	3.5859	3.3682	3.1277	2.9707	2.8474	2.7422	2.2503	.1170

Design (c): Central composite rotatable in 5-x variables (k = 21)

Replicates	n \ $P_{\tilde{A}}$	.005	.01	.02	.03	.04	.05	.10	$\max P_{\tilde{A}}(C) = P_{\tilde{A}}(0)$
r = 1	32	3.6680	3.4655	3.2481	3.1131	3.0119	2.9300	2.6459	.3134
2	64	3.5927	3.3765	3.1374	2.9817	2.8595	2.7548	2.2651	.1167
3	96	3.5859	3.3660	3.1193	2.9531	2.8177	2.6936		.0827

Table 3.1.5.

Values of G(C) and H(C)

For each n , the upper section gives values of G(C) , while the lower section gives values of 10,000 H(C). The values of C are not indicated, but they range from zero to 3.8 in steps of 0.2 as one proceeds along rows. The values of n are 15(1)30, 32, 40(8)96, 128.

n = 15	.29630	.29630	.29630	.29630	.29629
	.29611	.29472	.28878	.27296	.24361
	.20251	.15622	.11219	.07547	.14787
	.02879	.01649	.00901	.00471	.00236
	502.64450	502.64450	502.64440	502.64400	502.61530
	502.09800	498.19540	482.58590	444.01490	378.36120
	294.77760	209.91640	137.65080	83.96522	48.11387
	26.10965	13.49897	6.67670	3.16762	1.44379

n = 16	.28434	.28434	.28434	.28434	.28434
	.28422	.28319	.27837	.26461	.23786
	.19911	.15449	.11141	.07516	.04776
	.02876	.01648	.00901	.00471	.00236
	477.10520	477.10520	477.10520	477.10500	477.08930
	476.75250	473.86550	461.18950	427.67910	367.83400
	289.06350	207.27050	136.58680	83.58704	47.99330
	26.07478	13.48975	6.67445	3.16712	1.44369

n = 17	.27345	.27345	.27345	.27345	.27345
	.27337	.27261	.26869	.25673	.23233
	.19581	.15279	.11065	.07486	.04765
	.02872	.01646	.00901	.00471	.00236
	454.09390	454.09390	454.09390	454.09380	454.08520
	453.86560	451.72790	441.42860	412.30660	357.74870
	283.50930	204.67140	135.53410	83.21118	47.87315
	26.03998	13.48053	6.67221	3.16662	1.44358

	.26348	.26348	.26348	.26348	.26348
	.26343	.26286	.25967	.24947	.22702
	.19259	.15111	.10989	.07456	.04754
	.02868	.01645	.00900	.00471	.00236
n = 18					
	433.24850	433.24850	433.24850	433.24850	433.24370
	433.10030	431.51600	423.14330	397.82790	348.08310
	278.10970	202.11800	134.49260	82.83761	47.75339
	26.00523	13.47133	6.66997	3.16611	1.44348
	.25431	.25431	.25431	.25431	.25431
	.25438	.25386	.25126	.24222	.22191
	.18945	.14947	.10914	.07425	.04744
	.02865	.01644	.00900	.00471	.00236
n = 19					
	414.27330	414.27330	414.27330	414.27320	414.27060
	414.17680	413.00150	406.19150	384.17910	338.81650
	272.85980	199.60940	133.46200	82.46632	47.63404
	25.97055	13.46213	6.66773	3.16561	1.44338
	.24584	.24584	.24584	.24584	.24584
	.24582	.24551	.24340	.23553	.21701
	.18640	.14785	.10840	.07396	.04733
	.02861	.01643	.00900	.00471	.00236
n = 20					
	396.92440	396.92440	396.92440	396.92440	396.92300
	396.86160	395.98890	390.44700	371.30130	329.92920
	267.75470	197.14480	132.44240	82.09729	47.51510
	25.93593	13.45294	6.66549	3.16511	1.44327
	.23800	.23800	.23800	.23800	.23800
	.23799	.23776	.23603	.22919	.21229
	.18342	.14626	.10767	.07366	.04722
	.02858	.01642	.00900	.00471	.00236
n = 21					
	380.99920	380.99920	380.99920	380.99920	380.99840
	380.95810	380.30960	375.79740	359.14050	321.40270
	262.78960	194.72320	131.43350	81.73051	47.39655
	25.90137	13.44376	6.66325	3.16461	1.44317
	.23071	.23071	.23071	.23071	.23071
	.23070	.23053	.22913	.22317	.20774
	.18053	.14469	.10694	.07336	.04711
	.02854	.01641	.00899	.00471	.00236
n = 22					
	366.32730	366.32730	366.32730	366.32720	366.32680
	366.30030	365.81810	362.14240	347.64680	313.21930
	257.96020	192.34370	130.43530	81.36597	47.27841
	25.86688	13.43459	6.66102	3.16410	1.44306
	.22392	.22392	.22392	.22392	.22392
	.22391	.22378	.22264	.21745	.20337
	.17771	.14315	.10622	.07307	.04701
	.02851	.01640	.00899	.00471	.00236
n = 23					
	352.76470	352.76470	352.76470	352.76470	352.76450
	352.74710	352.38810	349.39240	336.77430	305.36250
	353.26210	190.00560	129.44750	81.00363	47.16066
	25.83244	13.42542	6.65878	3.16360	1.44296

	.21757	.21757	.21757	.21757	.21757
	.21756	.21747	.21653	.21201	.19916
	.17496	.14164	.10551	.07278	.04690
	.02848	.01639	.00899	.00471	.00236
n = 24					
	340.18900	340.18900	340.18900	340.18900	340.18880
	340.17740	339.91000	337.46730	326.48060	297.81640
	248.69100	187.70810	128.47020	80.64351	47.04330
	25.79807	13.41627	6.65655	3.16310	1.44285
	.19579	.19579	.19579	.19579	.19579
	.19579	.19576	.19534	.19273	.18380
	.16464	.13583	.10273	.07162	.04648
	.02834	.01635	.00898	.00470	.00236
n = 28					
	297.85710	297.85710	297.85710	297.85710	297.85700
	297.85490	297.77190	296.68700	290.35550	270.45300
	231.59860	178.90670	124.66200	79.22470	46.57782
	25.66119	13.37973	6.64762	3.16109	1.44244
	.17842	.17842	.17842	.17842	.17842
	.17842	.17841	.17823	.17671	.17049
	.15531	.13039	.10006	.07050	.04606
	.02820	.01631	.00897	.00470	.00236
n = 32					
	265.02510	265.02510	265.02510	265.02510	265.02510
	265.02470	264.99870	264.51340	260.84980	247.00160
	216.23980	170.69150	121.01020	77.83992	46.11856
	25.52528	13.34332	6.63870	3.15908	1.44202
	.15233	.15233	.15233	.15233	.15233
	.15233	.15233	.15229	.15178	.14874
	.13918	.12051	.09503	.06833	.04525
	.02793	.01623	.00894	.00470	.00236
n = 40					
	217.35110	217.35110	217.35110	217.35110	217.35110
	217.35110	217.34840	217.24950	216.00850	209.26260
	189.92310	155.84310	114.14660	75.16877	45.21831
	25.25637	13.27091	6.62092	3.15507	1.44119
	.13354	.13354	.13354	.13354	.13354
	.13354	.13354	.13353	.13335	.13186
	.12581	.11180	.09038	.06625	.14445
	.02766	.01615	.00892	.00469	.00236
n = 48					
	184.35400	184.35400	184.35400	184.35400	184.35400
	185.35400	184.35380	184.33310	183.90680	180.59470
	168.38890	142.84520	107.82510	72.62295	44.34183
	24.99126	13.19902	6.60321	3.15107	1.44035
	.11927	.11927	.11927	.11927	.11927
	.11927	.11927	.11927	.11921	.11847
	.11463	.10410	.08608	.06427	.04367
	.02739	.01607	.00890	.00469	.00235
n = 56					
	160.13250	160.13250	160.13250	160.13250	160.13250
	160.13250	160.13240	160.12810	159.97970	158.34140
	150.60880	131.43030	101.99660	70.19580	43.48841
	24.72990	13.12766	6.58556	3.14708	1.43952

	.10803	.10803	.10803	.10803	.10803
	.10803	.10803	.10803	.10801	.10764
	.10519	.09726	.08210	.06238	.04292
	.02713	.01599	.00888	.00468	.00235
n = 64					
	141.58260	141.58260	141.58260	141.58260	141.58260
	141.58260	141.58260	141.58160	141.52940	140.71340
	135.79670	121.37290	96.61691	67.88099	42.65736
	24.47222	13.05681	6.56797	3.14309	1.43869
	.09891	.09891	.09891	.09891	.09891
	.09891	.09891	.09891	.09891	.09872
	.09716	.09118	.07840	.06057	.04218
	.02687	.01591	.00885	.00467	.00235
n = 72					
	126.91350	126.91350	126.91350	126.91350	126.91350
	126.91350	126.91350	126.91330	126.89470	126.48560
	123.34840	112.48240	91.64064	65.67260	41.84800
	24.21817	12.98647	6.55044	3.13911	1.43786
	.09136	.09136	.09136	.09136	.09136
	.09136	.09136	.09136	.09135	.09126
	.09026	.08574	.07497	.05884	.04146
	.02662	.01583	.00883	.00467	.00235
n = 80					
	115.01820	115.01820	115.01820	115.01820	115.01820
	115.01820	115.01820	115.01810	115.01150	114.80520
	112.79650	104.59740	87.04788	63.56501	41.05971
	23.96769	12.91665	6.53298	3.13513	1.43703
	.08498	.08498	.08498	.08498	.08498
	.08498	.08498	.08498	.08498	.08493
	.08428	.08086	.07178	.05718	.04076
	.02637	.01575	.00881	.00466	.00235
n = 88					
	105.17510	105.17510	105.17510	105.17510	105.17510
	105.17510	105.17510	105.17510	105.17270	105.06810
	103.77790	97.58115	82.78977	61.55294	40.29185
	23.72072	12.84733	6.51557	3.13117	1.43620
	.07951	.07951	.07951	.07951	.07951
	.07951	.07951	.07951	.07951	.07949
	.07907	.07648	.06881	.05560	.04008
	.02612	.01567	.00879	.00466	.00235
n = 96					
	96.89367	96.89367	96.89367	96.89367	96.89367
	96.89367	96.89367	96.89367	96.89280	96.83947
	96.00817	91.31744	78.84215	59.63141	39.54383
	23.47721	12.77851	6.49823	3.12720	1.43537
	.06366	.06366	.06366	.06366	.06366
	.06366	.06366	.06366	.06366	.06366
	.06358	.06271	.05880	.04990	.03751
	.02516	.01537	.00870	.00464	.00234
n = 128					
	73.72770	73.72770	73.72770	73.72770	73.72770
	73.72770	73.72770	73.72770	73.72768	73.72393
	73.57659	72.01337	65.65812	52.76032	36.73863
	22.53666	12.50815	6.42949	3.11143	1.43206

$y_{16} = N(\underset{\sim}{x}'_{16}.\underset{\sim}{\theta} + 5;1)$ i.e. a=5, and $y_j = N(\underset{\sim}{x}'_j.\underset{\sim}{\theta},1)$, $j \neq 16$, with $\underset{\sim}{\theta}' = (.1,.2,.3,.4)$, and found that in all but 15 of the 100, the smallest and largest of the z_i' s correspond to the smallest and largest of the ω_i' s, in that the same y's gave rise to both, as in the example of Table 3.1.1. For larger n , briefer experiments with a=0 were performed and the above repeats itself when we generated data for the 2^3 , replicated 4 times (n = 32) and for the 2^3 , replicated 6 times (n = 48) , with the added result that the $\underset{\sim}{\omega}$ lie even closer to $\underset{\sim}{z}$ than in the example of Table 3.1.1.

Returning to the data of Table 3.1.1, we may now ask if any of the observations are suspect. We note that ω_5 is such that $\omega_5 = 2.771 > |\omega_j|$, all $j \neq 5$, so that y_5 is suspect. Suppose it had been decided to employ the A-rule with premium of 5%. From Table 3.1.2, we find (n = 16) that the relevant C is C=2.7949. Since $\omega_5 = 2.771 < C = 2.7949$, the A-rule calls for the estimator

$$\underset{\sim}{\hat{\theta}}_{\underset{\sim}{A}} = \underset{\sim}{\hat{\theta}} + \underset{\sim}{0} = \underset{\sim}{\hat{\theta}} = (.248,\ .303,\ .509,\ .224)'$$

(We note that if it had been decided to use the 10% rule, $\omega_5 > C = 2.4787$, so that, as the reader may easily verify,

$$\underset{\sim}{\hat{\theta}}_{\underset{\sim}{A}} = \underset{\sim}{\hat{\theta}} + \underset{\sim}{A} = \underset{\sim}{\hat{\theta}} + (-.188,.188,-.188,.188)'=(.060,.491,.321,.412)'$$

In addition to the constants given in Table 3.1.2, we also give the necessary C's for other designs, viz optimal regression designs of K . Smith (1918)(Table 3.1.3 (a-f)) and certian rotatable designs of Box and Hunter (1957). Of course, it often will be the case that an experimenter finds himself in a "full-rank design" situation not included in our small catalogue. The experimenter could, however, determine the necessary C required for a specific premium by utilizing the "do-it yourself kit" of Table 3.1.5, which tabulates the functions G and 10,000 H for values of C ranging from 0 to 3.8 in steps of 0.2 for various n. Hence, upon determining the characteristics of the design X needed, namely $\underset{\sim}{d}_i$ and M , permitting the calculation of $\underset{\sim}{d}_i'\underset{\sim}{d}_i$, m_{ii} and $\sum_{j\neq i} m_{ij}^2$, for i=1,...,n , then (3.17) may be evaluated by use of Table 3.1.5. Once calculated, (3.17) is divided by $tr(X'X)^{-1}$ to find the premium. Repeated back and forward use of this procedure will lead to the required C for a desired premium. Of course, if the experimenter has his "favorite" C , he may use the A-rule with that C , and if he wishes to inquire about the associated premium, Table 3.1.5 may be employed in the obvious way. (The authors' have tabulated G and H for $n = 5(1)100;128$ and $C = 0(.05)4$, and this is available on request.)

(ii) Premium of the S-rule.

Again, without loss of generality, we put the known value of $\sigma^2 = 1$. From (2.5) and (2.5a) we have that $\underset{\sim}{S} = \underset{\sim}{S}(\underset{\sim}{\omega}) = \underset{\sim}{S}(\underset{\sim}{z},\underset{\sim}{u})$, so that $\underset{\sim}{S}$ is independent of $\hat{\underset{\sim}{\theta}}$ in the null case. As $\hat{\underset{\sim}{\theta}}_{\underset{\sim}{S}} = \hat{\underset{\sim}{\theta}} + \underset{\sim}{S}$, we then have that

$$V(\hat{\underset{\sim}{\theta}}_{\underset{\sim}{S}}) = V(\hat{\underset{\sim}{\theta}}) + V(\underset{\sim}{S}) = (X'X)^{-1} + E(\underset{\sim}{S}\underset{\sim}{S}'). \tag{3.21}$$

since, as remarked earlier, $E(\underset{\sim}{S}) = 0$. This means that the premium of the S-rule may be written as

$$P_{\underset{\sim}{S}} = \mathrm{tr}\, E(\underset{\sim}{S}\underset{\sim}{S}')/\mathrm{tr}(X'X)^{-1} \tag{3.22}$$

and it remains to determine $\mathrm{tr}\, E(\underset{\sim}{S}\underset{\sim}{S}')$. From (2.5a), we have that

$$E(\underset{\sim}{S}\underset{\sim}{S}') = \sum_{i=1}^{n} X^{+} [\mathcal{E}_i^{+} + \mathcal{E}_i^{-}] X^{+'} \tag{3.23}$$

where the $(n \times n)$ matrix $\mathcal{E}_i^{+}$ is a matrix of zeroes, except for its $(i\text{-}i)^{th}$ element, which has value

$$e_{ii}^{+} = \int_{R_i^{+}} [C^2 - 2C\underset{\sim}{m}_{i}'.\underset{\sim}{\omega} + (\underset{\sim}{m}_{i}'.\underset{\sim}{\omega})^2][\prod_{j=1}^{n} \phi(\omega_j)][\prod_{j\neq i} d\omega_j]d\omega_i \tag{3.24}$$

with R_i^{+} defined as in (3.8a), and where the $(n \times n)$ matrix $\mathcal{E}_i^{-}$ is also a matrix of zeroes except for its $(i\text{-}i)^{th}$ element, which has value

$$e_{ii}^{-} = \int_{R_i^{-}} [C^2 + 2C\underset{\sim}{m}_{i}'.\underset{\sim}{\omega} + (\underset{\sim}{m}_{i}'.\underset{\sim}{\omega})^2][\prod_{j=1}^{n} \phi(\omega_j)][\prod_{j=i}^{n} d\omega_j]d\omega_i \tag{3.25}$$

with

$$R_i^{-} = \{\underset{\sim}{\omega} | \omega_i < -C,\ \omega_i < \omega_j < -\omega_i ,\quad \text{all} \quad j \neq i\} \tag{3.25a}$$

Hence, the $(n \times n)$ matrix $\mathcal{E}_i = \mathcal{E}_i^{+} + \mathcal{E}_i^{-}$ is a matrix of zeroes except for its $(i\text{-}i)^{th}$ element $e_{ii} = e_{ii}^{+} + e_{ii}^{-}$. Now it is easy to see that the leading terms of e_{ii}^{+} and e_{ii}^{-} sum to

$$C^2[P(R_i^{+}) + P(R_i^{-})] \tag{3.26}$$

and because of symmetry that this in turn equals

$$2C^2\left[\int_C \phi(\omega_i)b^{n-1}(\omega_i)d\omega_i\right] = C^2[1 - b^n(C)]/n \tag{3.26a}$$

that is, the first term of e_{ii} is given by (3.21a). Further, the second terms of e^+_{ii} and e^-_{ii} are such that their sum may be written as

$$-2C\{m_{ii}\int_C^{\infty}\omega_i\phi(\omega_i)b^{n-1}(\omega_i)d\omega_i + \sum_{j\neq i} m_{ij}\int_C^{\infty}\int_{-\omega_i}^{\omega_i}\omega_j\phi(\omega_j)b^{n-2}(\omega_i)d\omega_j d\omega_i \tag{3.27}$$

$$-m_{ii}\int_{-\infty}^{-C}\omega_i\phi(\omega_i)b^{n-1}(-\omega_i)d\omega_i - \sum_{j\neq i} m_{ij}\int_{-\infty}^{-C}\int_{\omega_i}^{-\omega_i}\omega_j\phi(\omega_j)b^{n-2}(-\omega_i)d\omega_j d\omega_i\} .$$

Now integrals of the form $\int_{-t}^{t}\omega\phi(\omega)d\omega = 0$, since $\phi(\omega) = \phi(-\omega)$, so that we may write (3.22) as

$$m_{ii}T(C) \tag{3.27a}$$

where

$$T(C) = -4C\int_C^{\infty}\omega\phi(\omega)b^{n-1}(\omega)d\omega \tag{3.27b}$$

That is, the second term of e_{ii} is given by (3.27a). Finally, we note that the third terms of e^+_{ii} and e^-_{ii} sum to

$$\left\{\int_{R_i^+} + \int_{R_i^-}\right\} [(\underset{\sim}{m}_i'\underset{\sim}{\omega})^2][\prod_{j=1}^{n}\phi(\omega_j)][\prod_{j\neq i} d\omega_j]d\omega_i \tag{3.28}$$

But, by paralleling the development in (3.9) - (3.11), we have that (3.28) may be written as

$$\left\{\int_{R_i^+} + \int_{R_i^-}\right\} [m_{ii}^2\omega_i^2 + \sum_{j\neq i} m_{ij}^2\omega_j^2][\prod_{j=1}^{n}\phi(\omega_i)][\prod_{j\neq i} d\omega_j]d\omega_i \tag{3.28a}$$

$$= 2\int_{R_i^+} [m_{ii}^2\omega_i^2 + \sum_{j\neq i} m_{ij}^2\omega_j^2][\prod_{j=1}^{n}\phi(\omega_j)][\prod_{j\neq i} d\omega_j]d\omega_i ,$$

for reasons of symmetry. But the above is simply $2e_i$, where e_i is defined in (3.10), and from (3.14) then, we have that (3.28) - (3.28a) may be written as

$$m_{ii}^2 G(C) + H(C) \sum_{j \neq i} m_{ij}^2 \tag{3.28b}$$

In summary, then, the $(n \times n)$ matrix $\mathcal{E}_i$ is a matrix of zeroes except for its $(i\text{-}i)^{th}$ element e_i, given by (from (3.26a), (3.27a) and (3.28b))

$$e_{ii} = C^2[1-b^n(C)]/n + m_{ii}T(C) + m_{ii}^2 G(C) + H(C) \sum_{j \neq i} m_{ij}^2 \tag{3.29}$$

Now, since $E(\underset{\sim}{S}\underset{\sim}{S}') = \sum_{i=1}^{n} X^+ \mathcal{E}_i X^{+'}$, and in view of the structure of $\mathcal{E}_i$ just mentioned, we have that

$$\text{tr } E(\underset{\sim}{S}\underset{\sim}{S}') = \sum_{i=1}^{n} e_{ii} \sum_{j=1}^{k} x_{ji}^{+2} \tag{3.30}$$

where

x_{ji}^{+}=j-th element of the i-th column of X^+, i=1,...,n, j=1,...,k (3.30a)

and e_{ii} is given by (3.29). Dividing (3.30) by $\text{tr}(X'X)^{-1}$ gives the premium of the S-rule. We note that if $k = 1$, $E(\underset{\sim}{y}) = \underset{\sim}{1}\mu$, then $\text{tr } E(\underset{\sim}{S}\underset{\sim}{S}')/\text{tr}(X'X)^{-1}$ reduces to the corresponding result found in Guttman (1971).

We also note that, as with the A-rule, $P_{\underset{\sim}{S}} = P_{\underset{\sim}{S}}(C)$. For the same small catalogue as given earlier, we give necessary C's for the .005, .01(.01).05, .10 premium S-rules in Tables 3.2.1 - 3.2.3. We also tabulate $T(C)$ in Table 3.2.4; this table, together with Table 3.1.5 may be used as a basis of a "do-it-yourself kit" to find necessary C's, for given X, that makes the S-rule of desired premium, or, for given X, to determine the premium for a given C. A more complete set of values of $T(C)$ for $n = 5(1)100$ 128 and $C = 0(.05)4$, is available on request from the authors.

To illustrate the S-rule, we again refer to the data of Table 3.1.1, which was generated using a replicated 2^3 design. We have seen that y_5 is suspect, because $\omega_5 > |\omega_j|$, all $j \neq 5$. If it had been decided to use the 5% premium S-rule, $C = .8334$, as given in Table 3.2.1. Since $\omega_5 = 2.771 > C = .8334$, the S-rule calls for the estimate

$$\hat{\underset{\sim}{\theta}}_5 = \hat{\underset{\sim}{\theta}} + \underset{\sim}{S} = X^+(y_1,\ldots,y_4,y_5 + C - \underset{\sim}{m}_5'.\underset{\sim}{\omega},y_6,\ldots,y_{16})' \tag{3.31}$$

and since $y_5 + C - \underset{\sim}{m}_5' \cdot \underset{\sim}{\omega} = y_5 - z_5 + C = 0.231 + .8334 = 1.0644$, (text continues page 94).

Table 3.2.1

Values of C corresponding to various values of $P_{\underset{\sim}{S}}$ for 2^3 factorial design with r replicates (k = 4).

r	n \ $P_{\underset{\sim}{S}}$	.005	.01	.02	.03	.04	.05	.10	$\max P_{\underset{\sim}{S}}(C) = P_{\underset{\sim}{S}}(0)$
2	16	2.4007	1.8883	1.3258	1.1041	.9537	.8334	.4069	.1689
3	24	2.1266	1.7457	1.3955	1.1879	1.0293	.8963	.4026	.1558
4	32	2.0821	1.7606	1.4262	1.2090	1.0371	.8903	.3351	.1395
5	40	2.0798	1.7734	1.4336	1.2032	1.0174	.8573	.2456	.1253
6	48	2.0825	1.7787	1.4284	1.1838	.9843	.8115	.1476	.1136
7	56	2.0848	1.7790	1.4154	1.1565	.9440	.7593	.0466	.1039
8	64	2.0858	1.7754	1.3976	1.1247	.8996	.7036		.0958
9	72	2.0855	1.7691	1.3765	1.0900	.8530	.6461		.0889
10	80	2.0839	1.7606	1.3531	1.0536	.8051	.5879		.0830
11	88	2.0813	1.7505	1.3283	1.0161	.7565	.5296		.0779
12	96	2.0776	1.7394	1.3025	.9780	.7078	.4714		.0734
16	128	2.0555	1.6867	1.1940	.8241	.5149	.2437		.0600

Table 3.2.2

Values of C corresponding to various values of $P_{\underset{\sim}{S}}$ for optimal regression designs in the interval $\{x \mid -1 \leq x \leq 1\}$ for fitting the polynomial of degree $d(d = k-1)$ (a) $d = 1$; $k = 2$ (See Table 3.1.3 (a)).

n \ $P_{\underset{\sim}{S}}$	.005	.01	.02	.03	.04	.05	.10	$\max P_{\underset{\sim}{S}}(C) = P_{\underset{\sim}{S}}(0)$
6	2.7001	2.3489	1.8331	1.2858	1.0246	.8844	.5226	.2466
8	2.4360	2.0393	1.5292	1.2569	1.0950	.9784	.6146	.2583
10	2.2742	1.8984	1.5027	1.2895	1.1439	1.0302	.6550	.2542
12	2.1936	1.8603	1.5158	1.3161	1.1735	1.0599	.6698	.2448
14	2.1573	1.8475	1.5289	1.3344	1.1918	1.0766	.6701	.2339
16	2.1423	1.8476	1.5388	1.3461	1.2025	1.0847	.6613	.2229
18	2.1376	1.8499	1.5462	1.3540	1.2080	1.0869	.6466	.2125
20	2.1364	1.8564	1.5520	1.3585	1.2094	1.0850	.6278	.2027
22	2.1369	1.8617	1.5567	1.3602	1.2078	1.0800	.6061	.1937
24	2.1381	1.8657	1.5592	1.3597	1.2040	1.0726	.5823	.1854
26	2.1396	1.8686	1.5602	1.3574	1.1983	1.0634	.5571	.1778
28	2.1411	1.8707	1.5599	1.3539	1.1912	1.0527	.5307	.1708
30	2.1424	1.8721	1.5586	1.3492	1.1830	1.0409	.5036	.1643
32	2.1436	1.8729	1.5565	1.3437	1.1738	1.0283	.4759	.1584
34	2.1446	1.8732	1.5536	1.3374	1.1639	1.0148	.4478	.1528
36	2.1454	1.8731	1.5502	1.3304	1.1533	1.0008	.4194	.1477
38	2.1461	1.8726	1.5465	1.3229	1.1422	.9864	.3908	.1430
40	2.1466	1.8718	1.5424	1.3150	1.1306	.9715	.3621	.1385

(Table 3.2.2 continued on next page.)

(Table 3.2.2 cont'd)

n \ P_S	.005	.01	.02	.03	.04	.05	.10	$\max P_S(C) = P_S(0)$
42	2.1470	1.8706	1.5387	1.3066	1.1187	.9563	.3334	.1344
44	2.1473	1.8691	1.5329	1.2979	1.1065	.9409	.3047	.1305
46	2.1474	1.8674	1.5276	1.2889	1.0941	.9252	.2760	.1268
48	2.1475	1.8655	1.5221	1.2796	1.0814	.9094	.2474	.1234
50	2.1474	1.8633	1.5163	1.2702	1.0685	.8935	.2189	.1201
52	2.1472	1.8609	1.5103	1.2605	1.0556	.8775	.1905	.1171
54	2.1470	1.8584	1.5042	1.2507	1.0424	.8614	.1623	.1142
56	2.1467	1.8557	1.4979	1.2408	1.0292	.8452	.1341	.1115
58	2.1463	1.8528	1.4914	1.2307	1.0159	.8290	.1061	.1089
60	2.1458	1.8499	1.4848	1.2206	1.0026	.8128	.0783	.1064
62	2.1452	1.8474	1.4782	1.2103	.9892	.7965	.0507	.1040
64	2.1446	1.8447	1.4714	1.2000	.9758	.7803	.0232	.1018
66	2.1439	1.8419	1.4645	1.1896	.9623	.7641		.0997
68	2.1431	1.8389	1.4575	1.1792	.9488	.7479		.0976
70	2.1422	1.8359	1.4504	1.1687	.9354	.7317		.0957
72	2.1413	1.8327	1.4433	1.1582	.9219	.7155		.0938
74	2.1403	1.8294	1.4362	1.1477	.9084	.6994		.0921
76	2.1392	1.8261	1.4289	1.1371	.8949	.6834		.0903
78	2.1381	1.8226	1.4217	1.1265	.8815	.6673		.0887
80	2.1368	1.8191	1.4144	1.1160	.8681	.6514		.0871
82	2.1355	1.8156	1.4070	1.1054	.8547	.6354		.0856
84	2.1342	1.8120	1.3997	1.0948	.8413	.6196		.0842
86	2.1328	1.8083	1.3923	1.0842	.8279	.6038		.0827
88	2.1313	1.8046	1.3848	1.0736	.8146	.5880		.0814
90	2.1297	1.8009	1.3774	1.0630	.8013	.5723		.0801
92	2.1281	1.7972	1.3699	1.0525	.7881	.5567		.0788
94	2.1264	1.7935	1.3625	1.0419	.7749	.5411		.0776
96	2.1247	1.7897	1.3550	1.0314	.7617	.5256		.0764
98	2.1229	1.7860	1.3475	1.0209	.7486	.5101		.0753
100	2.1210	1.7821	1.3400	1.0104	.7355	.4947		.0742
128	2.0930	1.7265	1.2351	.8658	.5569	.2860		.0618

<u>Table 3.2.2 (b)</u>

d = 2, k = 3

(See Table 3.1.3 (b))

n \ P_S	.005	.01	.02	.03	.04	.05	.10	$\max P_S(C) = P_S(\tilde{0})$
9	2.6923	2.3276	1.6466	1.0940	.9159	.7964	.4288	.1961
12	2.4184	1.9767	1.4085	1.1675	1.0155	.8993	.5076	.2028
15	2.2441	1.8343	1.4271	1.2163	1.0684	.9500	.5331	.1977
18	2.1602	1.7984	1.4480	1.2463	1.0972	.9749	.5324	.1890
21	2.1278	1.7967	1.4667	1.2647	1.1118	.9846	.5170	.1795
24	2.1121	1.7998	1.4788	1.2746	1.1173	.9849	.4928	.1703
27	2.1042	1.8077	1.4862	1.2787	1.1164	.9789	.4630	.1617
30	2.1028	1.8137	1.4904	1.2786	1.1111	.9684	.4295	.1537
33	2.1050	1.8178	1.4920	1.2753	1.1026	.9547	.3935	.1464
36	2.1081	1.8206	1.4915	1.2697	1.0916	.9386	.3558	.1398
39	2.1109	1.8223	1.4893	1.2621	1.0788	.9208	.3171	.1337

(Table 3.2.2 (b) continued on next page.)

(Table 3.2.2 (b) cont'd)

n \ P_S	.005	.01	.02	.03	.04	.05	.10	max $P_S(C)$ = $P_S(0)$
42	2.1132	1.8231	1.4858	1.2531	1.0645	.9061	.2776	.1282
45	2.1150	1.8231	1.4811	1.2430	1.0492	.8815	.2377	.1231
48	2.1163	1.8224	1.4754	1.2318	1.0329	.8606	.1976	.1184
51	2.1172	1.8211	1.4689	1.2199	1.0160	.8391	.1574	.1141
54	2.1176	1.8193	1.4618	1.2074	.9986	.8171	.1172	.1101
57	2.1177	1.8170	1.4541	1.1944	.9807	.7948	.0771	.1064
60	2.1175	1.8143	1.4460	1.1809	.9624	.7723	.0371	.1030
63	2.1170	1.8113	1.4375	1.1671	.9440	.7496		.0998
66	2.1163	1.8080	1.4286	1.1530	.9253	.7267		.0968
69	2.1153	1.8044	1.4194	1.1387	.9064	.7038		.0940
72	2.1140	1.8005	1.4100	1.1242	.8875	.6809		.0913
75	2.1126	1.7967	1.4003	1.1095	.8684	.6579		.0889
78	2.1109	1.7926	1.3905	1.0947	.8493	.6350		.0865
81	2.1091	1.7884	1.3805	1.0799	.8302	.6120		.0843
84	2.1071	1.7839	1.3703	1.0649	.8111	.5892		.0822
87	2.1050	1.7793	1.3601	1.0499	.7920	.5664		.0803
90	2.1027	1.7745	1.3497	1.0349	.7729	.5437		.0784
93	2.1002	1.7696	1.3393	1.0198	.7538	.5210		.0766
96	2.0986	1.7646	1.3287	1.0047	.7348	.4985		.0749
99	2.0969	1.7595	1.3182	.9896	.7158	.4760		.0733

Table 3.2.2 (c)

d = 3; k = 4

(See Table 3.1.3 (c))

n \ P_S	.005	.01	.02	.03	.04	.05	.10	max $P_S(C)$ = $P_S(0)$
12	2.6855	2.3057	1.3221	1.0044	.8486	.7324	.3420	.1647
16	2.4007	1.8883	1.3258	1.1041	.9537	.8334	.4069	.1689
20	2.2102	1.7615	1.3667	1.1582	1.0050	.8786	.4178	.1637
24	2.1266	1.7457	1.3955	1.1879	1.0293	.8963	.4026	.1558
28	2.0906	1.7499	1.4148	1.2031	1.0381	.8982	.3731	.1475
32	2.0821	1.7606	1.4262	1.2090	1.0371	.8903	.3351	.1395
36	2.0796	1.7682	1.4320	1.2084	1.0295	.8760	.2920	.1321
40	2.0798	1.7734	1.4336	1.2032	1.0174	.8573	.2456	.1253
44	2.0811	1.7768	1.4322	1.1947	1.0020	.8355	.1973	.1192
48	2.0825	1.7787	1.4284	1.1838	.9843	.8115	.1476	.1136
52	2.0838	1.7794	1.4227	1.1709	.9648	.7860	.0973	.1085
56	2.0848	1.7790	1.4154	1.1565	.9440	.7593	.0466	.1039
60	2.0855	1.7776	1.4070	1.1411	.9222	.7317		.0997
64	2.0858	1.7754	1.3976	1.1247	.8996	.7036		.0958
68	2.0858	1.7725	1.3873	1.1076	.8765	.6750		.0922
72	2.0855	1.7691	1.3765	1.0900	.8530	.6461		.0889
76	2.0849	1.7650	1.3650	1.0720	.8291	.6171		.0858
80	2.0839	1.7606	1.3531	1.0536	.8051	.5879		.0830
84	2.0827	1.7557	1.3409	1.0350	.7809	.5587		.0804
88	2.0813	1.7505	1.3283	1.0161	.7565	.5296		.0779
92	2.0795	1.7451	1.3155	.9971	.7322	.5004		.0756
96	2.0776	1.7394	1.3025	.9780	.7078	.4714		.0734
100	2.0754	1.7334	1.2892	.9588	.6834	.4424		.0714
128	2.0555	1.6867	1.1940	.8241	.5149	.2437		.0600

Table 3.2.3

Values of C for the designs of Table 3.1.4

Design (a)

Replicates	n \ $P_{\underset{\sim}{S}}$	.005	.01	.02	.03	.04	.05	.10	$\max P_{\underset{\sim}{S}}(C) = P_{\underset{\sim}{S}}(0)$
r = 2	26	2.4475	1.7912	1.2540	1.0249	.8552	.7141	.1948	.1241
3	39	2.0859	1.6768	1.3100	1.0744	.8868	.7261	.1163	.1117

Design (b)

Replicates	n \ $P_{\underset{\sim}{S}}$	.005	.01	.02	.03	.04	.05	$\max P_{\underset{\sim}{S}}(C) = P_{\underset{\sim}{S}}(0)$
r = 1	20	2.5330	1.4689	1.0485	.8013	.6070	.4416	.0834
2	40	1.9198	1.5189	1.1245	.8505	.6272	.4339	.0770

Design (c)

Replicates	n \ $P_{\underset{\sim}{S}}$	.005	.01	.02	.03	.04	$\max P_{\underset{\sim}{S}}(C) = P_{\underset{\sim}{S}}(0)$
r = 2	64	2.7485	1.1629	.7219	.4290	.1931	.0495
3	96	1.6751	1.2785	.8148	.4802	.2043	.0484

Table 3.2.4

Values of $-T(C)$. For each n , the table gives values of $-T(C)$. The values of C are not listed but they range from 0 to 3.8 in steps of 0.2 as one proceeds along the rows of the Table.

n = 15	0.00000	.05469	.10938	.16407	.21874
	.27305	.32470	.36619	.38499	.37119
	.32599	.26154	.19328	.13286	.08569
	.05223	.03024	.01668	.00879	.00443
n = 16	0.00000	.05193	.10386	.15579	.20771
	.25934	.30908	.35035	.37129	.36126
	.31994	.25839	.19185	.13228	.08549
	.05216	.03022	.01668	.00879	.00443
n = 17	0.00000	.04945	.09890	.14836	.19780
	.24710	.29489	.33572	.35839	.35174
	.31405	.25529	.19044	.13171	.08528
	.05210	.03020	.01667	.00879	.00443
n = 18	0.00000	.04722	.09443	.14165	.18886
	.23597	.28196	.32219	.34624	.34261
	.30832	.25224	.18904	.13115	.08508
	.05203	.03018	.01667	.00879	.00443

n = 19	0.00000	.04518	.09037	.13555	.18073
	.22585	.27013	.30966	.33479	.33386
	.30275	.24925	.18765	.13058	.08487
	.05196	.03016	.01666	.00879	.00443
n = 20	0.00000	.04333	.08666	.12999	.17332
	.21661	.25927	.29802	.32397	.32546
	.29733	.24630	.18628	.13002	.08467
	.05190	.03014	.01665	.00879	.00443
n = 21	0.00000	.04163	.08327	.12490	.16654
	.20814	.24928	.28719	.31376	.31740
	.29206	.24341	.18492	.12947	.08447
	.05183	.03012	.01665	.00878	.00443
n = 22	0.00000	.04007	.08015	.12022	.16029
	.20034	.24005	.27709	.30411	.30966
	.28693	.24057	.18358	.12892	.08426
	.05176	.03010	.01664	.00878	.00443
n = 23	0.00000	.03863	.07726	.11589	.15453
	.19314	.23150	.26767	.29497	.30223
	.28194	.23778	.18225	.12837	.08406
	.05170	.03008	.01664	.00878	.00443
n = 24	0.00000	.03730	.07459	.11189	.14919
	.18647	.22357	.25886	.28632	.29509
	.27708	.23503	.18094	.12782	.08386
	.05163	.03006	.01663	.00878	.00443
n = 28	0.00000	.03281	.06562	.09843	.13125
	.16406	.19681	.22873	.25593	.26916
	.25889	.22451	.17581	.12567	.08307
	.05137	.02998	.01661	.00877	.00443
n = 32	0.00000	.02934	.05868	.08802	.11735
	.14669	.17601	.20496	.23107	.24690
	.24253	.21468	.17089	.12356	.08228
	.05110	.02990	.01659	.00877	.00443
n = 40	0.00000	.02429	.04859	.07288	.09717
	.12147	.14576	.16997	.19318	.21098
	.21443	.19688	.16163	.11951	.08074
	.05059	.02974	.01655	.00876	.00443
n = 48	0.00000	.02079	.04158	.06238	.08317
	.10396	.12475	.14553	.16595	.18358
	.19136	.18126	.15310	.11563	.07924
	.05008	.02959	.01650	.00875	.00442
n = 56	0.00000	.01821	.03642	.05463	.07285
	.09106	.10927	.12748	.14556	.16221
	.17224	.16752	.14522	.11194	.07778
	.04957	.02943	.01646	.00874	.00442

n = 64	0.00000	.01623	.03245	.04868	.06490
	.08113	.09735	.11358	.12976	.14521
	.15626	.15538	.13793	.10842	.07635
	.04908	.02928	.01642	.00873	.00442
n = 72	0.00000	.01465	.02930	.04394	.05859
	.07324	.08789	.10253	.11717	.13143
	.14279	.14462	.13120	.10505	.07497
	.04895	.02913	.01638	.00872	.00442
n = 80	0.00000	.01336	.02672	.04009	.05345
	.06681	.08017	.09353	.10689	.12006
	.13132	.13506	.12495	.10184	.07361
	.04810	.02898	.01633	.00870	.00441
n = 88	0.00000	.01229	.02459	.03688	.04917
	.06146	.07376	.08605	.09834	.11054
	.12148	.12654	.11916	.09877	.07230
	.04763	.02883	.01629	.00869	.00441
n = 96	0.00000	.01139	.02278	.03417	.04556
	.05695	.06834	.07973	.09111	.10245
	.11298	.11890	.11379	.09583	.07101
	.04716	.02868	.01625	.00868	.00441
n = 128	0.00000	.00884	.01767	.02651	.03534
	.04418	.05302	.06185	.07069	.07952
	.08820	.09523	.08577	.08532	.06619
	.04534	.02809	.01608	.00864	.00440

$$\hat{\underset{\sim}{\theta}}_S = X^+(-.480,\ldots,-1.521,1.0644,-.290,\ldots,.291)' \\ = (.159,.392,.509,.224)' \tag{3.32}$$

(as compared with $\hat{\underset{\sim}{\theta}} = (.248,.303,.509,.224)'$.)

(iii) Premium of the W-rule.

Again, without loss of generality, we put the known value of $\sigma^2 = 1$. From (2.8) - (2.8a), we have that $\underset{\approx}{W} = \underset{\approx}{W}(\underset{\approx}{\omega}) = \underset{\approx}{W}(\underset{\approx}{z},\underset{\approx}{u})$, so that $\underset{\approx}{W}$ is independent of $\hat{\underset{\approx}{\theta}}$ in the null case. As $\hat{\underset{\approx}{\theta}}_{\underset{\approx}{W}} = \hat{\underset{\approx}{\theta}} + \underset{\approx}{W}$, we have that

$$V(\hat{\underset{\sim}{\theta}}_{\underset{\sim}{W}}) = V(\hat{\underset{\sim}{\theta}}) + V(\underset{\sim}{W}) = (X'X)^{-1} + E(\underset{\sim}{W}\underset{\sim}{W}') \tag{3.33}$$

since, as remarked earlier, $E(\underset{\sim}{W}) = \underset{\sim}{0}$. This means that the premium of the $\underset{\sim}{W}$-rule may be written as

$$P_{\underset{\sim}{W}} = \mathrm{tr}\, E(\underset{\sim}{W}\underset{\sim}{W}')/\mathrm{tr}(X'X)^{-1}, \tag{3.34}$$

and it remains to determine $\operatorname{tr} E(\underset{\sim}{W}\underset{\sim}{W}')$. From (2.8a) we have that

$$E(\underset{\sim}{W}\underset{\sim}{W}') = \sum_{j\neq i}^{n} \sum_{i=1}^{n} X^{+}\mathcal{E}_{i,j}X^{+'} \tag{3.35}$$

where the $(n \times n)$ matrix $\mathcal{E}_{i,j}$ is a matrix of zeroes, except for its i-th diagonal element e^{j}_{ii} which is given by

$$e^{j}_{ii} = \left\{ \int_{R^{+}_{ij}} + \int_{R^{-}_{ij}} \right\} ([\underset{\sim}{m}'_{j}. - \underset{\sim}{m}'_{i}.]\underset{\sim}{\omega})^{2} [\prod_{t=1}^{n} \phi(\omega_t)] [\prod_{t\neq j} d\omega_t] d\omega_i \tag{3.36}$$

with

$$R^{+}_{ij} = \{\underset{\sim}{\omega} | \omega_i > C;\ -\omega_i < \omega_t < \omega_j < \omega_i\ ,\ \text{all}\ \ t \neq j,i\}$$

and (3.36a)

$$R^{-}_{ij} = \{\underset{\sim}{\omega} | \omega_i < -C;\ \omega_i < \omega_j < \omega_t < -\omega_i\ ,\ \text{all}\ \ t \neq j,i\}$$

Hence

$$\operatorname{tr} E(\underset{\sim}{W}\underset{\sim}{W}') = \sum_{j\neq i}^{n} \sum_{i=1}^{n} \operatorname{tr} X^{+'}X^{+}\mathcal{E}_{ij} \tag{3.37}$$

when divided by $\operatorname{tr}(X'X)^{-1}$ gives the premium of the W-rule $P_{\underset{\sim}{W}}(C)$, that is

$$P_{\underset{\sim}{W}}(C) = \sum_{j\neq i} \sum_{i} \operatorname{tr} X^{+'}X^{+}\mathcal{E}_{ij}/\operatorname{tr}(X'X)^{-1} \tag{3.38}$$

The numerator of (3.38) may be further simplified due to the fact that $\mathcal{E}_{ij}$ is a matrix of zeroes except for the i-th diagonal element which has value e^{j}_{ii}. In fact it is easy to see that

$$P_{\underset{\sim}{W}}(C) = \sum_{j\neq i} \sum_{i} e^{j}_{ii} \sum_{r=1}^{k} x^{+2}_{ri}/\operatorname{tr}(X'X)^{-1} \tag{3.39}$$

where x^{+}_{ri} is the r-th element of the i-th column of X^{+}, $r = 1,\ldots,k$, $i = 1,\ldots,n$. It is interesting to note that if $E(\underset{\sim}{y}) = \underset{\sim}{1}\mu$, $V(\underset{\sim}{y}) = I$ and $\underset{\sim}{y} = N(\underset{\sim}{1}\mu, I)$, that is, $k = 1$, $X = \underset{\sim}{1}$ etc., (3.33) reduces to the formula for P_W given in Guttman(1971). We give C's necessary to give the W-rule (2.8) - (2.8a) the usual values of the premium for various n in Tables 3.3.1 - 3.3.3 for our small catalogue.

As an example, we again return to the data of Table 3.1.1. There $|\omega_s| > |\omega_j|$, all $j \neq s$ and $-\omega_5 < \omega_t < \omega_{15} < \omega_5$, all $t \neq 5,15$. Now $\omega_5 = 2.771 > C = 2.5573(P_{\underset{\sim}{\omega}} = .01)$, (from Table 3.3.1), so that the W-rule calls for the estimate

$$\hat{\underset{\sim}{\theta}}_{\underset{\sim}{W}} = \hat{\underset{\sim}{\theta}} + \underset{\sim}{W} = (.196,\ .355,\ .456,\ .277)' \ , \tag{3.40}$$

since we may write $\underset{\sim}{W} = X^{+}(0,0,0,0,z_{15}-z_5,0,\ldots,0)' = X^{+}(0,0,0,0,-.844,0,\ldots,0)'$, and $X^{+} = (X'X)^{-1}X' = \frac{1}{16}X'$ (see below (3.20) and Table 3.1.1).

4. The Protections of the rules - σ known; X full rank; mean-shift spurosity.

Throughout this section, we assume that the following specific non-null situation obtains: there is present in our sample of n independent observations, a spurious observation (we don't know beforehand which it will be) of the shifted mean type, that is

$$y_t = N(\underset{\sim}{x}_t'.\underset{\sim}{\theta} + a\sigma;\sigma^2);\ y_i = N(\underset{\sim}{x}_i'.\underset{\sim}{\theta};\sigma)\ ,\ \text{all}\quad i \neq t\ , \tag{4.1}$$

Table 3.3.1

Values of C corresponding to various values of P_W for a 2^3 factorial design with r replicates $(k = 4)$.

r	n \ $P_{\underset{\sim}{W}}$	.005	.01	$\max P_{\underset{\approx}{W}}(C) = P_{\underset{\approx}{W}}(\tilde{0})$
2	16	2.9339	2.5573	.0177
3	24	2.9029	2.4855	.0147
4	32	2.9002	2.3553	.0118
5	40	2.8870		.0096
6	48	2.8279		.0079
7	56	2.6569		.0066
8	64	2.3849		.0056
9	72			.0048
10	80			.0042

Table 3.3.2

Values of C corresponding to various values of $P_{\underset{\approx}{W}}$ for optimal regression designs in the interval $\{x|1- \leq x \leq 1\}$ for fitting the polynomial of degree $d(d = k-1)$ (a) $d = 1;\ k = 2$ (See Table 3.1.3 (a)).

(Table 3.3.2 continued on next page.)

(Table 3.3.2 cont'd)

n \ $P_{\underset{\sim}{W}}$	.005	.01	.02	.03	.04	.05	max $P_{\underset{\sim}{W}}(C) = P_{\underset{\sim}{W}}(\bar{0})$
6	3.2127	2.9393	2.6193	2.3946	2.2050	2.0259	.0677
8	3.2034	2.9200	2.5809	2.3340	2.1123	1.8758	.0595
10	3.1805	2.8852	2.5239	2.2456	1.9635		.0497
12	3.1485	2.8429	2.4570	2.1324	1.6686		.0416
14	3.1143	2.7990	2.3852	1.9793			.0354
16	3.0809	2.7582	2.3098	1.6571			.0306
18	3.0510	2.7222	2.2278				.0270
20	3.0278	2.6913	2.1293				.0241
22	3.0097	2.6649	1.9840				.0217
24	2.9967	2.6412					.0198
26	2.9879	2.6184					.0182
28	2.9818	2.5948					.0168
30	2.9774	2.5682					.0156
32	2.9739	2.5366					.0145
34	2.9705	2.4974					.0136
36	2.9667	2.4463					.0127
38	2.9618	2.3778					.0119
40	2.9556	2.2937					.0112
42	2.9475	2.1442					.0106
44	2.9370						.0100
46	2.9234						.0094
48	2.9061						.0090
50	2.8838						.0085
52	2.8556						.0081
54	2.8189						.0077
56	2.7723						.0073
58	2.7142						.0070
60	2.6470						.0067
62	2.5776						.0064
64	2.5115						.0061
66	2.4500						.0058
68	2.3900						.0056
70	2.3258						.0054
72	2.2453						.0052
74							.0050
76							.0048
78							.0046

Table 3.3.2 (b)

$d = 2,\ k = 3$

(See Table 3.1.3 (b))

n \ $P_{\underset{\sim}{W}}$	.005	.01	.02	max $P_{\underset{\sim}{W}}(C) = P_{\underset{\sim}{W}}(\bar{0})$
9	3.0649	2.7461	2.8266	.0286
12	3.0562	2.7272	2.2722	.0282
15	3.0242	2.6837	2.1665	.0249
18	2.9911	2.6404	2.0023	.0218
21	2.9670	2.6036		.0192
24	2.9521	2.5712		.0172
27	2.9446	2.5374		.0155

(Table 3.3.2 (b) continued on next page.)

(Table 3.3.2 (b) cont'd)

n \ P_W	.005	.01	.02	max $P_W(C)$ = $P_W(\tilde{0})$
30	2.9407	2.4956		.0140
33	2.9377	2.4362		.0128
36	2.9334	2.3445		.0117
39	2.9263	2.1845		.0107
42	2.9145			.0099
45	2.8964			.0091
48	2.8692			.0084
51	2.8293			.0078
54	2.7708			.0073
57	2.6877			.0068
60	2.5856			.0064
63	2.4841			.0060
66	2.3890			.0056
69	2.2840			.0053
72				.0050
75				.0047
78				.0045

Table 3.3.2 (c)

$d = 3$, $k = 4$

(See Table 3.1.3 (c))

n \ P_W	.005	.01	max $P_W(C)$ = $P_W(\tilde{0})$
16	2.9339	2.5573	.0177
20	2.9135	2.5223	.0163
24	2.9029	2.4855	.0147
28	2.9005	2.4370	.0132
32	2.9002	2.3553	.0118
36	2.8971	2.1848	.0107
40	2.8870		.0096
44	2.8659		.0087
48	2.8279		.0079
52	2.7629		.0072
56	2.6569		.0066
60	2.5194		.0061
64	2.3849		.0056
68	2.2255		.0052
72			.0048
76			.0045
80			.0042

Table 3.3.3

Values of C for the designs of Table 3.1.4.

Design (a)

Replicates	n \ P_W	.005	.01	max $P_W(C)$ = $P_W(\tilde{0})$
r = 2	26	2.8158	2.3151	.0122
3	39	2.8061		.0087

Design (b)

Replicates	n^{P_W}	.005	.01	max $P_W(C)$ = $P_W(0)$
r = 1	20	2.7151	2.3355	.0172
2	40	2.5601		.0066

Design (c)

Replicates	n^{P_W}	.005	.01	.02	max $P_W(C)$ = $P_W(0)$
r = 1	32	2.8210	2.4875	1.7457	.0205

$$\underset{\sim}{y} = N(X\underset{\sim}{\theta} + a\sigma \underset{\sim}{e}_t; \sigma^2 I) \ . \qquad (4.1a)$$

We note that when the non-null situation (4.1) obtains, the least squares estimator $\hat{\underset{\sim}{\theta}} = X^+\underset{\sim}{y}$ is no longer unbiased for $\underset{\sim}{\theta}$, for, as is easily verified, we have that

$$E(\hat{\underset{\sim}{\theta}}) = \underset{\sim}{\theta} + \underset{\sim}{\tau}_t, \text{ where } \underset{\sim}{\tau}_t = a\sigma(X'X)^{-1}X'\underset{\sim}{e}_t = a\sigma(X'X)^{-1}\underset{\sim}{x}_t. \quad (4.2)$$

It is also true that $\hat{\underset{\sim}{\theta}}_A$, $\hat{\underset{\sim}{\theta}}_S$ and $\hat{\underset{\sim}{\theta}}_W$ will no longer be unbiased for $\underset{\sim}{\theta}$ when (4.1) obtains. Hence, we will use as an overall measure of spread of any estimator, say $\hat{\underset{\sim}{\theta}}_R$, the trace of its mean square error matrix, $MS\mathcal{E}(\hat{\underset{\sim}{\theta}}_R)$, given by

$$\text{tr } MS\mathcal{E}(\hat{\underset{\sim}{\theta}}_R) = \text{tr } E[\hat{\underset{\sim}{\theta}}_R - \underset{\sim}{\theta})(\hat{\underset{\sim}{\theta}}_R - \underset{\sim}{\theta})'] \qquad (4.3)$$

At this point, we may now ask for the protection that is afforded when we use a rule, say R, with its intendent estimator, say $\hat{\underset{\sim}{\theta}}_R$, when indeed there is a spurious observation present. The protection is defined as the gain in using a rule R when spurosity exists, and in view of (3.4), and following Anscombe (1960), Tiao and Guttman (1967), Guttman and Smith (1969, 1971), and Guttman (1971), we define protection of a rule R as

$$\text{Prot}_{\underset{\sim}{R}} = [\text{tr } MS\mathcal{E}(\hat{\underset{\sim}{\theta}}) - \text{tr } MS\mathcal{E}(\hat{\underset{\sim}{\theta}}_R)]/\text{tr } MS\mathcal{E}(\hat{\underset{\sim}{\theta}}) \qquad (4.4)$$

where expectations are taken with respect to the non-null distributions of $\hat{\underset{\sim}{\theta}}$ and $\hat{\underset{\sim}{\theta}}_R$ induced by the non-null situation, in our case described by (4.1). (Note that when $a = 0$, and if $E(\hat{\underset{\sim}{\theta}}_R) = \underset{\sim}{\theta}$

when the null-situation obtains, as is the case with $\hat{\theta}_{\underset{\sim}{A}}$, $\hat{\theta}_{\underset{\sim}{W}}$, $\hat{\theta}_{\underset{\sim}{S}}$, then $\text{Prot}_{\underset{\sim}{R}}$ reduces to

$$\text{Prot}_{\underset{\sim}{R}} = -[\text{tr } V(\hat{\theta}_{\underset{\sim}{R}}) - \text{tr } V(\hat{\theta}_{\underset{\sim}{}})]/\text{tr } V(\hat{\theta}_{\underset{\sim}{}}) = -P_{\underset{\sim}{R}} \qquad (4.5)$$

that is, the protection when the null situation obtains is minus the premium, as should be, for then we are paying a premium for "no fires", that is, no spurious observations.)

We will need $\text{tr } MS\mathcal{E}(\hat{\theta}_{\underset{\sim}{}})$. We have, given (4.1a), that

$$MS\mathcal{E}(\hat{\theta}_{\underset{\sim}{}}) = E[(\hat{\theta}_{\underset{\sim}{}} - \theta_{\underset{\sim}{}} - \tau_{\underset{\sim}{t}} + \tau_{\underset{\sim}{t}})(\hat{\theta}_{\underset{\sim}{}} - \theta_{\underset{\sim}{}} - \tau_{\underset{\sim}{t}} + \tau_{\underset{\sim}{t}})']$$

$$= E[(\hat{\theta}_{\underset{\sim}{}} - \theta_{\underset{\sim}{}} - \tau_{\underset{\sim}{t}})(\hat{\theta}_{\underset{\sim}{}} - \theta_{\underset{\sim}{}} - \tau_{\underset{\sim}{t}})'] + \tau_{\underset{\sim}{t}}\tau_{\underset{\sim}{t}}' \qquad (4.6)$$

$$= V(\hat{\theta}_{\underset{\sim}{}}) + \tau_{\underset{\sim}{t}}\tau_{\underset{\sim}{t}}' ,$$

since, from (4.2), $E(\hat{\theta}_{\underset{\sim}{}} - \theta_{\underset{\sim}{}} - \tau_{\underset{\sim}{t}}) = 0$. Hence

$$\text{tr } MS\mathcal{E}(\hat{\theta}_{\underset{\sim}{}}) = \sigma^2 \text{tr}(X'X)^{-1} + \tau_{\underset{\sim}{t}}'\tau_{\underset{\sim}{t}} \qquad (4.7)$$

We now investigate the protections for the rules of section 2.

(i)   Protection of the A-rule.

Without loss of generality, we put $\sigma^2 = 1$. Now, under the assumption (4.1), we have that

$$MS\mathcal{E}(\hat{\theta}_{\underset{\sim}{A}}) = E[(\hat{\theta}_{\underset{\sim}{A}} - \theta_{\underset{\sim}{}})(\hat{\theta}_{\underset{\sim}{A}} - \theta_{\underset{\sim}{}})'] = E[(\hat{\theta}_{\underset{\sim}{}} + A_{\underset{\sim}{}} - \theta_{\underset{\sim}{}})(\hat{\theta}_{\underset{\sim}{}} + A_{\underset{\sim}{}} - \theta_{\underset{\sim}{}})'] , \qquad (4.8)$$

which may be written as

$$MS\mathcal{E}(\hat{\theta}_{\underset{\sim}{A}}) = E[\{(\hat{\theta}_{\underset{\sim}{}} - \theta_{\underset{\sim}{}} - \tau_{\underset{\sim}{t}}) + (A_{\underset{\sim}{}} + \tau_{\underset{\sim}{t}})\}\{(\hat{\theta}_{\underset{\sim}{}} - \theta_{\underset{\sim}{}} - \tau_{\underset{\sim}{t}}) + (A_{\underset{\sim}{}} + \tau_{\underset{\sim}{t}})\}']$$

$$(4.8a)$$

$$= V(\hat{\theta}_{\underset{\sim}{}}) + E[A_{\underset{\sim}{}} + \tau_{\underset{\sim}{t}})(A_{\underset{\sim}{}} + \tau_{\underset{\sim}{t}})'] ,$$

since the cross product vanishes because $A_{\underset{\sim}{}}$ is independent of $\hat{\theta}_{\underset{\sim}{}}$ and $E(\hat{\theta}_{\underset{\sim}{}} - \theta_{\underset{\sim}{}} - \tau_{\underset{\sim}{t}}) = 0$. (Here, $\tau_{\underset{\sim}{t}} = aX^+e_{\underset{\sim}{t}} = a(X'X)^{-1}x_{\underset{\sim}{t.}}$, since $\sigma = 1$.) Hence,

$$\text{tr } MS\mathcal{E}(\hat{\theta}_{\underset{\sim}{A}}) = \text{tr}(X'X)^{-1} + E[(A_{\underset{\sim}{}} + \tau_{\underset{\sim}{t}})'(A_{\underset{\sim}{}} + \tau_{\underset{\sim}{t}})]$$

$$(4.9)$$

$$= \sum_{i=1}^{k} C_{ii} + E[(\underset{\sim}{A} + \underset{\sim}{\tau}_t)'(\underset{\sim}{A} + \underset{\sim}{\tau}_t)] ,$$

where C_{ii} = (i-i)th element of $(X'X)^{-1}$. Now from (4.7), with $\sigma = 1$, we have

$$\text{tr } MS\mathcal{E}(\hat{\underset{\sim}{\theta}}) = \sum_{i=1}^{k} C_{ii} + \underset{\sim}{\tau}_t'\underset{\sim}{\tau}_t,\ \underset{\sim}{\tau}_t = a(X'X)^{-1}\underset{\sim}{x}_{i\cdot} \tag{4.10}$$

Substituting (4.9) and (4.10) in (4.4), we have that the protection afforded by use of the A-rule, when the jth observation is spurious of the shifted mean type, is given by

$$\text{Prot.}_{t,\underset{\approx}{A}} = [\underset{\sim}{\tau}_t'\underset{\sim}{\tau}_t - E(\underset{\sim}{A}+\underset{\sim}{\tau}_t)'(\underset{\sim}{A}+\underset{\sim}{\tau}_t)]/[\sum_{i=1}^{k} C_{ii} + \underset{\sim}{\tau}_t'\underset{\sim}{\tau}_t] \tag{4.11}$$

It remains to find a formula for $E(\underset{\sim}{A} + \underset{\sim}{\tau}_t)'(\underset{\sim}{A} + \underset{\sim}{\tau}_t)$. For this, we need the distribution of $\underset{\sim}{\omega}$ when (4.1) obtains, where $\underset{\sim}{\omega} = \underset{\sim}{z} + X\underset{\sim}{Pu}$. As $\underset{\sim}{z} = M\underset{\sim}{y}$, we have that

$$E(\underset{\sim}{z}) = M[X\underset{\sim}{\theta} + a\underset{\sim}{e}_t] = aM\underset{\sim}{e}_t = a\underset{\sim}{m}_{t\cdot} , \tag{4.12}$$

since $MX = 0$, and $M\underset{\sim}{e}_t = \underset{\sim}{m}_{\cdot t}$, the t-th column of M . But since M is symmetric, $\underset{\sim}{m}_{\cdot t} = \underset{\sim}{m}_{t\cdot}$, where $\underset{\sim}{m}_{t\cdot}'$ = t-th row of M . Further, $V(\underset{\sim}{z}) = MIM' = M$, since M is symmetric and idempotent. As $\underset{\sim}{y} = N(X\underset{\sim}{\theta} + a\underset{\sim}{e}_t, I)$, we thus have

$$\underset{\sim}{z} = N(a\underset{\sim}{m}_{t\cdot}, M) . \tag{4.13}$$

Because $\underset{\sim}{u} = N(\underset{\sim}{0}, I)$ is independent of $\underset{\sim}{y}$ and hence $\underset{\sim}{z}$, we have then that

$$\underset{\sim}{\omega} = N(a\underset{\sim}{m}_{t\cdot}, I) , \tag{4.14}$$

that is, the ω_r are independent $N(am_{tr}, 1)$ variables, $r=1,\ldots,n$.

Recalling the definitions of $\underset{\sim}{A}$, R_i^+ and R_i^- given in (2.2) with $\sigma = 1$, (3.8a) and (3.20a), respectively, then it is easy to see that

$$E(\underset{\sim}{A}+\underset{\sim}{\tau}_t)'(\underset{\sim}{A}+\underset{\sim}{\tau}_t) = \underset{\sim}{\tau}_t'\underset{\sim}{\tau}_t \int_{-C}^{C}\cdots\int_{-C}^{C} [\prod_{r=1}^{n} \phi(\omega_r - am_{tr})] \prod_{r=1}^{n} d\omega_r$$

$$+ \sum_{i=1}^{n} \left\{\int_{R_i^+} + \int_{R_i^-}\right\}(d_i\underset{\sim}{m}_{i\cdot}'\underset{\sim}{\omega}-\underset{\sim}{\tau}_t)'(d_i\underset{\sim}{m}_{i\cdot}'\underset{\sim}{\omega}-\underset{\sim}{\tau}_t)[\prod_{r=1}^{n} \phi(\omega_r - am_{tr})][\prod_{r\neq i} d\omega_r]d\omega_i \tag{4.15}$$

Thus, the numerator of $\mathrm{Prot}_{t,A}$ in (4.11) may be written as

$$\underset{\sim}{\tau}_t'\underset{\sim}{\tau}_t[1-\prod_{r=1}^{n} b(C;am_{tr})]-\sum_{i=1}^{n}\left\{\int_{R_i^+}+\int_{R_i^-}\right\}(\underset{\sim}{d}_i\underset{\sim}{m}_i'.\underset{\sim}{\omega}-\underset{\sim}{\tau}_t)'(\underset{\sim}{d}_i\underset{\sim}{m}_i'.\underset{\sim}{\omega}-\underset{\sim}{\tau}_t)$$

$$[\prod_{r=1}^{n}\phi(\omega_r-am_{tr})][\prod_{r\neq i}d\omega_r]d\omega_i\ , \qquad (4.16)$$

where

$$b(C,d) = \Phi(C+d)+\Phi(C-d)-1 \qquad (4.16a)$$

has the property that

$$b(C,0) = 2\Phi(C)-1 = b(C)\ . \qquad (4.16b)$$

Dividing (4.16) by $[\sum_{i=1}^{k} C_{ii}+\underset{\sim}{\tau}_t'\underset{\sim}{\tau}_t]$ yields the protection afforded by the A-rule, when y_t is spurious. Of course, (4.16) is capable of further development, but since different X matrices yield different M and $\underset{\sim}{d}_i$, we leave (4.16) in its present form. Actual calculation for the small catalogue of the designs mentioned in Section 3, as well as curves of the protection as functions of "a", for the 5 and 1% premium A-rules, are underway and will be given in a later paper.

(ii) <u>Protection of the S-rule</u>.

Again, with $\sigma^2 = 1$, and following the development of the previous subsection, we are quickly led to the protection of the S-rule, when the t-th observation is as in (4.1), viz

$$\mathrm{Prot}_{t,\underset{\sim}{S}} = [\underset{\sim}{\tau}_t'\underset{\sim}{\tau}_t-E(\underset{\sim}{S}+\underset{\sim}{\tau}_t)'(\underset{\sim}{S}+\underset{\sim}{\tau}_t)]/[\sum_{i=1}^{k} C_{ii}+\underset{\sim}{\tau}_t'\underset{\sim}{\tau}_t]\ . \qquad (4.17)$$

It remains to find a convenient formula for $E[(\underset{\sim}{S}+\underset{\sim}{\tau}_t)'(\underset{\sim}{S}+\underset{\sim}{\tau}_t)]$. But from the definition of $\underset{\sim}{S}$ given in (2.5a), with $\sigma = 1$, we have, on using (4.14), that

$$E[(\underset{\sim}{S}+\underset{\sim}{\tau}_t)'(\underset{\sim}{S}+\underset{\sim}{\tau}_t)] = \underset{\sim}{\tau}_t'\underset{\sim}{\tau}_t[\prod_{r=1}^{n} b(C;am_{tr})] \qquad (4.18)$$

$$+\sum_{i=1}^{n}\int_{R_i^+}[(X^+\hat{\underset{\sim}{\delta}}_i+\underset{\sim}{\tau}_t)'(X^+\hat{\underset{\sim}{\delta}}_i+\underset{\sim}{\tau}_t)][\prod_{r=1}^{n}\phi(\omega_r-am_{tr})][\prod_{r\neq i}d\omega_r]d\omega_i$$

$$+ \sum_{i=1}^{n} \int_{R_i^-} [X^+\hat{\underset{\sim}{\gamma}}_i + \underset{\sim}{\tau}_t)'(X^+\hat{\underset{\sim}{\gamma}}_i + \underset{\sim}{\tau}_t)][\prod_{r=1}^{n} \phi(\omega_r - am_{tr})][\prod_{r\neq i} d\omega_r]d\omega_i$$

Since $\hat{\underset{\sim}{\delta}}_i$ and $\hat{\underset{\sim}{\gamma}}_i$ are $(n \times 1)$ vectors of zeroes, except for their i^{th} components which are $(C - \underset{\sim}{m}_i'\underset{\sim}{\omega})$ and $(C + \underset{\sim}{m}_i'\underset{\sim}{\omega})$ respectively, we have that $X^+\hat{\underset{\sim}{\delta}}_i$ and $X^+\hat{\underset{\sim}{\gamma}}_i$ can be written as $(C - \underset{\sim}{m}_i'\underset{\sim}{\omega})\underset{\sim}{x}_{\cdot i}^+$ and $-(C + \underset{\sim}{m}_i'\underset{\sim}{\omega})\underset{\sim}{x}_{\cdot i}^+$, where $\underset{\sim}{x}_{\cdot i}^+$ is the i-th column of X^+. Hence, re-writing (4.18) to take account of that fact, and substituting in (4.17), we find that the numerator of (4.17) is

$$\underset{\sim}{\tau}_t'\underset{\sim}{\tau}_t[1-\prod_{r=1}^{n} b(C;am_{tr})] - \sum_{i=1}^{n} \int_{R_i^+} [((C-\underset{\sim}{m}_i'\underset{\sim}{\omega})^2)\underset{\sim}{x}_{\cdot i}^{+\prime}\underset{\sim}{x}_{\cdot i}^+) + 2\underset{\sim}{\tau}_t'\underset{\sim}{x}_{\cdot i}^+(C-\underset{\sim}{m}_i'\underset{\sim}{\omega})$$

$$+ \underset{\sim}{\tau}_t'\underset{\sim}{\tau}_t][\prod_{r=1}^{n} \phi(\omega_r - am_{tr})][\prod_{r\neq i} d\omega_r]d\omega_i - \sum_{i=1}^{n} \int_{R_i^-} [(C+\underset{\sim}{m}_i'\underset{\sim}{\omega})^2(\underset{\sim}{x}_{\cdot i}^{+\prime}\underset{\sim}{x}_{\cdot i}^+) - 2\underset{\sim}{\tau}_t'\underset{\sim}{x}_{\cdot i}^+$$

$$(C + \underset{\sim}{m}_i'\underset{\sim}{\omega}) + \underset{\sim}{\tau}_t'\underset{\sim}{\tau}_t][\prod_{r=1}^{n} \phi(\omega_r - am_{tr})][\prod_{r\neq i} d\omega_r]d\omega_i \; . \qquad (4.19)$$

The expression (4.19) is capable of further development, but we leave it in this form to be illustrated for the catalogue of section 3 in a later paper. Actual computations for this catalogue, of the protection afforded by the S-rule (obtained by dividing (4.19) by $[\sum_{i=1}^{k} C_{ii} + \underset{\sim}{\tau}_i'\underset{\sim}{\tau}_i])$ and graphs, as a function of a, are in preparation.

(iii) Protection of the W-rule.

When the non-null situation of (4.1) obtains, and if $\sigma^2 = 1$, we find first that the protection afforded by the W-rule when the t-th observation is spurious, may be written as

$$\text{Prot.}_{t,W} = \left\{\underset{\sim}{\tau}_t'\underset{\sim}{\tau}_t - E[(\underset{\sim}{W}+\underset{\sim}{\tau}_t)'(\underset{\sim}{W}+\underset{\sim}{\tau}_t)]\right\}/[\sum_{i=1}^{k} C_{ii} + \underset{\sim}{\tau}_t'\underset{\sim}{\tau}_t] \qquad (4.20)$$

From (2.8) - (2.8a), we see that we may write the numerator of (4.19) as

$$\underset{\sim}{\tau}_t'\underset{\sim}{\tau}_t[1-\prod_{r=1}^{n} b(C;am_{tr})] - \sum_{j\neq i}^{n} \sum_{i=1}^{n} \left\{\int_{R_{ij}^+} + \int_{R_{ij}^-}\right\} [(\underset{\sim}{\ell}_{ij}'\underset{\sim}{\omega})^2(\underset{\sim}{x}_{\cdot i}^{+\prime}\underset{\sim}{x}_{\cdot i}^+) + \qquad (4.21)$$

$$2(\underset{\sim}{\tau}_j'\underset{\sim}{x}_{\cdot i}^+)(\underset{\sim}{\ell}_{ij}'\underset{\sim}{\omega}) + \underset{\sim}{\tau}_j'\underset{\sim}{\tau}_j][\prod_{r=1}^{n} \phi(\omega_r - am_{tr})][\prod_{r\neq i,j} d\omega_r]d\omega_j d\omega_i \; ,$$

where $\underset{\sim}{\ell}'_{ij} = (\underset{\sim}{m}_{\cdot j} - \underset{\sim}{m}_{\cdot i})'$. The expression (4.21) is capable of further development, but we leave it in this form and illustrate its use in a paper now in preparation for our small catalogue of section 3. For this catalogue, we will also provide graphs of the protection afforded by the 5% and 1% premium W-rules. The protection is obtained by dividing (4.21) by $[\sum_{i=1}^{k} C_{ii} + \underset{\sim}{\tau}'_i\underset{\sim}{\tau}_i]$.

5. σ^2 unknown, X of full rank.

Because of "independence" considerations, we conduct the discussion of this section under the assumption, often met in practice that there exists an independent estimator of σ^2 , say s^2 , which is such that $s^2 = \sigma^2\chi^2_\nu/\nu$. We will also need adjusted residuals ω_i^{**} obtained as follows.

Firstly, suppose that after generating y_j , $j=1,\ldots,n$, where $E(\underset{\sim}{y})$ is thought to be $\underset{\sim}{X}$, we select a point in our factor space, say $\underset{\sim}{x}$ and generate a further $2k$ observations v_i , $i=1,\ldots,2k$, where the $(2k \times 1)$ vector $\underset{\sim}{v}$ is such that it is independent of $\underset{\sim}{y}$ and

$$\underset{\sim}{v} = N[(\underset{\sim}{x}'\underset{\sim}{\theta})\underset{\sim}{1},\ \sigma^2 I_{2k}] \tag{5.1}$$

Then, if we form the k differences $t_1 = (v_2-v_1)/\sqrt{2}$, $t_2 = (v_4-v_3)/\sqrt{2},\ldots,t_k = (v_{2k}-v_{2k-1})/\sqrt{2}$, we have that

$$\underset{\sim}{t} = N[\underset{\sim}{0},\sigma^2 I_k] \tag{5.2}$$

Now let us define new adjusted residuals given by

$$\underset{\sim}{\omega}^{**} = \underset{\sim}{z} + XP\ \underset{\sim}{t} \tag{5.3}$$

where here $\underset{\sim}{z} = M\underset{\sim}{y}$ is $N(\underset{\sim}{0},\sigma^2 M)$, and independent of $\underset{\sim}{t}$. Thus

$$\underset{\sim}{\omega}^{**} = N(\underset{\sim}{0},\sigma^2 I) \tag{5.4}$$

We incorporate these residuals into definitions of A*, S* and W*-rules by using the definitions (2.1) - (2.2), (2.5) - (2.5a), and (2.8 - 2.8a), with σ of these rules replaced by s , and double-astericking the ω's . Denoting the premiums of the A-rule, S-rule and W-rule by P_A, P_S and P_W respectively, which are given by (3.6) - (3.17), (3.22) - (3.30) and (3.34) - (3.39), respectively, and denoting $\underset{\sim}{A}$, $\underset{\sim}{S}$ and $\underset{\sim}{W}$ generally by $\underset{\sim}{R}$, we have that for given s, the premium of the A*, S* and W*-rules are (denoting the latter generally by R*) equal to $P_{R*}(Cs/\sigma)$. (The divisor σ is needed to make the distribution of $\underset{\sim}{\omega}^{**}/\sigma = N(\underset{\sim}{0}, I)$, which is the same as $\underset{\sim}{\omega}$, and makes the premiums of R*-rules, given s, of the same form as R-rules, except that C is replaced by Cs/σ.) Hence, the unconditional premium, that is the premium of the R* rules is given by

$$P_{\underset{\sim}{R}*}(C) = \int_0^\infty P_{\underset{\sim}{R}}(Cs/\sigma)\ p(s|\sigma)ds \tag{5.5}$$

where $p(s|\sigma)$ is the probability density function of s, which, using the fact that $s = \sigma\chi_v/\sqrt{v}$ is given by

$$p(s|\sigma) = v^{v/2}[\sigma^{v/2}2^{(v-2)/2}\Gamma(\tfrac{v}{2})]^{-1}s^{v-1}\exp[-vs^2/2\sigma^2] \tag{5.6}$$

if $s > o$, and zero otherwise. If we now make the transformation $s/\sigma = x\sqrt{v}$, in (5.5), we find

$$P_{\underset{\sim}{R}*}(C) = \int_0^\infty P_{\underset{\sim}{R}}(Cx/\sqrt{v})\ p(x)dx \tag{5.7}$$

and

$$p(x) = [2^{(v-2)/2}\Gamma(v/2)]^{-1}\ x^{v-1}e^{-x^2/2} \tag{5.7a}$$

In the same way, it is easy to see that

$$\text{Prot.}_{t,R*}(C;a) = \int_0^\infty \text{Prot.}_{t,R}(Cx/\sqrt{v};a)\ p(x)dx \tag{5.8}$$

We will illustrate (5.7) and (5.8) for the designs discussed in section 3, in a paper now in preparation. At this point we wish to point out that we may generate the v_i of (5.1) by carefully replicating at a "safe" selected point of the design (e.g. a suitable "center" point, etc.), and reserving the 2k observations made there to calculate $\underset{\approx}{\omega}^{**}$, put the rule we are using (i.e., one of the A*, S* or W* rules) into operation, with the first n observations made at the other points of the design used to calculate $\hat{\underset{\approx}{\theta}}$ or one of $\hat{\underset{\approx}{\theta}}_{A*}$, $\hat{\underset{\approx}{\theta}}_{S*}$, $\hat{\underset{\approx}{\theta}}_{W*}$ etc.

References

1. Andrews, D. F., (1971), "Significance tests based on residuals," Biometrika, 58, 139.

2. Anscombe, F. G., (1960), "Rejection of outliers," Technometrics, 2, 123.

3. Box, G. E. P. and J. S. Hunter, (1957), "Multifactor experimental designs;" Annals of Mathematical Statistics, 28, 195.

4. Guttman, Irwin and Dennis E. Smith, (1969), "Investigation of rules for dealing with outliers in small samples from the normal distribution: I: Estimation of the Mean," Technometrics, 11, 527.

5. Guttman, Irwin, and Dennis E. Smith, (1971)," ________ II. Estimation of the variance," Technometrics, 13, 101.

6. Guttman, Irwin, (1971), "Premium and Protection of Several Procedures for dealing with outliers when sample sizes are moderate to large," Technical Report #88 of the Centre de Recherches Mathématiques, Université de Montréal; to be published in Technometrics, 1973, Volume 15.

7. Rand Corporation, (1955), "A Million Random Digits with 100,000 Normal Deviates;" The Free Press, New York.

8. Smith, K., (1918), "On the standard deviations of adjusted and interpolated values of an observed polynomial function and its constants, and the guidance they give towards a proper choice of the distribution of observations;" Biometrika, 12, 1.

9. Tiao, G. C. and Irwin Guttman, (1967), "Analysis of outliers with adjusted residuals," Technometrics, 9, 541.

Appendix

As we have mentioned in Section 2, we wish to prove in this appendix the following lemma.

Lemma: Using previous notation, if $X_i = \tilde{X} - \mathbf{e}_i \mathbf{x}'_{i\cdot}$ is such that $r(\tilde{X}_i) = r(X) = k$, then

$$\tilde{X}_i^+ = X^+ - \mathbf{d}_i \mathbf{m}'_{i\cdot}. \tag{A.1}$$

Proof: To prove (A.1), it is equivalent to show that

$$[(X-\mathbf{e}_i\mathbf{x}'_{i\cdot})'(X-\mathbf{e}_i\mathbf{x}'_{i\cdot})]^{-1}(X-\mathbf{e}_i\mathbf{x}'_{i\cdot})' - (X'X)^{-1}X' = -\frac{1}{m_{ii}}(X'X)^{-1}\mathbf{x}_{i\cdot}\mathbf{m}'_{i\cdot}. \tag{A.2}$$

But since $\mathbf{m}'_{i\cdot} = \mathbf{e}'_i - \mathbf{x}'_{i\cdot}(X'X)^{-1}X'$ $m_{ii} = 1-\mathbf{x}'_{i\cdot}(X'X)^{-1}\mathbf{x}_{i\cdot}$, and

$e_i'X = x_{i.}'$ (or $X'e_i = x_{i.}$), we have that the left-hand side of (A.2) is, as easily seen,

$$[X'X - x_{i.}x_{i.}']^{-1}(X' - x_{i.}e_i') - (X'X)^{-1}X'$$
$$= (X'X)^{-1}\left\{[I - x_{i.}x_{i.}'(X'X)^{-1}]^{-1}(X' - x_{i.}e_i') - X'\right\}, \qquad (A.3)$$

while the right-hand side of (A.2) is

$$(X'X)^{-1}\left\{\frac{1}{m_{ii}}\, x_{i.}[x_{i.}'(X'X)^{-1}X' - e_i']\right\} \qquad (A.4)$$

Now the quantity inside the braces { } of line 2 of (A.3) is, on using the Sherman-Morrison formula

$$[I + x_{i.}(m_{ii})^{-1}\, x_{i.}'(X'X)^{-1}](X' - x_{i.}e_{i.}') - X' \qquad (A.5)$$

which after some algebra may be seen to be equal to

$$\frac{1}{m_{ii}}\,[x_{i.}x_{i.}'(X'X)^{-1}X' - x_{i.}e_i'] \qquad (A.6)$$

But (A.6) is equal to the quantity inside the braces { } of (A.4), so we have, on pre-multiplication by $(X'X)^{-1}$, shown that the left-hand side of (A.2) is equal to the right-hand side of (A.2), so that (A.1) is proved.

(A Comment is in order. In a technical report to be issued shortly, Styan and Guttman have shown that (A.1) holds under any assumptions on the ranks of $\tilde{X}_i$ and X. Specifically, they prove the following Lemma.

Lemma. If $\tilde{X}_i = X - e_i x_{i.}'$, $d_i = \frac{1}{m_{ii}}\, x_{.i}^{+}$ where $m_{ii} = 1 - x_{i.}'X^{+}e_i \neq 0$, then

$$\tilde{X}_i^{+} = X^{+} - d_i m_{i.}' ,$$

with $m_{i.}' = e_i' - x_{i.}'X^{+}$, where, in general, A^{+} denotes the Moore-Penrose inverse of A. It is easy to see that if we assume $r(\tilde{X}_i) = r(X) = k$, that the above Lemma reduces to the lemma stated and proved for that special case given at the beginning of the Appendix.)

Gene H. Golub was supported in part by the Atomic Energy Commission. This work was started at the Fondacion Bariloche, San Carlos de Bariloche, Argentina. The research work was carried out by Irwin Guttman when he was at the Centre de Recherches Mathématiques,

Université de Montréal.

Gene H. Golub
Computer Science Department
Stanford University
Stanford, California
U.S.A.

Irwin Guttman
Department of Mathematics
University of Toronto
Toronto, Ontario
Canada

Rudolf Dutter
Centre de Recherches Mathématiques
Université de Montréal
Montreal, Quebec
Canada

ADEQUACY OF DISCRIMINANT FUNCTIONS AND THEIR APPLICATIONS

R. D. Gupta and R. P. Gupta*

1. Introduction. The problem of association between two vectors $x(p\times 1)$ and $y(q\times 1,\ p\leq q)$ arises in regression analysis, multivariate analysis of variance, discriminant analysis, analysis of association in a contingency table, stochastic processes, homogeniety of samples, etc. This relationship has different interpretations and implications in these fields, but in each case it can be expressed in terms of canonical correlations $(\rho_1, \rho_2, \ldots, \rho_s)$ and canonical variables. If all the ρ_i's $(i=1,2,\ldots,p)$ are zero, there is no association between x and y and under the normality assumption, this is tested by using the following criterion:

$$\text{Wilks's } \Lambda \text{ criterion} \quad ; \quad \Lambda = \frac{|W|}{|B+W|} \tag{1.1}$$

or

$$\text{Pillai's criterion} \quad ; \quad \gamma = n\ \mathrm{tr}B(B+W)^{-1} \tag{1.2}$$

or

$$\text{Hotteling-Lawley criterion} \quad ; \quad \xi = n\ \mathrm{tr}\ BW^{-1}\ , \tag{1.3}$$

where

$$B = C_{xy}C_{yy}^{-1}C_{yx}\ , \quad W = C_{xx} - C_{xy}C_{yy}^{-1}C_{yx}\ , \tag{1.4}$$

and

$$\begin{bmatrix} C_{xx} & C_{xy} \\ C_{yx} & C_{yy} \end{bmatrix}$$

is the matrix of the corrected s.s. and s.p. of observations on x and y, based on n d.f. In case the hypothesis that ρ_i's are zero is rejected, one needs further analysis. If $\rho_1 \geq \cdots \geq \rho_s$ are non-null and $\rho_{s+1} = \rho_{s+2} = \cdots = \rho_p = 0$, then we say that the association between x and y is of rank s. In such a case, the entire association may be described by first s canonical variates corresponding to $\rho_1, \ldots, \rho_s$. Suppose at this stage one is interested in testing the goodness-of-fit of Γx, where Γ is $s\times p$ and of rank s. Testing of Γx amounts to the testing of the direction and collinearity aspects of Γx. Radcliffe [1966] and Williams [1952] have factorized Wilks's Λ criterion as

*Research supported by NRC Grant A-5290.

$$\Lambda_1 = \frac{|\Gamma W \Gamma'|}{|\Gamma (W+B) \Gamma'|} , \tag{1.5}$$

$$\Lambda_2 = \frac{|\Gamma B (B+W)^{-1} W \Gamma'|}{|\Gamma B \Gamma'|} , \tag{1.6}$$

$$\Lambda_3 = \Lambda / \Lambda_1 \Lambda_2 , \tag{1.7}$$

here Λ_1 is the direction factor and Λ_2 is the collinearity factor. Wilks's Λ may also be factorized as

$$\Lambda_4 = \frac{\Lambda |\Gamma B W^{-1} (B+W) \Gamma'|}{|\Gamma B \Gamma'|} , \tag{1.8}$$

$$\Lambda_5 = \Lambda / \Lambda_1 \Lambda_4 . \tag{1.9}$$

Under the null hypothesis, the distributions of Λ_2, Λ_3, Λ_4, Λ_5 are $\Lambda(n-s, s, p-s)$, $\Lambda(n-2s, p-s, q-s)$, $\Lambda(n-s, p-s, q-s)$, $\Lambda(n-q, p-s, s)$, respectively.

Kshirsagar [1969] factorizes γ and ξ in the case Γ is $1 \times p$. Following his theory, we extend his factorization for the case Γ is $s \times p$, as follows: Let x^*, y^* be the vector of the population canonical variables and let the relationship between x^* and x be

$$x^* = H'x , \tag{1.10}$$

where columns of H are canonical vectors in the x-space corresponding to $\rho_1, \ldots, \rho_p$. Therefore, x^* and y^* have I_p and I_q as their variance-covariance matrices, respectively, and except $\rho_1, \rho_2, \ldots, \rho_s$ all other correlations are zero. Define

$$W^* = H'WH , \quad B^* = H'BH , \quad C^*C^{*'} = W^* + B^* , \tag{1.11}$$

where C^* is a lower triangular matrix. Then, the density of

$$L^* = C^{*-1} B^* C^{*'-1} , \tag{1.12}$$

when $y^{*'}$ is fixed, is shown to be (Radcliff 1968)

$$K \, \phi(L^*_{11}, \rho_1, \ldots, \rho_s) |L^*|^{(q-p-1)/2} |I_p - L^*|^{(n-q-p-1)/2} , \tag{1.13}$$

where $\phi(L^*_{11}, \rho_1, \ldots, \rho_s)$ is a function of $L_{11}, \rho_1, \ldots, \rho_s$, and

$$L = \begin{bmatrix} L_{11} & L_{12} \\ L_{21} & L_{22} \end{bmatrix} \begin{matrix} s \\ p-s \end{matrix} \qquad (1.14)$$
$$\qquad s \qquad p-s$$

Kshirsagar [1969] has shown, for large n, $|I_p - L^*|^{(n-q-p-1)/2}$ can be replaced by

$$\exp\{-\tfrac{1}{2}\operatorname{tr} Z^*\}\ , \qquad (1.15)$$

where $Z^* = nL^*$ and so, Z^* will have a non-central Wishart density for large n. Make a further transformation

$$Z^* = nL^* = S^*S^{*\prime}\ , \qquad (1.16)$$

where

$$S^* = \begin{bmatrix} S^*_{11} & 0 \\ S^*_{12} & S^*_{22} \end{bmatrix} \begin{matrix} s \\ p-s \end{matrix} \qquad (1.17)$$
$$\qquad s \qquad p-s$$

is a lower triangular matrix. Then it can be readily seen that, for large n, $\operatorname{tr} S^*_{11}S^{*\prime}_{11}$ is a non-central χ^2, s^{*2}_{ii} is a χ^2 with $(q+1-i)$ d.f. $(i=s+1,\ldots,p)$, $s^*_{ij}\,(i=s+1,\ldots,p;\ j=1,2,\ldots,p)$ is normally distributed with mean zero and variance one and all these variables are independent. The overall criterion for testing the independence of x and y is $n\operatorname{tr}B(B+W)^{-1}$, which is the same as $\operatorname{tr}Z^*$ on account of (1.7) and

$$\begin{aligned} \operatorname{tr}Z^* &= \operatorname{tr} S^*_{11}S^{*\prime}_{11} + \operatorname{tr} S^*_{12}S^{*\prime}_{12} + \operatorname{tr} S^*_{22}S^{*\prime}_{22} \\ &= \gamma_1 + \gamma_2 + \gamma_3\ ,\text{(say)}. \end{aligned} \qquad (1.18)$$

Then γ_1 contains the entire non-centrality; γ_2 is a χ^2 with $s(p-s)$ d.f. and γ_3 is a χ^2 with $(p-s)(p-s)$ d.f.

Under the null hypothesis (H_o) of the goodness-of-fit of Γx, Γx are the first s population canonical variates. By little algebra, one can find that

$$\left.\begin{aligned} \gamma_1 &= n\operatorname{tr}\Gamma B\Gamma'(\Gamma(B+W)\Gamma')^{-1} \\ \gamma_2 &= n\operatorname{tr}(\Gamma B(B+W)^{-1}B\Gamma)(\Gamma B\Gamma')^{-1} - \gamma_1 \\ \gamma_3 &= n\operatorname{tr}B(B+W)^{-1} - \gamma_1 - \gamma_2\ . \end{aligned}\right\} \qquad (1.19)$$

The overall test of null hypothesis is given by $\gamma_2+\gamma_3$ and γ_2, γ_3 are the direction and collinearity parts of γ.

In exactly a similar manner, we can show that, for the other criterion ξ, the partitioning is

$$\xi = n \text{ tr } BW^{-1} = \xi_1 + \xi_2 + \xi_3 \quad , \tag{1.20}$$

where

$$\left.\begin{aligned} \xi_1 &= n \text{ tr } \Gamma B\Gamma'(\Gamma W\Gamma')^{-1} \\ \xi_2 &= n \text{ tr}(\Gamma BW^{-1}B\Gamma')(\Gamma B\Gamma')^{-1}-\xi_1 \\ \xi_3 &= n \text{ tr } BW^{-1}-\xi_1-\xi_2 \end{aligned}\right\} \tag{1.21}$$

ξ_2 is a χ^2 with $s(p-s)$ d.f. and ξ_3 is a χ^2 with $(p-s)(p-s)$ d.f. in large samples and these are respectively the direction and collinearity parts and can be used to test these aspects of the null hypothesis.

Kshirsagar [1970] applies the above results for goodness-of-fit of an assigned set of scores for the analysis of association in a contingency table. In section 3, we extend his results for s assigned sets of scores and apply these results in markov chains and in homogeneity of samples, respectively, in section 4 and 5.

2. Elimination of irrelevant variables. In some situations we may be interested in studying the relationship between — not x and y — but between residual variates z and w, where the latter are obtained from the former by eliminating the first t sample canonical variables which are assumed to be irrelevant to the problem under consideration and are therefore excluded. Assume that L_1x and M_1y, where

and

$$\left.\begin{aligned} L_1 \text{ , } t\times p \text{ , } &= [\ell_1, \ell_2, \ldots, \ell_t]' \\ M_1 \text{ , } t\times q \text{ , } &= [m_1, m_2, \ldots, m_t]' \end{aligned}\right\} \tag{2.1}$$

are the first t canonical variables. Since the column vectors ℓ_i, m_i satisfy the equation

$$\begin{bmatrix} -r_iC_{xx} & C_{xy} \\ C_{yx} & -r_iC_{yy} \end{bmatrix} \begin{bmatrix} \ell_i \\ m_i \end{bmatrix} = 0 \quad , \tag{2.2}$$

where γ_i is the i-th sample canonical correlation, we have

$$C_{xx}L_1'R = C_{xy}M_1' , \tag{2.3}$$

where R is the $t \times t$ diagonal matrix of γ_i^2, $i=1,2,\ldots,t$. It follows by using (2.2) that

$$\left.\begin{aligned} BL_1' &= (B+W)L_1'R \\ \text{and} \qquad WL_1' &= -(B+W)L_1'(R-I_t) \end{aligned}\right\} \tag{2.4}$$

Further let L_2 be a $(p-t)\times p$ matrix and M_2 a $(q-t)\times q$ matrix such that

$$L_2C_{xx}L_1' = 0 \quad , \quad M_2C_{yy}M_1' = 0 \quad , \tag{2.5}$$

i.e., L_1x and L_2x are uncorrelated and so also M_1y and M_2y. From (2.4) and (2.5), it is seen that

$$L_2BL_1' = 0 \quad , \quad L_2WL_1' = 0 \ . \tag{2.6}$$

We may now take

$$z = L_2x \quad , \quad w = M_2y \ , \tag{2.7}$$

as our residual variables after eliminating L_1x and M_1y. We define

$$B_z = C_{zw}C_{ww}^{-1}C_{wz} \quad , \quad W_z = C_{zz}-C_{zw}C_{ww}^{-1}C_{wz} \ , \tag{2.8}$$

where

$$\begin{bmatrix} C_{zz} & C_{zw} \\ C_{wz} & C_{ww} \end{bmatrix} \tag{2.9}$$

is the matrix of corrected s.s. and s.p. of observations on z and w. To test the independence of z and w, we may use any one of the three criteria; (1.1), (1.2) or (1.3), by replacing W and B by W_z and B_z, respectively. We must also replace n by $n-t$, p by $p-t$, q by $q-t$ as t variables are eliminated from x and y. However, we wish to express these test statistics in terms of our old matrices W and B. This is done as follows. From (2.3) and (2.5), we observe

$$L_2C_{xy}M_1' = 0 \ , \tag{2.10}$$

and hence we find that

$$\begin{aligned} B_z &= L_2 C_{xw} C_{ww}^{-1} C_{wx} L_2' \\ &= L_2 [C_{xy} C_{yy}^{-1} C_{yx} - C_{xy} M_1' (M_1 C_{yy} M_1')^{-1} M_1 C_{yx}] L_2' \qquad (2.11) \\ &= L_2 B L_2' \ , \end{aligned}$$

on account of (2.10).
Also we may notice that

$$W_z + B_z = C_{zz} = L_2 (B+W) L_2' \qquad (2.12)$$

By writing

$$L = [L_1' \ , \ L_2']' \qquad (2.13)$$

we may note that

$$\begin{aligned} (B+W)^{-1} &= L' (L C_{xx} L')^{-1} L \\ &= \sum_{i=1}^{2} L_i' (L_i C_{xx} L_i')^{-1} L_i \ . \end{aligned} \qquad (2.14)$$

It now follows that

$$\begin{aligned} \Omega B_z (B_z + W_z)^{-1} B_z \Omega' &= \Omega L_2 B L_2' (L_2 C_{xx} L_2')^{-1} L_2 B L_2' \Omega' \\ &= \Gamma B [(B+W)^{-1} - L_1' (L_1 C_{xx} L_1')^{-1} L_1] B \Gamma' \qquad (2.15) \\ &= \Gamma B (B+W)^{-1} B \Gamma \ , \end{aligned}$$

where $\Omega L_2 = \Gamma$.

In exactly the same way, we may show that

$$\Omega B_z W_z^{-1} B_z \Omega' = \Gamma B W^{-1} B \Gamma' \ . \qquad (2.16)$$

Further, by using (2.4) we may prove that

$$\begin{aligned} = \frac{|W|}{|B+W|} &= \frac{|L_1 W L_1'| \, |L_2 W L_2'|}{|L_1 (B+W) L_1'| \, |L_2 (B+W) L_2'|} \\ &= \prod_{i=1}^{t} (1 - r_i^2) \cdot \Lambda_z \ , \end{aligned} \qquad (2.17)$$

$$\Omega B_z \Omega' = \Omega L_2 B L_2' \Omega' = \Gamma B \Gamma' \ , \qquad (2.18)$$

and that

$$\Omega W_z \Omega' = \Omega L_2 W L_2' \Omega' = \Gamma W \Gamma' \quad . \tag{2.19}$$

By using these identities, the new criterion for testing independence of z and w can be written as

$$\text{Wilks's } \Lambda \text{ criterion;} \quad \Lambda_z = \frac{|W_z|}{|B_z+W_z|} = \prod_{i=t+1}^{p} (1-r_i^2), \tag{2.20}$$

$$\text{Pillai's criterion;} \quad \gamma_z = n \text{ tr } B_z(B_z+W_z)^{-1} = n \sum_{i=t+1}^{p} r_i^2, \tag{2.21}$$

$$\text{Hotelling-Lawley criterion;} \quad \xi_z = n \text{ tr } B_z W_z^{-1} = n \sum_{i=t+1}^{p} \frac{r_i^2}{(1-r_i^2)} \tag{2.22}$$

Under the assumption that only s true canonical correlations are different from zero, we wish to test the goodness-of-fit of s assigned functions x for the relationship between z and w. Note that the assigned functions must be so chosen that they are uncorrelated with eliminated $L_1 x$. The following will be our new direction and collinearity factors:

$$\Lambda_{1z} = \frac{|\Omega W_z \Omega'|}{|\Omega(B_z+W_z)\Omega'|} = \frac{|\Gamma W \Gamma'|}{|\Gamma(B+W)\Gamma'|}, \tag{2.23}$$

$$\Lambda_{2z} = \frac{|\Omega B_z (B_z+W_z)^{-1} W_z \Omega'|}{|\Omega B_z \Omega'|} = \frac{|\Gamma B (B+W)^{-1} W \Gamma'|}{|\Gamma B \Gamma'|}, \tag{2.24}$$

$$\Lambda_{3z} = \frac{\Lambda_z}{\Lambda_{1z} \cdot \Lambda_{2z}}, \tag{2.25}$$

$$\Lambda_{4z} = \frac{\Lambda_z |\Omega B_z W_z^{-1} (B_z+W_z)\Omega'|}{|\Omega B_z \Omega'|}$$

$$= \frac{\Lambda_z |\Gamma B W^{-1} (B+W)\Gamma'|}{|\Gamma B \Gamma'|}, \tag{2.26}$$

$$\Lambda_{5z} = \frac{\Lambda_z}{\Lambda_{1z} \cdot \Lambda_{4z}}, \tag{2.27}$$

$$\gamma_{1z} = n \text{ tr } \Omega B_z \Omega' (\Omega(B_z+W_z)\Omega')^{-1}$$

$$= n \text{ tr } \Gamma B \Gamma' (\Gamma(B+W)\Gamma')^{-1} \quad , \tag{2.28}$$

$$\gamma_{2z} = n \text{ tr}[\Omega B_z (B_z+W_z)^{-1} B_z \Omega'](\Omega B_z \Omega')^{-1} - \gamma_{1z}$$

$$= n \text{ tr}[\Gamma B(B+W)^{-1} B \Gamma'](\Gamma B \Gamma')^{-1} - \gamma_{1z} \quad , \tag{2.29}$$

$$\gamma_{3z} = \gamma_z - \gamma_{1z} - \gamma_{2z} \quad , \tag{2.30}$$

$$\xi_{1z} = n \text{ tr } \Omega B_z \Omega' (\Omega W_z \Omega')^{-1}$$

$$= n \text{ tr } \Gamma B \Gamma' (\Gamma W \Gamma')^{-1} \quad , \tag{2.31}$$

$$\xi_{2z} = n \text{ tr}(\Omega B_z W_z^{-1} B_z \Omega')(\Omega B_z \Omega')^{-1} - \xi_{1z}$$

$$= n \text{ tr}(\Gamma B W^{-1} B \Gamma')(\Gamma B \Gamma')^{-1} - \xi_{1z} \quad , \tag{2.32}$$

$$\xi_{3z} = \xi_z - \xi_{1z} - \xi_{2z} \quad . \tag{2.33}$$

Hence we found that these new direction, collinearity factors and their distributions are the same as the old ones except that Λ must be changed to Λ_z , γ to γ_z , ξ to ξ_z , n to $n-t$, p to $p-t$ and q to $q-t$.

3. Association between two attributes. Consider a $p \times q$ contingency table with the rows corresponding to p categories $a_1, a_2, \ldots, a_p$ of an attribute a and columns to q categories $b_1, b_2, \ldots, b_q$ of another attribute b . Let n_{ij} , $i=1,2,\ldots,p$; $j=1,2,\ldots,q$ be the be the frequency in the (i,j)-th cell. Let $n_{i\cdot}$, $n_{\cdot j}$ be the totals of i-th row and j-th column, respectively, and $\sum_{i=1}^{p} n_{i\cdot} = \sum_{j=1}^{q} n_{\cdot j} = n$. We define

$$\left.\begin{array}{ll} N = [n_{ij}] \quad , & D_1 = \text{diag.}[n_{1\cdot}, n_{2\cdot}, \ldots, n_{p\cdot}] \\ & D_2 = \text{diag.}[n_{\cdot 1}, n_{\cdot 2}, \ldots, n_{\cdot q}] . \end{array}\right\} \tag{3.1}$$

Let p_{ij} be the probability that an individual belongs to the category a_i of a and the category b_j of $b(i=1,2,\ldots,p$;

$j=1,2,\ldots,q$). Then $p_{i\cdot} = \sum_{j=1}^{q} p_{ij}$ is the probability of an individual belonging to a_i and $p_{\cdot j} = \sum_{i=1}^{p} p_{ij}$ is the probability of belonging to b_j, while $\sum_{i=1}^{p} \sum_{j=1}^{q} p_{ij} = 1$. In order to study the relationship between these attributes, on the basis of this qualitative data, we define two sets of variables x_i $(i=1,2,\ldots,p)$, y_j $(j=1,2,\ldots,q)$ as follows:

$$\left.\begin{aligned} x_i &= 1 , \text{ if an individual belongs to } a_i \\ &= 0 , \quad \text{otherwise } (i=1,2,\ldots,p). \\ y_j &= 1 , \text{ if an individual belongs to } b_j \\ &= 0 , \quad \text{otherwise } (j=1,2,\ldots,q). \end{aligned}\right\} \qquad (3.2)$$

Note that all x_i's and all y_j's are not independent as $\sum_{i=1}^{p} p_{i\cdot} = \sum_{j=1}^{q} p_{\cdot j} = 1$. Let x denote the $p\times 1$ column vector of the x_i's and y denote the $q\times 1$ column vector of the y_j's. The relationship between a and b is then the relationship between x and y, and we can, therefore, use the theory of canonical vectors and canonical correlations. By writing the values of x_i and y_j for all the n individuals it is easy to see that the matrices of the corrected s.s. and s.p. of x and y are

$$C_{xx} = D_1 , \quad C_{xy} = N , \quad C_{yy} = D_2 , \qquad (3.3)$$

and hence

$$\left.\begin{aligned} B &= N D_2^{-1} N' = \left[\sum_{j=1}^{q} \frac{n_{ij} n_{hj}}{n_{\cdot j}}\right] , \; i,h=1,2,\ldots,p , \\ W &= D_1 - N D_2^{-1} N' . \end{aligned}\right\} \qquad (3.4)$$

The sample canonical correlations $\gamma_1, \gamma_2, \ldots, \gamma_p$ are the roots of the equation

$$\begin{bmatrix} r^2 C_{xx} & C_{xy} \\ C_{yx} & C_{yy} \end{bmatrix} = 0 \qquad (3.5)$$

and the canonical vectors ℓ_i and m_i, vectors of optimal scores corresponding to a and b, are the solutions of

$$\begin{bmatrix} r^2 C_{xx} & C_{xy} \\ C_{xy} & C_{yy} \end{bmatrix} \begin{bmatrix} \ell \\ m \end{bmatrix} = 0 \ . \tag{3.6}$$

Let e be a column vector $(p\times 1)$ of unities everywhere. It is easy to observe that

$$We = 0 \quad , \tag{3.7}$$

i.e., $r^2 = 1$ is a canonical correlation between x and y, the corresponding variates being $x_1+x_2+\ldots+x_p$ and $y_1+\ldots+y_q$. Obviously, these are irrelevant because it assigns the same value to all the categories and fails to differentiate between them. We must therefore eliminate these variates and study the residual variates z and w. By using the procedure given in section 2 for the calculations of z and w, we find that

$$\left.\begin{aligned} z &= L_2 x \ , \text{ i.e., } z_i = x_i - n_{i\cdot}(x_1+\ldots+x_p)/n \\ \text{and} \quad w &= M_2 y \ , \text{ i.e., } w_j = y_j - n_{\cdot j}(y_1+\ldots+y_q)/n \ . \end{aligned}\right\} \tag{3.8}$$

We may also easily calculate C_{zz}, C_{zw}, C_{ww}, B_z and W_z and find

$$W_z = W_o \quad , \quad B_z = B_o - d_o d_o'/n \quad , \tag{3.9}$$

where W_o and B_o are matrices obtained from W and B by deleting the last row and the last column, and

$$d_o = [n_{1\cdot}, n_{2\cdot}, \ldots, n_{p-1\cdot}]' \ , \ d = [d_o', n_{p\cdot}]' \ . \tag{3.10}$$

Note that

$$|W_z+B_z| = |D_1^o - d_o d_o'/n| = (\prod_{i=1}^{p} n_{i\cdot})/n \quad , \tag{3.11}$$

where D_1^o is obtained from D_1 by deleting the last row and last column.

The independence of z and w i.e., no association between two attributes a and b, may be tested by using the criterion given in the following table

Table

Test Statistic	Distribution for large n
$\Lambda_z = \dfrac{n\|W_o\|}{\prod_{i=1}^{p} n_{i\cdot}}$	$\Lambda(n-1,\ p-1,\ q-1)$
$\gamma_z = n[\mathrm{tr}\ B(B+W)^{-1}-1] = n[(\sum_{i=1}^{p}\sum_{j=1}^{q}\dfrac{n_{ij}^2}{n_{i\cdot}\,n_{\cdot j}})-1]$	χ^2 with $(p-1)(q-1)$ d.f.
$\xi_z = n\ \mathrm{tr}\ B_z W_z^{-1}$	χ^2 with $(p-1)(q-1)$ d.f.

Note that the classical test of independence of a and b uses the statistic

$$U = \sum_{i=1}^{p}\sum_{j=1}^{q}[n_{ij} - \frac{n_{i\cdot}n_{\cdot j}}{n}]^2/[\frac{n_{i\cdot}n_{\cdot j}}{n}]$$

$$= n[\sum_{i=1}^{p}\sum_{j=1}^{q}\frac{n_{ij}^2}{n_{i\cdot}n_{\cdot j}} - 1] \ . \tag{3.12}$$

Under the null hypothesis, U has χ^2 distribution with $(p-1)(q-1)$ d.f., provided n and n_{ij} are not small. Hence the classical χ^2 test is the same as the test based on Pillai's γ_z, even though the assumption of normality does not hold.

If the null hypothesis is rejected, one needs further analysis. Suppose only s sample canonical correlations are significant then s sets of scores are adequate. Suppose one is interested in testing the goodness-of-fit of a set of hypothetical scores Γ, $s\times p$ and of rank s, for the rows. The null hypothesis of goodness -of-fit comprises two aspects, (i) the association between a's and b's is of rank s, and (ii) the true scores corresponding to non-zero canonical correlations are Γ. The part (i) is the collinearity part and (ii) is the direction part of our null hypothesis.

Since we have eliminated $\sum_{i=1}^{p} x_i$, the assigned functions Γx must, as noticed in section 2, be uncorrelated with $\sum_{i=1}^{p} x_i$, i.e.,

$$d'\Gamma' = 0 \ . \tag{3.13}$$

and hence Γ must satisfy $\Gamma = \Omega L_2$, L_2 as defined in (3.8), and find that

$$\Omega' = \Gamma_o - e_o\Gamma_p \ , \tag{3.14}$$

where $\Gamma' = [\Gamma_o' \ , \ \Gamma_p]'$ and $e' = [e_o \ , \ 1]'$.
We have observed in (3.7) that W is a singular matrix. Thus we cannot obtain the direction and collinearity factors straightforward as they involve inverse of the matrix W . We must obtain them by using B_z and W_z . This is done as follows: On partitioning B and W as

$$B = \underset{\quad p-1 \qquad 1}{\begin{bmatrix} B_o & b \\ b' & b_{pp} \end{bmatrix}}\begin{matrix} p-1 \\ 1 \end{matrix} \ , \qquad W = \underset{\quad p-1 \qquad 1}{\begin{bmatrix} W_o & -b \\ -b' & W_{pp} \end{bmatrix}}\begin{matrix} p-1 \\ 1 \end{matrix} \tag{3.15}$$

we may observe with the help of (3.7) that

$$W_oe_o = b \ . \tag{3.16}$$

Now we set

$$B\Gamma' = F = \underset{s}{\begin{bmatrix} F_o \\ F_p \end{bmatrix}}\begin{matrix} p-1 \\ 1 \end{matrix} \ , \tag{3.17}$$

and by using (3.13) we find that

$$e'F = e'B\Gamma' = d'\Gamma' = 0 \ . \tag{3.18}$$

The equations

$$WG = F \ , \tag{3.19}$$

in sp unknowns, $G = \underset{s}{\begin{bmatrix} G_o \\ G_p \end{bmatrix}}\begin{matrix} p-1 \\ 1 \end{matrix}$, are solvable. A solution is

$$G = W^-F \ , \tag{3.20}$$

where W^- is the Moore-Penrose inverse of W , see Rao and Mitra [1971]. However, (3.19) and (3.20) yield

$$W_oG_o - b\ G_p = F_o$$

or

$$G_o - W_o^{-1}\ b\ G_p = W_o^{-1}F_o \ ,$$

or $$G_o - e_o G_p = W_o^{-1} F_o \ . \quad (3.21)$$

Using (3.17) we may prove that

$$B_z \Omega' = (B_o - \frac{d_o d_o'}{n})(\Gamma_o - e_o \Gamma_p) = B_o \Gamma_o + b\Gamma_p = F_o \ . \quad (3.22)$$

Thus it follows that

$$\Omega B_z W_z^{-1} B_z \Omega' = F_o' W_z^{-1} F_o = F_o'(G_o - e_o G_p)$$

$$= F'G = \Gamma B W^- B \Gamma' \ . \quad (3.23)$$

The direction and collinearity factors, in terms of data, and thier distributions are given in the following table.

Table

Factor	Distribution for large n
$\Lambda_{2z} = \dfrac{\lvert\Gamma(I_p - D_1^{-1} N D_2 N')\Gamma'\rvert \lvert\Gamma D_1 \Gamma'\rvert}{\lvert\Gamma F\rvert \lvert\Gamma(D_1 - N D_2^{-1} N')\Gamma'\rvert}$	$\Lambda(n-s-1, s, p-s-1)$
$\Lambda_{3z} = \dfrac{\lvert W_z\rvert}{\lvert B_z + W_z\rvert} \cdot \dfrac{\lvert\Gamma F\rvert}{\lvert F'(I_p - D_1^{-1} N D_2 N')\Gamma'\rvert}$	$\Lambda(n-2s-1, p-s-1, q-s-1)$
$\Lambda_{4z} = \dfrac{\lvert W_z\rvert}{\lvert B_z + W_z\rvert} \cdot \dfrac{\lvert\Gamma F + F'G\rvert}{\lvert\Gamma F\rvert}$	$\Lambda(n-s-1, p-s-1, q-s-1)$
$\Lambda_{5z} = \dfrac{\lvert\Gamma D_1 \Gamma'\rvert \lvert\Gamma F\rvert}{\lvert\Gamma(D_1 - N D_2^{-1} N')\Gamma'\rvert \lvert\Gamma F + F'G\rvert}$	$\Lambda(n-q, s, p-s-1)$
$\gamma_{2z} = n\ \mathrm{tr}[(F' D_1^{-1} F)(\Gamma F)^{-1}] - n\ \mathrm{tr}[\Gamma N D_2^{-1} N' \Gamma' (\Gamma D_1^{-1} \Gamma')^{-1}]$	χ^2 with $s(p-s-1)$ d.f.
$\gamma_{3z} = n[(\sum_{i=1}^{p} \sum_{j=1}^{q} \frac{n_{ij}^2}{n_{i\cdot} n_{\cdot j}}) - 1] - n\ \mathrm{tr}[(F' D_1^{-1} F)(\Gamma F)^{-1}]$	χ^2 with $(p-s-1)(q-s-1)$ d.f.
$\xi_{2z} = n\ \mathrm{tr}[F'(D - N D_2^{-1} N')^{-1} F (\Gamma F)^{-1}] - n\ \mathrm{tr}\{(\Gamma F)[\Gamma(D_1 - N D_2^{-1} N')\Gamma']^{-1}\}$	χ^2 with $s(p-s-1)$ d.f.
$\xi_{3z} = n\ \mathrm{tr}\ B_z W_z^{-1} - n\mathrm{tr}[F'(D_1 - N D_2^{-1} N')^{-1} F (\Gamma F)^{-1}]$	χ^2 with $(p-s-1)(q-s-1)$ d.f.

4. Association between the states of finite Markov chains. We consider a one-step Markov chain in u states. Let $p_i(m+1)$ be the probability that the chain is in state i at stage $m+1$, $i=1,\ldots,u$, and $p_{ij}(m+1)$ be the probability that the chain is in state j at stage $m+1$, given that it was in state i at stage m, $j=1,2,\ldots,u$. Let $p' = (p_1(m),\ldots,p_u(m))$ and $q' = (p_1(m+1),\ldots,p_u(m+1))$, then we have

$$q = p'P , \tag{4.1}$$

where $P = [p_{ij}(m+1)]$, $i,j=1,2,\ldots,u$, is the one step transition probability matrix. In order to study the relationship between these states, we define two sets of variables $x_i\,(i=1,\ldots,u)$ and $y_j\,(j=1,2,\ldots,u)$ as follows:

$$\begin{aligned} Y_j &= 1 \text{ , if chain is in state } j \text{ at stage } m+1 \\ &= 0 \text{ , otherwise.} \\ x_i &= 1 \text{ , if chain is in state } i \text{ at stage } m \\ &= 0 \text{ , otherwise.} \end{aligned}$$

Note that all x's and all y's are not independent as $\sum_{i=1}^{u} p_i(m)=1$ for all m (at each stage).

Let x denote the $u\times 1$ column vector of x_i's and y denote the $u\times 1$ column vector of y_j's. The relationship among the states is then the relationship between x and y, and we can, therefore, use the theory of canonical vectors and canonical correlations. Suppose x represents states along the rows and y along the columns. Further, let $n_{ij}\,(i,j=1,2,\ldots,u)$ be the frequency in the (i,j)-th cell. Let $n_{i\cdot}$, $n_{\cdot j}$ be the respective totals of the i-th row and j-th column, and $\sum_{i=1}^{u} n_{i\cdot} = \sum_{j=1}^{u} n_{\cdot j} = n$.

The corrected s.s. and s.p. of x and y are

$$\left.\begin{aligned} C_{xx} &= D_1 = \operatorname{diag}(n_{1\cdot},n_{2\cdot},\ldots,n_{u\cdot}) , \\ C_{yy} &= D_2 = \operatorname{diag}(n_{\cdot 1},n_{\cdot 2},\ldots,n_{\cdot u}) , \\ C_{xy} &= D_1 N , \text{ where } N = [n_{ij}] , \end{aligned}\right\} \tag{4.2}$$

and hence

$$\left.\begin{aligned} B &= D_1ND_2^{-1}N'D_1 \quad , \\ W &= D_1-D_1ND_2^{-1}N'D_1 \; . \end{aligned}\right\} \tag{4.3}$$

The canonical correlations $\gamma_1,\gamma_2,\ldots,\gamma_u$ in sample, are the roots of the equation

$$\begin{bmatrix} r^2C_{xx} & C_{xy} \\ C_{yx} & C_{yy} \end{bmatrix} = 0 \quad . \tag{4.4}$$

and the canonical vectors ℓ and m , vectors of optimal scores corresponding to states at stage m and $m+1$, are the solution of

$$\begin{bmatrix} r^2C_{xx} & C_{xy} \\ C_{yx} & C_{yy} \end{bmatrix} \begin{bmatrix} \ell \\ m \end{bmatrix} = 0 \quad . \tag{4.5}$$

To test the null hypothesis $p_{ij}(m+1) = p_j(m+1)(i,j=1,2,\ldots,u)$, i.e., all the true canonical correlations between x and y are zero may be tested by using criterion given in the following table

Table

Test Statistic	Distribution for large n
$\Lambda = \lvert I_u-ND_2^{-1}N'D_1\rvert$	$\Lambda(n, u, u)$
$\gamma = n \text{ tr } D_1ND_2^{-1}N'$	χ^2 with u^2 d.f.
$\xi = n \text{ tr } D_1ND_2^{-1}N'(I_u-D_1ND_2^{-1}N')^{-1}$	χ^2 with u^2 d.f.

If the null hypothesis is rejected, one needs further analysis. Suppose only s sample canonical correlations are significant then s sets of scores are adequate. Suppose at this stage one is interested in testing the goodness-of-fit of a set of hypothetical scores Γ , $s\times p$ and of rank s , for the states at stage m . The null hypothesis of goodness-of-fit comprises two aspects, (i) the association among the states at stages m and $m+1$ is of rank s , (ii) the true scores corresponding to non-zero canonical correlations are Γ . The part (i) is the collinearity part and (ii) is the direction part of our null hypothesis. The direction and collinearity factors, in terms of data, and their distribution under null hypothesis are given in the following table.

Table

Factor	Distribution for large n
$\Lambda_2 = \dfrac{\lvert \Gamma D_1 N D_2^{-1} N'(D_1 - D_1 N D_2^{-1} N' D_1)\Gamma' \rvert}{\lvert \Gamma D_1 N D_2^{-1} N' D_1 \Gamma' \rvert \lvert \Gamma (D_1 - D_1 N D_2^{-1} N' D_1)\Gamma' \rvert}$	$\Lambda(n-s, s, u-s)$
$\Lambda_3 = \dfrac{\lvert I_u - N D_2^{-1} N' D_1 \rvert \lvert \Gamma D_1 N D_2^{-1} N' D_1 \Gamma' \rvert}{\lvert \Gamma D_1 N D_2^{-1} N'(D_1 - D_1 N D_2^{-1} N' D_1)\Gamma' \rvert}$	$\Lambda(n-2s, u-s, u-s)$
$\Lambda_4 = \dfrac{\lvert I_u - N D_2^{-1} N' D_1 \rvert \lvert \Gamma D_1 N D_2^{-1} N' D_1 (I_u - N D_2^{-1} N' D_1)^{-1} \Gamma' \rvert}{\lvert \Gamma D_1 N D_2^{-1} N' D_1 \Gamma' \rvert}$	$\Lambda(n-s, u-s, u-s)$
$\Lambda_5 = \dfrac{\lvert \Gamma D_1 N D_2^{-1} N' D_1 \Gamma' \rvert \lvert \Gamma D_1 \Gamma' \rvert}{\lvert \Gamma D_1 N D_2^{-1} N' D_1 (I_u - N D_2^{-1} N' D_1)^{-1} \Gamma' \rvert \lvert \Gamma (D_1 - D_1 N D_2^{-1} N' D_1)\Gamma' \rvert}$	$\Lambda(n-u, u-s, u-s)$
$\gamma_2 = n\, \mathrm{tr}[\Gamma D_1 N D_2^{-1} N' D_1 N D_2^{-1} N' D_1 \Gamma' (\Gamma D_1 N D_2^{-1} N' D_1 \Gamma')^{-1} - \Gamma D_1 N D_2^{-1} N' D_1 \Gamma' (\Gamma D_1 \Gamma')^{-1}]$	χ^2 with $s(u-s)$ d.f.
$\gamma_3 = n\, \mathrm{tr}[D_1 N D_2^{-1} N' - \Gamma D_1 N D_2^{-1} N' D_1 N D_2^{-1} N' D_1 \Gamma' (\Gamma D_1 N D_2^{-1} N' D_1 \Gamma')^{-1}]$	χ^2 with $(u-s)^2$ d.f.
$\xi_2 = n\, \mathrm{tr}[\Gamma D_1 N D_2^{-1} N'(I_u - D_1 N D_2^{-1} N')^{-1} D_1 N D_2^{-1} N' D_1 \Gamma' (\Gamma D_1 N D_2^{-1} N' D_1 \Gamma')^{-1} - \Gamma D_1 N D_2^{-1} N' D_1 \Gamma' (\Gamma (D_1 - D_1 N D_2^{-1} N' D_1)\Gamma')^{-1}]$	χ^2 with $s(u-s)$ d.f.
$\xi_3 = n\, \mathrm{tr}[D_1 N D_2^{-1} N'(I_u - D_1 N D_2^{-1} N')^{-1} - \Gamma D_1 N D_2^{-1} N'(I_u - D_1 N D_2^{-1} N')^{-1} D_1 N D_2^{-1} N' D_1 \Gamma' (\Gamma D_1 N D_2^{-1} N' D_1 \Gamma')^{-1}$	χ^2 with $(u-s)^2$ d.f.

If Markov chain has stationary distribution, i.e.,

$$p_i(m+1) = p_i(m) \quad , \quad i=1,2,\dots,u \ ; \ m=0,1,2,\dots\dots ,$$

then equation (4.1) becomes

$$p' = p'P \tag{4.6}$$

The variance-covariance matrices of x and y are

$$\left.\begin{aligned} \Sigma_{mm} &= Vr(x) = Vr(y) = D_u - pp' = \Sigma_{m+1\ m+1} , \\ \Sigma_{m\ m+1} &= Cov(x,y) = D_u P - pp' , \end{aligned}\right\} \tag{4.7}$$

where $D_u = diag(p_1(m),p_2(m),\dots,p_u(m))$.

The population canonical correlations between x and y are the roots of the equation

$$\begin{vmatrix} \rho^2\Sigma_{mm} & \Sigma_{m\ m+1} \\ \Sigma'_{m\ m+1} & \Sigma_{mm} \end{vmatrix} = 0 . \tag{4.8}$$

We note that Σ_{mm} is of rank $u-1$, however its generalized inverse see Rao [1971], may be easily calculated and find that $\Sigma^{-}_{mm} = D_u^{-1}$. So the roots of the equation (4.8) are the same as the roots of the matrix

$$\Sigma^{-}_{mm}\,\Sigma_{m\ m+1} = P - ep' . \tag{4.9}$$

Let $\alpha_1,\alpha_2,\dots,\alpha_u$ be the corresponding canonical vectors, then

$$P\alpha_i - ep'\alpha_i = \rho_i\alpha_i \quad , \quad i=1,2,\dots,u . \tag{4.10}$$

On premultiplying (4.10) by p', we find that

$$p'\alpha_i - p'\alpha_i = \rho_i p'\alpha_i \quad , \quad i=1,2,\dots,u \tag{4.11}$$

which, since $\rho_i \neq 0$, implies that $p'\alpha_i = 0$, and hence (4.10) reduces to

$$P\alpha_i = \rho_i\alpha_i \quad , \quad i=1,2,\dots,u . \tag{4.12}$$

Note that unity is a root of P, as it is well-known, and the corresponding canonical variate is irrelevant to study the relationship among the states at stage m and $m+1$. For testing the null hypothesis of no association and further analysis, testing the goodness-of-fit of assigned set of scores, we may write the same theory given in section 3.

5. Homogeneity of samples. Fisher [1940] analysed the data on twelve samples of human blood tested with twelve different sera. The reactions obtained were represented by the five symbols: -, ?, w, (+), + .

Fisher's method of assigning optimal scores to these symbols consisted of maximizing the ration of the determinant of between s.s. and s.p. matrix to the determinent of total s.s. and s.p. matrix. Later the same data was analysed by Bartlett [1941] who brought forth the analogy of Fisher's procedure to that determining a discriminant function in case of several groups.

The general problem which is akin to the homogeneity of a series of samples may be stated as follows: Suppose there are k independent samples with $n_{\cdot 1}, n_{\cdot 2}, \ldots, n_{\cdot k}$ observations respectively. These observations are classified into m mutually exclusive classes, such that n_{ij} is the observed frequency of the j-th sample in the i-th class. We note that the two-way table of frequencies n_{ij} is neither a bivariate frequency table nor a contingency table. However, we may quantify the given table by two vectors $x' = (x_1, x_2, \ldots, x_m)$ and $y' = (y_1, \ldots, y_k)$ such that

$$\left.\begin{aligned} &P_r[x_i=1] = p_i \;,\; P_r[x_i=0] = 1-p_i \;,\; p_1+p_2+\ldots+p_m=1, \\ &\quad y_i=1 \;,\; \text{if observation comes from i-th sample} \\ &\qquad =0 \;,\; \text{otherwise.} \end{aligned}\right\} \qquad (5.1)$$

Let p_{ij} denote the probability of an observation of the j-th sample falling in the i-th class. The null hypothesis of homogeneity of samples, i.e., $p_{ij} = p_i$, falling of an observation in the i-th class does not depend on the sample it came from. Note the y being fixed does not effect our approximation theory as long as x has an asymptotic multivariate normal distribution. Let $\sum_{j=1}^{k} n_{\cdot j} = n$, $\sum_{j=1}^{k} n_{ij} = n_{i\cdot}$ is the total of the i-th class, and

$\sum_{i=1}^{m} n_{ij} = n_{\cdot j}$ is the total number of observations in the j-th sample.

By writing down the values of x_i and y_j for all the n observations it is easy to see that the matrices of the corrected s.s. and s.p. of x and y are

$$\left.\begin{aligned} C_{xx} &= D_1 = \mathrm{diag}(n_{1\cdot}, n_{2\cdot}, \ldots, n_{m\cdot}) , \\ C_{xy} &= N = [n_{ij}] , \\ C_{yy} &= D_2 = \mathrm{diag}(n_{\cdot 1}, n_{\cdot 2}, \ldots, n_{\cdot k}) , \end{aligned}\right\} \qquad (5.2)$$

and hence

$$\left.\begin{aligned} B &= N D_2^{-1} N' \\ W &= D_1 - N D_2^{-1} N' . \end{aligned}\right\} \qquad (5.3)$$

The sample canonical correlations are the roots of the equation

$$\begin{vmatrix} r^2 C_{xx} & C_{xy} \\ C_{yx} & C_{yy} \end{vmatrix} = 0 . \qquad (5.4)$$

It is readily observed from (5.4) that unity is a canonical correlation between x and y , the corresponding variates are irrelevant. We must therefore eliminate these variates and study the residual variates z and w . For testing the null hypothesis and further analysis, testing assigned set of scores, we may write down word by word theory given in section 3.

References

[1] Fisher, R. A. (1940). "Statistical Methods for Research Workers," 11-th ed., Edinburgh, Oliver and Boyd.

[2] Gupta, R. P. and Kabe, D. G. (1971). "Distributions of certain factors useful in discriminant analysis," Ann. Inst. Stat. Math., 23, 97.

[3] Kshirsagar, A. M. (1964). "Distributions of direction and collinearity factors in discriminant analysis," Proc. Camb. Phil. Soc., 60, 217.

[4] Kshirsagar, A. M. (1969). "Correlation between two vector variables," J. Roy. Stats. Soc., B, 31, 477.

[5] Kshirsagar, A. M. (1970). "Goodness-of-fit of an assigned set of scores for the analysis of association in a contingency table," Ann. Inst. Stat. Math., 22, 295.

[6] Radcliffe, J. (1966). "Factorizations of the residual likelihood criterion in discriminant analysis," Proc. Camb. Phil. Soc., 62, 743.

[7] Radcliffe, J. (1968). "The distribution of certain factors occurring in discriminant analysis," Proc. Camb. Phil. Soc., 64, 731.

[8] Rao, C. R. and Mitra, S. K. (1971). "Generalized Inverse of Matrices and its Applications," John Wiley, New York.

[9] Williams, E. J. (1952). "Use of scores for the analysis of association in contingency tables," Biometrika, 39, 274.

[10] Williams, E. J. (1961). "Tests for discriminant functions," J. Austral. Math. Soc., 2, 243.

[11] Williams, E. J. (1967). "The analysis of association among many variates," J. Roy. Stat. Soc., B, 20, 199.

[12] Bartlett, M. S. (1951), "The goodness-of-fit of a single hypothetical function in the case of several groups," Ann. Eugen., 16, 199.

R. D. Gupta
Department of Mathematics
Dalhousie University
Halifax, Nova Scotia
Canada

R. P. Gupta
Department of Mathematics
Dalhousie University
Halifax, Nova Scotia
Canada

ON RAO'S GENERALIZED U STATISTIC

D. G. Kabe

When a $p \times N$ random matrix Y has a pN variate normal distribution with mean BX and covariance matrix $\Sigma \otimes I$, and Σ and B are partitioned as in (3) below, then the nonnull distribution of the likelihood ratio statistic for testing the hypothesis that $B_2 - \Sigma_{21}\Sigma_{11}^{-1} B_1$ has a specified value is investigated.

1. Introduction and Summary

Let the joint density of a $(p_1+p_2) \times (p_1+p_2)$ positive definite symmetric matrix S, and a $(p_1+p_2) \times q$ matrix $\hat{B}$ be

$$(1) \quad g(S,\hat{B}) = C_1 \exp\{-\tfrac{1}{2}\mathrm{tr}\ \Sigma^{-1}[S + (\hat{B}-B)XX'(\hat{B}-B)']\} \ |S|^{\frac{1}{2}(N-p_1-p_2-1)} ,$$

where

$$(2) \quad C_1 = (2\pi)^{-\frac{1}{2}p(N+q)} \ |\Sigma|^{-\frac{1}{2}(N+q)} \ 2^{-p} \prod_{i=1}^{p} C(N-p+i) \ |XX'|^{\frac{1}{2}p} ,$$

XX' is a $q \times q$ matrix of rank $q < p_1$, $p_1+p_2 = p$, and $C(N)$ represents the surface area of a unit N dimensional sphere. Let $\hat{B}$ be partitioned into two parts $\hat{B}_1$ and $\hat{B}_2$, $\hat{B}_1$ $p_1 \times q$, $\hat{B}' = (\hat{B}_1' \ \hat{B}_2')'$, and let the corresponding partitions of Σ, Σ^{-1}, S, and B be

$$(3) \quad S = \begin{pmatrix} S_{11} & S_{12} \\ S_{21} & S_{22} \end{pmatrix}, \quad \Sigma = \begin{pmatrix} \Sigma_{11} & \Sigma_{12} \\ \Sigma_{21} & \Sigma_{22} \end{pmatrix}, \quad \Sigma^{-1} = \begin{pmatrix} \Sigma^{11} & \Sigma^{12} \\ \Sigma^{21} & \Sigma^{22} \end{pmatrix},$$

and $B' = (B_1' \ B_2')'$. Then Rao's generalized U statistic is

$$(4) \quad U = \frac{|S|}{|S + \hat{B}XX'\hat{B}'|} \ \frac{|S_{11} + \hat{B}_1 XX' \hat{B}_1'|}{|S_{11}|}$$

The U statistic is used for testing the hypothesis that $B_2 = 0$, against $B_2 \neq 0$, when $B_1 = 0$ is given. Under the null hypothesis U is distributed as $U_{N-p_1,\, q,\, p_2}$, see Anderson ([1],pp. 191-193). Khatri [9] obtains the nonnull distribution of U when B_2 is of rank unity, i.e., he obtains the distribution of Rao's U statistic, see Kabe [7]. We give here the nonnull distribution of U when B_2 is of full rank $q \leq p_1$, and we derive the results directly and straightforwardly in terms of original variates, unlike Khatri who resorts to triangular transformations of the original variates.

The general structure of the distribution problem of U is important in statistical literature, see e.g., Gleser and Olkin [3], [4], who are interested in obtaining the nonnull distribution of U, when S and $\hat{B}$ have the density (1). They show that the testing problem and the distribution of U remain invariant under a group of triangular matrix transformations. This means that the maximal invariant in the parameter space and its counterpart in the sample space may be transformed to their canonical forms and hence the derivations may be simplified to some extent. However a considerable amount of matrix algebra required for these canonical transformations may obscure the statistical properties of the original problem, and thus the derivations in terms of original variates may be preferable.

The derivation given here exactly parallels the one given by Kabe [7], when $q = 1$. Some results which are found useful in the sequel are stated in the next section, and the main result of the paper is given in section 3. We assume that all the integrals occurring in this paper are evaluated over the appropriate ranges of the variables of integration.

2. Some Useful Results

Let Y be a $p \times N$ matrix, $-\infty < Y < \infty$, D a given $q \times N$ matrix of rank $q(<N)$, μ a $p \times N$ matrix of constant terms, A an $N \times N$ positive definite symmetric matrix, $N \geq p + q$. Then Kabe [8] proves that

$$(5)\quad \int_{(Y-\mu)A(Y-\mu)' = G,\ DY' = V'} f((Y-\mu)A(Y-\mu)')\,dY$$

$$= 2^{-p} \prod_{i=1}^{p} C(N-p-q+i)\,|DA^{-1}D'|^{-\frac{1}{2}p}|A|^{-\frac{1}{2}p} f(G)$$

$$|G - (V-\mu D')(DA^{-1}D')^{-1}(V-\mu D')'|^{\frac{1}{2}(N-p-q-1)} .$$

Here G is $p \times p$, V is $p \times q$. When f is a suitable density function, then, obviously, the right hand side of (5) represents the joint density of G and V.

If Y is $p \times N$, $-\infty < Y < \infty$, then we have that

$$(6)\quad \int \exp\{-\tfrac{1}{2}\mathrm{tr}\Sigma^{-1}[YAY' + (Y-\mu)\Phi(Y-\mu)']\}dY$$

$$= (2\pi)^{\frac{1}{2}pN}|\Sigma|^{\frac{1}{2}N}|A + \Phi|^{-\frac{1}{2}p}\exp\{-\tfrac{1}{2}\mathrm{tr}\Sigma^{-1}\mu\Phi(A + \Phi)^{-1}A\mu'\}$$

Let S be $p \times p$, Y $p \times N$, A $N \times N$, then we know that

(7) $$(S + YAY')^{-1} = S^{-1} - S^{-1} Y(A^{-1} + Y'SY)^{-1} Y'S^{-1} ,$$

and hence we deduce that

(8) $$Y'S^{-1} Y = Y'\rho^{-1} Y(I - AY'\rho^{-1}Y)^{-1} , \quad \rho = S + YAY' .$$

Kabe [6], and Hyakawa [5] evaluate the following integral which occurs in the theory of noncentral multivariate beta distributions

(9) $$\int \exp \{-\tfrac{1}{2}\mathrm{tr}\Sigma_{11}^{-1}S_{11} + \mathrm{tr}(XX'B_1'\Sigma_{11}^{-1} S_{11}\Sigma_{11}^{-1} B_1 XX')^{\frac{1}{2}}z\}$$
$$|S_{11}|^{\frac{1}{2}(N-p_1+q-1)} |M_{11} - zz'|^{\frac{1}{2}(p_1-2q-1)} dz\, dS_{11}$$
$$= C_6 |M_{11}|^{\frac{1}{2}(p_1-q-1)} {}_1F_1(\tfrac{1}{2}N + \tfrac{1}{2}q, \tfrac{1}{2}N, XX'B_1' \Sigma_{11}^{-1} B_1 XX'M_{11}) ,$$

where z is $q \times q$ nonsymmetric matrix and M_{11} is $q \times q$ symmetric matrix, and

(10) $$C_6 = 2^{p_1} |\Sigma_{11}|^{\frac{1}{2}(N+q)} (2\pi)^{\frac{1}{2}Np_1} \frac{\Gamma_{p_1}(\tfrac{1}{2}q+\tfrac{1}{2}N)}{\Gamma_{p_1}(\tfrac{1}{2}N) \prod_{i=1}^{p_1} C(N-p_1-q+i)} .$$

Here

(11) $$\Gamma_m(u) = \pi^{\frac{1}{2}m(m-1)} \prod_{i=1}^{m} \Gamma[u-\tfrac{1}{2}(i-1)]$$, and the hypergeometric series ${}_1F_1$ of matrix argument is defined by Constantine [2]. The Jacobian of transformation $J(M_{11}:M_{11}^*)$ defined by the relation

$$M_{11} = M_{11}^* (I + M_{11}^* XX')^{-1}$$ is

(12) $$J(M_{11}:M_{11}^*) = |I + M_{11}^* XX'|^{-(q+1)}$$

Now we proceed to derive the density of U .

3. Distribution of U

By using (1) we write the joint density of S_{11}, S_{21}, D_{22}, $\hat{B}_1$, and z_2 as

(13) $$g(S_{11}, S_{21}, D_{22}, \hat{B}_1, z_2)$$
$$= C_1 \exp \{-\tfrac{1}{2}\mathrm{tr}\Sigma_{11}^{-1}S_{11} - \tfrac{1}{2}\mathrm{tr}\Sigma^{22}D_{22}$$
$$-\tfrac{1}{2}\mathrm{tr}\Sigma^{22} (S_{21} - \Sigma_{21}\Sigma_{11}^{-1}S_{11}) S_{11}^{-1} (S_{21} - \Sigma_{21}\Sigma_{11}^{-1}S_{11})'$$
$$-\tfrac{1}{2}\mathrm{tr}\Sigma_{11}^{-1} (\hat{B}_1 - B_1) XX' (\hat{B}_1 - B_1)'$$

$$-\tfrac{1}{2}\mathrm{tr}\Sigma^{22}(z_2-\eta_2+S_{21}S_{11}^{-1}\hat{B}_1-\Sigma_{21}\Sigma_{11}^{-1}\hat{B}_1)XX'(z_2-\eta_2$$
$$+S_{21}S_{11}^{-1}\hat{B}_1-\Sigma_{21}\Sigma_{11}^{-1}\hat{B}_1)'\}|S_{11}|^{\frac{1}{2}(N-p_1-p_2-1)}$$
$$|D_{22}|^{\frac{1}{2}(N-p_1-p_2-1)},$$

where

(14) $D_{22} = S_{22} - S_{11}^{-1}S_{12}$, $z_2 = \hat{B}_2 - S_{21}S_{11}^{-1}\hat{B}_1$, $\eta_2 = B_2 - \Sigma_{21}\Sigma_{11}^{-1}B_1$.

By using (5) and assuming $p_1 > p_2$, we integrate out S_{21} from (13) over the region

(15) $(S_{21} - \Sigma_{21}\Sigma_{11}^{-1}S_{11})S_{11}^{-1}(S_{21} - \Sigma_{21}\Sigma_{11}^{-1}S_{11})' = G$, $\hat{B}_1'S_{11}^{-1}S_{21}' = V'$,

and we find the joint density of G, V, $\hat{B}_1$, Z_2, D_{22}, S_{11} to be

(16) $$g(G, V, S_{11}, \hat{B}_1, Z_2, D_{22})$$
$$= C_1C_2 \exp\{-\tfrac{1}{2}\mathrm{tr}\Sigma_{11}^{-1}S_{11} - \tfrac{1}{2}\mathrm{tr}\Sigma_{11}^{-1}(\hat{B}_1 - B_1)XX'(\hat{B}_1 - B_1)'$$
$$-\tfrac{1}{2}\mathrm{tr}\Sigma^{22}(G + D_{22})$$
$$-\tfrac{1}{2}\mathrm{tr}\Sigma^{22}(Z_2 - \eta_2+V - \Sigma_{21}\Sigma_{11}^{-1}\hat{B}_1)XX'(Z_2-\eta_2+V-\Sigma_{21}\Sigma_{11}^{-1}\hat{B}_1)'\}$$
$$|S_{11}|^{\frac{1}{2}(N-p_1-1)}|D_{22}|^{\frac{1}{2}(N-p_1-p_2-1)}|\hat{B}_1'S_{11}^{-1}\hat{B}_1|^{-\frac{1}{2}p_2}$$
$$|G-(V-\Sigma_{21}\Sigma_{11}^{-1}\hat{B}_1)(\hat{B}_1S_{11}^{-1}\hat{B}_1)^{-1}(V-\Sigma_{21}\Sigma_{11}^{-1}\hat{B}_1)'|^{\frac{1}{2}(p_1-p_2-q-1)},$$

where

(17) $$C_2 = 2^{-p_2}\prod_{i=1}^{p_2} C(p_1 - p_2 - q + i) .$$

Now we integrate out G from (16) by setting

(18) $$G_1 = G-(V-\Sigma_{21}\Sigma_{11}^{-1}\hat{B}_1)(\hat{B}_1'S_{11}^{-1}\hat{B}_1)^{-1}(V-\Sigma_{21}\Sigma_{11}^{-1}\hat{B}_1)' ,$$

the integral with respect to G_1 is a known integral, see Anderson ([1], p.177, example 6), and we find the joint density of V, S_{11}, D_{22}, $\hat{B}_1$, Z_2 . We integrate out V from this joint density by using (6), and we have that

(19) $$g(S_{11}, D_{22}, \hat{B}_1, Z_2) = C_1C_2C_3C_4 \exp\{-\tfrac{1}{2}\mathrm{tr}\Sigma_{11}^{-1}S_{11}$$
$$-\tfrac{1}{2}\mathrm{tr}\Sigma_{11}^{-1}(\hat{B}_1 - B_1)XX'(\hat{B}_1 - B_1)' - \tfrac{1}{2}\mathrm{tr}\Sigma^{22}D_{22}$$

$$-\tfrac{1}{2}\mathrm{tr}\Sigma^{22}(Z_2 - \eta_2)XX'(I + \hat{B}_1'S_{11}^{-1}\hat{B}_1XX')^{-1}(Z_2 - \eta_2)'\}$$

$$|S_{11}|^{\frac{1}{2}(N-p_1-1)}|D_{22}|^{\frac{1}{2}(N-p_1-p_2-1)}|I + \hat{B}_1'S_{11}^{-1}\hat{B}_1 XX'|^{-\frac{1}{2}p_2} .$$

where

$$(20) \quad C_3 = \pi^{\frac{1}{4}p_2(p_2-1)}2^{\frac{1}{2}p_2(p_1-q)}|\Sigma^{22}|^{-\frac{1}{2}(p_1-q)}\prod_{i=1}^{p_2}\Gamma(\tfrac{1}{2}[p_1-q+1-i]) ,$$

and

$$(21) \quad C_4 = (2\pi)^{\frac{1}{2}p_2q}|\Sigma^{22}|^{-\frac{1}{2}q} .$$

Now we set $S_{11} + \hat{B}_1 XX'\hat{B}_1' = \rho$, and consider the integral

$$(22) \quad \int d\rho \int_{\hat{B}_1'\rho^{-1}\hat{B}_1 = M_{11}} \exp\{-\tfrac{1}{2}\mathrm{tr}\Sigma_{11}^{-1}\rho + \mathrm{tr}\, XX' B_1'\Sigma_{11}^{-1}\hat{B}_1\}$$

$$|\rho|^{\frac{1}{2}(N-p_1-1)}|(XX')^{-1} - \hat{B}_1'\rho^{-1}\hat{B}_1|^{\frac{1}{2}(N-p_1-1)} d\hat{B}_1 ,$$

which is evaluated by using (5), and the density of M_{11} is

$$(23) \quad g(M_{11}) = C_5|I - M_{11}XX'|^{\frac{1}{2}(N-p_1-1)}$$

$$\int \exp\{-\tfrac{1}{2}\mathrm{tr}\Sigma_{11}^{-1}\rho + \mathrm{tr}(XX' B_1'\Sigma_{11}^{-1}\rho\Sigma_{11}^{-1}B_1 XX')^{\frac{1}{2}} Z\}$$

$$|\rho|^{\frac{1}{2}(N-p_1+q-1)}|M_{11} - ZZ'|^{\frac{1}{2}(p_1-2q-1)} d\rho\, dZ$$

$$= C_5C_6|I - M_{11}XX'|^{\frac{1}{2}(N-p_1-1)}|M_{11}|^{\frac{1}{2}(p_1-q-1)}$$
$${}_1F_1[\tfrac{1}{2}N + \tfrac{1}{2}q; \tfrac{1}{2}N; XX' B_1'\Sigma_{11}^{-1}B_1 XX' M_{11}] ,$$

where

$$(24) \quad C_5 = 2^{-q}\prod_{i=1}^{q}C(p_1-2q+i) ,$$ and C_6 is given by (10) .

By using (12) and (23) we obtain the density of $\hat{B}_1'S_{11}^{-1}\hat{B}_1 = M_{11}^*$.
Thus the joint density of the variates M_{11}^* , D_{22}, and Z_2 is

$$(25) \quad g(M_{11}^*, D_{22}, Z_2) = C_1C_2C_3C_4C_5C_6 \exp\{-\tfrac{1}{2}\mathrm{tr}\Sigma_{11}^{-1}B_1 XX' B_1'$$

$$-\tfrac{1}{2}\mathrm{tr}\Sigma^{22}(D_{22} + Z_2 XX'(I + M_{11}^* XX')^{-1}Z_2')$$

$$+ \mathrm{tr}\Sigma^{22}\eta_2 XX'(I + M_{11}^* XX')^{-1}Z_2'$$

$$-\tfrac{1}{2}\mathrm{tr}\Sigma^{22}\eta_2(I + M_{11}^* XX')^{-1}\eta_2'\}$$

$$|M_{11}^*|^{\frac{1}{2}(p_1-q-1)}|I + M_{11}^* XX'|^{-\frac{1}{2}(N+q+p_2+1)}|D_{22}|^{\frac{1}{2}(N-p_1-p_2-1)}$$

$${}_1F_1\left[\tfrac{1}{2}N+\tfrac{1}{2}q; \tfrac{1}{2}N; XX' B_1'\Sigma_{11}^{-1}B_1 XX' M_{11}^*(I + M_{11}^* XX')^{-1}\right] .$$

Now from (4) we note that

$$(26)\quad U = |I + M_{22}^* XX' (I + M_{11}^* XX')^{-1}|^{-1} ,$$

where $M_{22}^* = Z_2' D_{22}^{-1} Z_2$. Thus for given M_{11}^* it remains to find the density of M_{22}^* from the joint, conditional on M_{11}^* , density of D_{22} and Z_2 . We note that M_{22}^* has a noncentral multivariate beta density of the second kind. The density of M_{22}^* is found exactly on same lines as that of M_{11}^* , and the joint density of M_{11}^*, M_{22}^* , the maximal invariants of the sample space, is

$$\begin{aligned}(27)\quad g(M_{11}^*, M_{22}^*) &= C_1C_2C_3C_4C_5C_6C_7C_8 \\ &\exp\{-\tfrac{1}{2}\mathrm{tr}\Sigma_{11}^{-1} B_1 XX' B_1' - \tfrac{1}{2}\mathrm{tr}\Sigma^{22}\eta_2 XX' (I + M_{11}^* XX')^{-1}\eta_2\} \\ &|M_{11}^*|^{\frac{1}{2}(p_1-q-1)} |I + M_{11}^* XX'|^{-\frac{1}{2}(N+q+p_2+1)} \\ &|M_{22}^*|^{\frac{1}{2}(p_2-q-1)} |I + M_{22}^* XX' (I + M_{11}^* XX')^{-1}|^{-\frac{1}{2}(N-p_1+q+1)} \\ &{}_1F_1\left[\tfrac{1}{2}N + \tfrac{1}{2}q; \tfrac{1}{2}N; XX' B_1' \Sigma_{11}^{-1} B_1 XX' M_{11}^* (I + M_{11}^* XX')^{-1}\right] \\ &{}_1F_1\left[\tfrac{1}{2}(N-p_1+q); \tfrac{1}{2}(N-p_1); XX' (I + M_{11}^* XX')^{-1} \eta_2\Sigma^{22}\eta_2\right. \\ &\left.(I + M_{11}^* XX')^{-1} M_{22}^* (I + M_{22}^*)^{-1}\right] ,\end{aligned}$$

where

$$(28)\quad C_7 = 2^{-q} \prod_{i=1}^{q} C(p_2-2q+i)$$

and

$$(29)\quad C_8 = \frac{2^{p_2}\,\Gamma_{p_2}(\tfrac{1}{2}[N-p_1+g])}{\Gamma_{p_2}(\tfrac{1}{2}[N-p_1])}\,(2\pi)^{\frac{1}{2}(N-p_1)p_2}\,\frac{|\Sigma^{22}|^{-\frac{1}{2}(N-p_1+q)}}{\prod_{i=1}^{p_2} C(N-p_2-q+i)} .$$

It appears to be difficult to obtain the marginal density of U in the nonnull case. However, the moments of U may be obtained in a complicated form.

In case $m < (p_1+p_2)$, then we may use the criterion

$$(30)\quad U_1 = \frac{|S|}{|S_{11}|}\,\frac{|S_{11} + Y_1Y_1'|}{|S + YY'|} ,$$

where $E(Y) = BX$ and Y is (p_1+p_2) xm, Y_1 is p_1 xm. U_1 is distributed, under null hypothesis, as $U_{N-p_1,m,\,p_2}$.

REFERENCES

[1] Anderson, T. W. (1958), An Introduction to Multivariate Statistical Analysis, John Wiley, New York.

[2] Constantine, A. G. (1963), Some noncentral distribution problems in multivariate analysis, Ann. Math. Statist. 34, 1270.

[3] Gleser, L. J. and Olkin, I. (1966), A k-sample regression model with covariance. Multivariate Analysis, Ed. P. R. Krishnaiah, Academic Press, New York.

[4] Gleser, L. J. and Olkin, I. (1970), Linear models in multivariate analysis. Essays in Probability and Statistics, Ed. Bose, R. C. et. al. The University of North Carolina Press.

[5] Hayakawa, Takesi (1967), On the distribution of the maximum latent root of a positive definite symmetric random matrix. Ann. Inst. Statist. Math. 19, 1.

[6] Kabe, D. G. (1972), On the noncentral distribution of a certain correlation matrix. S. Afr. Statist. J. 6, 27.

[7] Kabe, D. G. (1965), On the noncentral distribution of Rao's U Statistic, Ann. Inst. Statist. Math. 17, 75.

[8] Kabe, D. G. (1965), Generalization of Sverdrup's lemma and its applications to multivariate distribution theory, Ann. Math. Statist. 36, 671.

[9] Khatri C. G. (1965), Nonnull distribution of the likelihood ratio statistic for a particular kind of multicollinearity. J. Ind. Statist. Ass. 3, 212.

D. G. Kabe
Deaprtment of Mathematics
St. Mary's University
Halifax, Nova Scotia
Canada

DISTRIBUTION OF TWO CORRELATED HOTELLING'S T^2'S

A. M. Kshirsagar and John C. Young

1. Introduction

Let X be a $p \times n$ matrix representing a random sample of size n from a p-variate normal population with mean vector $\underline{\mu}$ and variance-covariance matrix Σ. While considering multivariate statistical outliers, Wilks (1963) proposed the statistics

$$W_r = |A_r|/|A| \ , \ r = 1, 2, \ldots, n \tag{1.1}$$

where A is the matrix of the corrected sum of squares and products of all the n sample observations and A_r is the same matrix, when the rth observation (i.e., the rth column of X) is omitted. He considered some low order moments of these statistics and has stated that their joint distribution is very complicated. We shall first examine any two of these W's say W_{n-1} and W_n only, in order to consider their joint distribution. Let $\underline{x}_r$ denote the rth column of X. Consider the orthogonal transformation, as given by a Helmert's orthogonal matrix.

$$\begin{aligned}
\underline{y}_1 &= \frac{1}{\sqrt{(1)(2)}} (\underline{x}_1 - \underline{x}_2) \\
\underline{y}_2 &= \frac{1}{\sqrt{(2)(3)}} (\underline{x}_1 + \underline{x}_2 - 2\underline{x}_3) \\
&\vdots \\
\underline{y}_{n-2} &= \frac{1}{\sqrt{(n-2)(n-1)}} (\underline{x}_1 + \ldots + \underline{x}_{n-2} - (n-2)x_{n-1}) \\
\underline{y}_{n-1} &= \frac{1}{\sqrt{(n-1)n}} (\underline{x}_1 + \ldots + \underline{x}_{n-1} - (n-1)x_n) \\
\underline{y}_n &= \frac{1}{\sqrt{n}} (\underline{x}_1 + \ldots + \underline{x}_n) \ .
\end{aligned} \tag{1.2}$$

Also define

$$\begin{aligned}
z_{n-2} &= \frac{1}{\sqrt{(n-2)(n-1)}} (\underline{x}_1 + \ldots + \underline{x}_{n-2} - (n-2)\underline{x}_n) \\
z_{n-1} &= \frac{1}{\sqrt{(n-1)n}} (\underline{x}_1 + \ldots + \underline{x}_{n-2} + \underline{x}_n - (n-1)\underline{x}_{n-1}) \ .
\end{aligned} \tag{1.3}$$

Let E_{ab} denote an $a \times b$ matrix of all unit elements. Then it is easy to verify that

$$A = X(I - \frac{1}{n} E_{nn})X' ,$$

$$= \sum_{r=1}^{n-1} \underline{y}_r \underline{y}_r' \quad (1.4)$$

$$A_n = \sum_{r=1}^{n-2} \underline{y}_r \underline{y}_r' \quad . \quad (1.5)$$

Also observe that

$$y_{n-2} + \{(n-2)n\}^{\frac{1}{2}} \underline{y}_{n-1} = (n-1)\underline{z}_{n-2} \quad (1.6)$$

and that

$$A_{n-1} = \sum_{r=1}^{n-3} \underline{y}_r \underline{y}_r' + \underline{z}_{n-2}\underline{z}_{n-2}' \quad . \quad (1.7)$$

Then

$$W_n = |\sum_{r=1}^{n-2} \underline{y}_r \underline{y}_r'| / |\sum_{r=1}^{n-1} \underline{y}_r \underline{y}_r'| \quad (1.8)$$

$$W_{n-1} = |\sum_{r=1}^{n-3} \underline{y}_r \underline{y}_r' + \underline{z}_{n-2}\underline{z}_{n-2}'| / |\sum_{r=1}^{n-1} \underline{y}_r \underline{y}_r'| \quad . \quad (1.9)$$

Since the transformation (1.2) is orthogonal, it can be readily proved that $\underline{y}_r$ $(r = 1,2,\ldots,n-1)$ are independent, each having a p-variate normal distribution with zero means and variance-covariance matrix Σ . Even when W_n , W_{n-1} are expressed in this comparatively simpler form, the problem of the joint distribution of W_n , W_{n-1} appears to be difficult. However, if instead of the corrected sum of squares and products matrix A , the matrix of s.s. and s.p. of observations about the true means is considered, a little algebra will show that, we will get statistics of the form

$$t_1^2 = |\sum_{r=1}^{n-2} y_r y_r'| / |\sum_{r=1}^{n-1} \underline{y}_r \underline{y}_r'| \quad (1.10)$$

and

$$t_2^2 = |\sum_{r=1}^{n-3} \underline{y}_r \underline{y}_r' + \underline{y}_{n-1}\underline{y}_{n-1}'| / |\sum_{r=1}^{n-1} \underline{y}_r \underline{y}_r'| \quad . \quad (1.11)$$

Each is related to a Hotelling's T^2 distribution and the two are correlated. The joint distribution of these statistics is derived in this paper. Since the distribution will not involve Σ , it is assumed that $\Sigma = I$ and $\underline{\mu} = \underline{0}$.

2. Joint Distribution of t_1^2 , t_2^2

Let $B = \sum_{r=1}^{n-3} \underline{y}_r \underline{y}_r'$. Then the joint distribution of B, $\underline{y}_{n-2}$,

$\underline{y}_{n-1}$ is easily seen to be

$$W_p(B|I|f)dB\cdot\frac{1}{(2\pi)^p}\exp[-\tfrac{1}{2}\underline{y}'_{n-2}\underline{y}_{n-2}-\tfrac{1}{2}\underline{y}'_{n-1}\underline{y}_{n-1}]d\underline{y}_{n-2}\underline{y}_{n-1}\cdots \tag{2.1}$$

where

$$W_p(B|I|f) = C_{pf}|B|^{(f-p-1)/2}\,e^{-(1/2)\mathrm{tr}\,B}, \tag{2.2}$$

$$C_{pf}^{-1} = 2^{fp/2}\,\pi^{p(p-1)/4}\prod_{i=1}^{p}\Gamma\left(\frac{f+1-i}{2}\right) \tag{2.3}$$

$$f = n-3 .$$

In the distribution (2.1), make the following transformation

$$D = B + \underline{y}_{n-2}\underline{y}'_{n-2} + \underline{y}_{n-1}\underline{y}'_{n-1} \tag{2.4}$$

$$\underline{u} = D^{-1/2}\underline{y}_{n-2}$$

$$\underline{v} = D^{-1/2}\underline{y}_{n-1}$$

where $D^{-1/2}$ is any matrix such that $(D^{-1/2})^2 = D^{-1}$. The Jacobian of transformation from B to D is 1 and that of $\underline{y}_{n-2}$ to $\underline{u}$ or $\underline{y}_{n-1}$ to $\underline{v}$ is $|D|^{1/2}$. Hence the joint distribution of D, $\underline{u}$, $\underline{v}$ is

$$C_{pf}(2\pi)^{-p}|D|^{(f-p+1)/2}\exp(-\mathrm{tr}D/2)\cdot|I-\underline{u}\underline{u}'-\underline{v}\underline{v}'|^{(f-p-1)/2}\,dDd\underline{u}d\underline{v}, \tag{2.5}$$

as

$$|B| = |D-D^{1/2}\underline{u}\underline{u}'D^{1/2} - D^{1/2}\underline{v}\underline{v}'D^{1/2}| = |D||I-\underline{u}\underline{u}'-\underline{v}\underline{v}'| . \tag{2.6}$$

This shows that D, $\underline{u}$, $\underline{v}$ are independent. Adjusting the constant in (2.5) suitably, the joint distribution of $\underline{u}$, $\underline{v}$ is

$$\frac{\Gamma(f+1)}{(2\pi)^p\,\Gamma(f-p+1)}|I-\underline{u}\underline{u}'-\underline{v}\underline{v}'|^{(f-p-1)/2}d\underline{u}d\underline{v} . \tag{2.7}$$

Observe that the statistics t_1^2, t_2^2 of (1.10) and (1.11) are

$$t_1^2 = \frac{|B + \underline{y}_{n-2}\underline{y}'_{n-2}|}{|D|} = \frac{|D - \underline{y}_{n-1}\underline{y}'_{n-1}|}{|D|}$$

$$= |I - \underline{u}\underline{u}'| = 1 - \underline{u}'\underline{u} \tag{2.8}$$

and similarly

$$t_2^2 = 1 - \underline{v}'\underline{v} \tag{2.9}$$

Also observe that in (2.7),

$$|I - \underline{u}\underline{u}' - \underline{v}\underline{v}'| = (1 - \underline{u}'\underline{u})(1 - \underline{v}'\underline{v}) - (\underline{u}'\underline{v})^2$$

$$= t_1^2\, t_2^2 - (\underline{u}'\underline{v})^2 . \tag{2.10}$$

In (2.7), transform from $\underline{v}$ to $\underline{\xi} = [\xi_1,\ldots,\xi_p]'$ by an orthogonal transformation

$$\underline{\xi} = L\,\underline{v} \tag{2.11}$$

where L is a $p \times p$ orthogonal matrix, with $\underline{u}'/(\underline{u}'\underline{u})^{1/2}$ as its last row. The Jacobian of this transformation is $|L| = 1$ and $\underline{v}'\underline{v} = \underline{\xi}'\underline{\xi} = 1 - t_2^2$. Also,

$$\underline{u}'\underline{v} = \underline{u}'L'L\underline{v} = [0,\ldots,0,(\underline{u}'\underline{u})^{1/2}]\underline{\xi} = (1 - t_1^2)^{1/2} \cdot \xi_p . \tag{2.12}$$

The joint distribution of $\underline{u}$ and $\underline{\xi}$ is therefore

$$\frac{\Gamma(f+1)}{(2\pi)^p\Gamma(f-p+1)} \{t_1^2\, t_2^2 - (1 - t_1^2)\xi_p^2\}^{(f-p-1)/2} d\underline{u}d\underline{\xi} . \tag{2.13}$$

From $\underline{u}$, transform to $t_1^2 = 1-\underline{u}'\underline{u}$ and $p-1$ other variables $\phi_1, \phi_2,\ldots,\phi_{p-1}$ by the transformation

$$u_1 = (1 - t_1^2)^{1/2} \cos\phi_1 \cos\phi_2 \cdots \cos\phi_{p-1}$$

$$u_j = (1 - t_1^2)^{1/2} \cos\phi_1 \cos\phi_2 \cdots \cos\phi_{p-j} \sin\phi_{p-j+1}$$

$$j = 2,3,\ldots,p . \tag{2.14}$$

Similarly transform from $\underline{\xi}$ to $t_2^2 = 1 - \underline{\xi}'\underline{\xi}$ and $p-1$ other variables $\theta_1, \theta_2,\ldots,\theta_{p-1}$ by

$$\xi_1 = (1 - t_2^2)^{1/2} \cos\theta_1 \cos\theta_2 \cdots \cos\theta_{p-1}$$

$$\xi_j = (1 - t_2^2)^{1/2} \cos\theta_1 \cos\theta_2 \cdots \cos\theta_{p-j} \sin\theta_{p-j+1}$$

$$(j = 2,3,\ldots,p) \tag{2.15}$$

The Jacobian of the transformations are

$$\tfrac{1}{2}(1 - t_1^2)^{\frac{1}{2}p-1} \prod_{i=1}^{p-2} \cos^{p-i-1}\phi_i$$

and

$$\tfrac{1}{2}(1 - t_2^2)^{\frac{1}{2}p-1} \prod_{i=1}^{p-2} \cos^{p-i-1}\theta_i .$$

Note that θ_{p-1} and ϕ_{p-1} vary from 0 to 2π; the other θ's and ϕ's vary from $-\pi/2$ to $\pi/2$ while t_1^2 , t_2^2 vary from 0 to 1 .

Integrating out all the ϕ's and θ's except θ_1, the joint distribution of t_1^2, t_2^2, θ, comes out as

$$\frac{\Gamma(f+1)}{4\pi\Gamma(p-1)\Gamma(f-p+1)} \{t_1^2 t_2^2 - (1 - t_1^2)(1 - t_2^2)\sin^2\theta_1\}^{(f-p-1)/2} \cos^{p-2}\theta_1 \, dt_1^2 dt_2^2 d\theta_1 . \quad (2.16)$$

To integrate out θ_1, one must expand the density in (2.16) in powers of $\sin^2\theta_1$ and then integrate it out. This will then yield the joint distribution of t_1^2 and t_2^2.

3. Moments of t_1^2, t_2^2

Only the product moment of t_1^2 and t_2^2 is difficult. The mean and variance of t_1^2 (or of t_2^2) can be easily obtained from the marginal distributions of t_1^2, t_2^2 which are related to Hotelling's T^2's. Thus

$$t_1^2 = \frac{1}{1 + T_1^2/(f+1)}$$

where T_1^2 is Hotelling's T^2 based on $f+1$ d.f.

In the joint distribution of $\underline{u}$ and $\underline{v}$, given by (2.7), transform from $\underline{v}$ to $\underline{\eta}$ by

$$\underline{v} = (I - \underline{u}\underline{u}')^{1/2}\,\underline{\eta} . \quad (3.1)$$

Then we shall find that $\underline{u}$ and $\underline{\eta}$ are independently distributed as $K(\underline{u}|f)d\underline{u}$ and $K(\underline{\eta}|f-1)d\underline{\eta}$ respectively, where

$$K(\underline{u}|f)d\underline{u} = \frac{f}{\pi^{p/2}(f-p)} \cdot \frac{\Gamma(f/2)}{\Gamma\{(f-p)/2\}} |I - \underline{u}\underline{u}'|^{(f-p)/2} d\underline{u} . \quad (3.2)$$

From (3.2), one can easily show that

$$E(t_1^2)^h = E(1 - \underline{u}'\underline{u})^h = E|I - \underline{u}\underline{u}'|^h$$

$$= \frac{f(f + 2h - p)}{(f-p)(f+2h)} \cdot \frac{\Gamma\left(\frac{f-p}{2} + h\right)\Gamma\left(\frac{f}{2}\right)}{\Gamma\left(\frac{f}{2} + h\right)\Gamma\left(\frac{f-p}{2}\right)} . \quad (3.3)$$

This will also be the hth moment of t_2^2 by symmetry. This leads to

$$E(t_1^2) = \frac{f-p+2}{f+2} , \quad (3.4)$$

$$V(t_1^2) = \frac{2p(f-p+2)}{(f+2)^2(f+4)} . \quad (3.5)$$

Now

$$\begin{aligned}\mathrm{Cov}(t_1^2, t_2^2) &= E\{(1-\underline{u}'\underline{u})(1-\underline{v}'\underline{v})\} - E(t_1^2)E(t_2^2)\\ &= E\{(1-\underline{u}'\underline{u})[1-\underline{\eta}'(I-\underline{u}\underline{u}')\underline{\eta}]\} - \{E(t_1^2)\}^2\\ &= E(t_1^2) - E\{t_1^2[\underline{\eta}'\underline{\eta} - (\underline{\eta}'\underline{u})^2]\} - \{E(t_1^2)\}^2\\ &= E(t_1^2) - E(t_1^2)\,E(\underline{\eta}'\underline{\eta}) + E\{t_1^2(\underline{\eta}'\underline{u})^2\} - \{E(t_1^2)\}^2 \qquad (3.6)\end{aligned}$$

as t_1^2 and η are independent. Since $\underline{\eta}$ has the same distribution as $\underline{u}$ with f changed to $f-1$,

$$\begin{aligned}E(\underline{\eta}'\underline{\eta}) &= 1 - E(1-\underline{u}'\underline{u}) \quad \text{with } f \text{ replaced by } f-1\\ &= p/(f+1) \qquad (3.7)\end{aligned}$$

Hence (3.6) reduces to

$$\mathrm{Cov}(t_1^2, t_2^2) = \frac{-p(f-p+2)}{(f+1)(f+2)^2} + E\{t_1^2(\underline{\eta}'\underline{u})^2\} \quad . \qquad (3.8)$$

Now

$$E\{t_1^2(\underline{\eta}'\underline{u})^2\} = \int (1-u'u)(\underline{\eta}'\underline{u})^2\ K(\underline{u}|f)K(\underline{\eta}|f-1)\,d\underline{u}\,d\underline{\eta} \ \ldots \qquad (3.9)$$

where the integration is over the range of values of $\underline{u}$ and $\underline{\eta}$ such that $\underline{u}'\underline{u} \leq 1$, $\underline{\eta}'\underline{\eta} \leq 1$. Transform from $\underline{\eta}$ to $\underline{g} = [g_1,\ldots,g_p]$ by the transformation

$$\underline{g} = L\,\underline{\eta}$$

where L is already defined as the orthogonal matrix with $\underline{u}'/(\underline{u}'\underline{u})^{1/2}$ as its last row. Then

$$\underline{\eta}'\underline{u} = \underline{\eta}'L'L\underline{u} = \underline{g}'L\underline{u} = g_p(\underline{u}'\underline{u})^{1/2} = (1-t_1^2)^{1/2}\ g_p \ .$$

Hence (3.9) reduces to

$$\begin{aligned}&\int t_1^2(1-t_1^2)\ K(\underline{u}|f)\,d\underline{u} \cdot \int g_p^2\ K(\underline{g}|f-1)\,d\underline{g}\\ &= E(t_1^2 - t_1^4) \cdot \frac{1}{p}\,E(\underline{g}'\underline{g}) \ , \qquad (3.10)\end{aligned}$$

due to the symmetry of the distribution of $\underline{g}$. But $\underline{g}$ has the same distribution as $\underline{u}$ with f replaced by $f-1$ and hence (3.10) reduces to

$$\frac{p(f-p+2)}{(f+4)(f+2)(f+1)} \ .$$

The covariance between t_1^2 and t_2^2, therefore comes out as (from (3.6))

$$\frac{-2p(f-p+2)}{(f+1)(f+2)^2(f+4)} \ . \qquad (3.11)$$

Reference

[1] Wilks, S. S. (1963). "Multivariate Statistical Outliers," Sankhya B, 25, 407.

A. M. Kshirsagar
Institute of Statistics
Texas A & M University
College Station, Texas
U.S.A.

John C. Young
Department of Statistics
Southern Methodist University
Dallas, Texas
U.S.A.

REMARKS ON MULTIVARIATE MODELS BASED ON THE NOTION OF STRUCTURE

Andre G. Laurent

I. <u>Introduction</u> During the last fifty years, there has been, for worse or better, a progressive mathematization of Statistics, which, originally classified among the social sciences, is considered by many, nowadays, a branch of mathematics rather than a discipline using mathematics as a tool; that is, one has witnessed a shift of emphasis from the end to the means. Mathematics is logico-deductive and is characterized by the fact that its objects have no empirical referents. Statistics, by contrast, is essentially concerned with, or does, at least, mostly refer to empirical objects. The very expression "statistical inference" is not too meaningful without reference to the empirical world, even though the case may be made that Statistics is the study of measurable functions, associated with a very special class of measures. This is an old controversy. The nature and the diversity of the arguments that are raised seem to call for the creation of an anthropology and a sociology of Statistics and statisticians. This is a religious dispute and the quasireligious overtones of some of these discussions can be traced to the introductory statement of the statistical Bible "At the beginning was the Sample Space." The latter, in far more than six days, has been structured out of chaos according to many <u>a priori</u> probability models that have been forced upon it by some <u>Deus ex machina</u> or, then, stemmed from some unconscious remembrance of Plato's realm of ideas. It has been ministered ever since by a closed and powerful priesthood. The point that I want to make is that the emphasis on empirical data as a starting point of conceptual thought has shifted to an emphasis on a priori models, so that the statistic is viewed as the fleeting and uncertain shadow of the parameter and the frequency as a degraded and profane desecration of the probability rather than both parameters and probabilities as approximations of statistics and frequencies, or their abstract limits, or their non empirical idealizations. The "quest" of the statistician for the parameter or the probability is a mystical quest in the space of samples and statistics toward mythological, necessarily elusive, objects that belong to another space; and statistical methods are replica of the "Spiritual Exercises," while statistical theory is the theology induced by the mythology already mentioned.

The "old school" of statisticians has been frequently under attack for concentrating on such "meaningless" problems as to whether some statistics be "better" than others, or whether there be "best" ways of summarizing samples, without reference to any model; in other words, for raising the problem of descriptive Statistics with reference to empirical knowledge or theoretical but not probabilistically framed knowledge of empirical data rather than with reference to preconceived probability models--a view that is generally nowadays considered obsolete, as it has become axiomatic that a statistic can be assessed only within the context of estimation or testing hypotheses theory, that is, as estimators or test functions. The latter view is shared as well by the right wing of mythology fundamentalists and the left wing of decision theory agnostics. Being as it may, I feel that a strong case can be made for a new approach to descriptive Statistics and for descriptive Statistics as a starting point toward Probability Theory and, thus, for the re-examination of the concept of "physical probability" whether subjective or objective. I feel that very much is lost in a theory of measurements that chooses to have no reference to the measured quantity, and in a theory that deals with events as sets without reference to the members (states of nature) that constitute them (the crucial point for a theory of physical probability). Model building on the basis of empirical knowledge with the guidance of general paradigms should not be reconsidered; there may be great benefit in following the path opened by Herschel, Gauss, Maxwell, who, indeed, <u>start with the properties of the statistic and search for the model that be consistent with them, or approximate them, or embody them after a limiting process</u>. The recent development of, and the increased interest in, Structuralism [33] have suggested, here, the concept of "<u>structure</u>" and the notion of <u>entropy</u> (in a not too technical sense) as paradigmatic for the building and the interpretation, through a quasi "integral geometry," of multivariate models, as tentatively examplified in the following paragraphs. This current of thought seems germane, even if in a remote way, to that of Fraser [9], Good [10], Jaynes [11], and Kullback [16], to name only a few, even though it developed somewhat independently. There is some evidence that there exists in the statistical community a scattered interest in a synthetic theory that would unify and blend information theoretic, group theoretic, integral geometry theoretic, utility theoretic, and statistical concepts, and which, someday, will be apprehended.

II. Laplace-Gaussian Model, Spherical Distributions, Spherical Structure. The distribution of a sample M_n of n observations of a random variable X with a univariate Laplace-Gaussian distribution (with zero mean) provides an example of one distribution of the class of spherical distributions, that is, of distributions that remain invariant under the group of orthogonal transformations and whose density functions depend only on the radius vector--a sufficient statistic for the entire class. It is well known that, under mild conditions, the property, for a family of the class, of factoring into a product of n univariate distributions is a characteristic of the Laplace-Gaussian model. Note also that the latter is maximum entropy under the constraint: $E[X^2] = \sigma^2$. The spherical class and its properties have been surveyed in a noteworthy paper by Lord [30]; many other properties belong to the folklore of the subject and are scattered here and there, or have been summarily discussed apropos other subjects in the literature; recently, Thomas [37] has presented some results about them, outlined a new operational calculus to discuss them, with several statistical applications. There is, therefore, no point in discussing this class in detail. (For a detailed treatment see [27].) The important characteristics of the spherical class are 1) the class is projective, that is, indexed by n, it is closed under projection; 2) the characteristic function of a spherical distribution is formally invariant under projection as a function of its radial argument; 3) the characteristic function is reducible to a univariate Hankel's transform; 4) the distributions of the class induce a conditional uniform superficial distribution on the sphere, which makes the radius vector R not only a sufficient statistic but an "absolutely sufficient" (A.S.) statistic in that sense that this conditional distribution does not depend on which family of the class is considered, but only on the nature of the statistic itself; 5) this conditional distribution is maximum entropy and, under normalization of the X_i's by R into the ancillary statistic $\{\phi_i = X_i/R\}$, it is parameter free and uniform on the unit sphere; 6) the latter is marginal as well as conditional (that is, independent from R) and, expressed in terms of $X_1,\ldots,X_{n-1}$ (one of its representations), it is a (n-1)-dimensional Thompson's distribution, whose "projection" on the k dimensional space (conditional distribution of $X_1,\ldots,X_k$, given R) is again a k-dimensional Thompson distribution [18] (Thompson [38] studied the case $k = 1$, in reduced form, and as a marginal distribution; he was unaware of its conditional and minimum variance estimator properties); 7) the projection just mentioned provides the minimum variance unbiased estimator of the joint distribution of $X_1,\ldots,X_k$ (a fact underlined and used in the case $k = 1$ and for the univariate Gaussian distribution by Lieberman and Reskinoff [29] among many others from the S.R.G.C.U. [35] to Tate [36]--the latter in a different context, and discussed by Laurent for any k [18] or, with applications to bombing problems, for the multivariate Gaussian distribution [19] to [26]); 8) mixtures of these Thompson's distributions--that are characteristics of the spherical structure--generate the whole class of spherical distributions; 9) the joint distribution of $M_k = (X_1,\ldots,X_k)$ and R, derived either from Thompson's distribution or from the Sverdrup-Kabe lemma [12], [13], provides the kernel of a univariate transform relating the p.d.f. of M_k to that of M_n, considered respectively as functions of the k and the n-dimensional radius vectors R_k and R_n, namely, the first p.d.f. is a Weyl's integral transform of order $(n-k)/2$ of the second--this way a projection operator from the n dimensional to the k dimensional class is defined, whose inversion, when possible, allows for the antiprojection of k into n-dimensional spherical distributions (or, possible, distributions in the sense of Schwartz), and the generalization of univariate symmetric into multivariate distributions (Thomas [37] studies this question from the viewpoint of an operational calculus induced by Mellin's transform); 10) the behavior of the coordinates of a point, uniformly distributed on the sphere of radius $\sqrt{n}\,\theta$, is asymptotically Laplace-Gaussian, a fact used by Maxwell in one of his several derivations of the law of velocities of particles in a gas with constant energy [32], and leading Emile Borel [3] and Paul Levy [28], through a limit process reasoning, to an interpretation of the normal distribution as induced by a uniform distribution on a sphere in the space L_2 --interpretation that geometrizes Gaussian phenomena, reduces their mechanisms to spherical structure properties, and throws a new light on the central limit theorems whose information theoretic aspect is thus suggested (for the information theoretic aspect of central limit theorems one may refer to Renyi [34]). The important conclusion is that the class of spherical distributions is a class whose construction and interpretation can be entirely based on, and are reducible to, notions of maximum uniformity and spherical structure. The attempt is made in the following paragraphs to generalize this approach and interpretation to other structures and other classes of distributions.

III. The Negative Exponential Model. Simplicial Distributions. Simplicial Structure. Results very similar to those obtained for the Laplace-Gaussian family of distributions and the class of spherical distributions generated by the corresponding Thompson's kernel or, equivalently, induced by the conditional maximum entropy distribution on the sphere--defined by the "absolutely sufficient" statistic $T_n = \sum_1^n X_1^2$ --can be obtained, using as a starting point the negative exponential model, which generalizes similarly to a class of "simplicial" distributions through the mixture of an exponential-Thompson kernel or through the consideration of maximum entropy distribution on the main facet of the E_n simplex--defined by the "absolutely sufficient" statistic $T_n = \sum_1^n X_i$. This case will be given a more detailed treatment, at the cost of occasional triviality and redundancy, because its extreme simplicity makes it a good starting point for further generalizations. Let $(X_1,\ldots,X_n)$ be a random sample of n observations, with arithmetic mean $\overline{X}_n$, of a random variable X with the negative exponential distribution

$$f(x;\theta) = (1/\theta)\exp(-x/\theta) \ , \quad x \geq 0$$
$$= 0 \ , \qquad x < 0 \ .$$

It should be noted that $f(x;\theta)$ is maximum entropy under the restriction $E[X] = \theta$, therefore the "closest" to Lebesgue measure density under that restriction. The sample likelihood depends on X through $n\overline{X}_n$ and induces the same sufficient partition of the sample space as $n\overline{X}_n$, which is sufficient and complete. The family $\{n\overline{X}_n\}$ is projective, that is, the distribution of M_k , $k < n$, depends only on $k\overline{X}_k$, and the conditional distribution of M_n , given $n\overline{X}_n = t$, expressed as the p.d.f. of $n-1$ coordinates (a "subsample" M_{n-1} of X) is:

$$f(M_{n-1}|n\overline{X}_n = t)\, dM_{n-1} = (n-1)!\,(1/t)^{n-1}, \quad \sum_1^{n-1} x_i \leq t \ , \ x_i \geq 0 \ ,$$

and, in terms of the elementary probability with respect to the elementary area $dS_{t,n}$ on $n\overline{X}_n = t$ is:

$$g(M_n|nX_n = t)\,dS_{t,n} = dS_{t,n}|S_{t,n} \ ,$$

where $S_{t,n}$ is the area of the main facet of the simplex defined by $n\overline{X}_n = t$, $X_i \geq 0$. Hence the conditional superficial density of M_n on the main facet of the simplex is constant, and maximum entropy under the restriction $n\overline{X}_n = t$; such is also the marginal density of the coordinates of M_n on the facet. This well-known fact is of great interest in connection with problems arising in connection with the random division of an interval and the study of conditional homogeneous Poisson's processes (see for example [6] and [15]). The statistic $n\bar{X}_n$ is "absolutely sufficient" in that the distribution of M_n given $n\bar{X}_n = t$ is independent from the distribution of M_n and any vector M_n with a p.d.f. with structure

$$f(\underset{\sim}{x}) = g_n(n\bar{x}_n) \quad , \quad x_i \geq 0$$
$$= 0 \quad , \quad x_i < 0 \quad ,$$

will enjoy identical properties, which are implied by the nature of the absolutely sufficient statistic. Such distributions will be called "simplicial." The "Heaviside simplicial" is

$$f(\underset{\sim}{x}) = n!/\theta^n I(\bar{x}_n,\theta) \ ,$$

where $I(\bar{x}_n,\theta)$ denotes the indicator of the set $0 \leq \bar{x}_n \leq \theta$; and the "Dirac simplicial" is

$$f(\underset{\sim}{x}) = (n-1)!\Delta(\bar{x}_n - \theta)/\theta^{n-1} \quad .$$

The projection of a simplicial distribution is simplicial, that is, simplicial distributions are projective. For all simplicial distributions the conditional p.d.f. of a "subsample" $M_k = (X_1,\ldots,X_k)$, given $n\bar{X}_n = t$, is $f_k(M_k|n\bar{X}_n = t) = w_k(k\bar{x}_k;t)$ $= (n-1)^{[k]}(1 - k\bar{x}_k/t)^{n-k-1}/t^k$, $k\bar{x}_k \leq t$, $x_i \geq 0$, with c.d.f. $G_k(k\bar{x}_k;t)$. Also $G_k(k\bar{x}_k, n\bar{X}_n)$ is the minimum variance unbiased estimator of the marginal c.d.f. of M_k . This density will be called the "$n\bar{X}_n$-Thompson's generator of order k ."
The p.d.f. of $n\bar{X}_n$ is

$$h_n(t) = t^{n-1}g_n(t)/(n-1)! \ , \quad t \geq 0 \ ;$$

the joint p.d.f. of M_k and $n\bar{X}_n$ is

$$f(M_k,t) = (t - k\bar{x}_k)^{n-k-1}g_n(t)/(n-k-1)! \ ;$$

hence, the marginal p.d.f. of M_k is

$$f(M_k) = g_k(k\bar{x}_k) = \int_{k\bar{x}_k}^{+\infty} (t-k\bar{x}_k)^{n-k-1}g_n(t)/(n-k-1)!dt \ ;$$

that is, $f(M_k)$ is the Weyl integral transform of order $n-k$ of $g_n(t)$, or, equivalently, the $(n-k)$ repeated integral of $g_n(t)$. The kernel involved will be called the "$n\bar{X}_n$-projector of order k ," as the Weyl operator is here a projection operator. The "antiprojection operator" is, up to $(-1)^{n-k}$, the $n-k$ differentiation operator; and, under conditions of differentiability, it makes it possible to generalize to n dimensions a k-dimensional simplicial distribution, and, in case $k = 1$, to provide multivariate simplicial generalizations of univariate nonnegative random variables. The family of all sample sizes likelihoods of the negative exponential model is, of course, closed under projection and cannot lead to nontrivial generalizations of that model. One can also consider the p.d.f. of M_k as the mixture (under necessary conditions of convergence) of the $n\bar{X}_n$-Thompson's generator by arbitrary p.d.f. $h_n(t)$. Mixers are not required to be p.d.f., as exemplified by the case of Dirac's mixer $\Delta(\bar{X}_n - \theta)$, which generates the p.d.f.

$$f(M_k) = (n-1)^{[k-1]}(1 - k\bar{x}_k/n\theta)^{n-k-1}/(n\theta)^k ,$$

to be called the "$n\bar{X}_n$-Thompson distribution of order k," (also the projection of the n-simplicial Dirac distribution).

Conversely, <u>we may start with the notion of "simplicial structure</u>" and construct all these distributions on the basis of the consideration of the projective family $\{n\bar{X}_n\}$, $X_i \geq 0$, and a principle of maximum entropy: the maximum entropy induced conditional superficial distribution under restriction $n\bar{x}_n = t$, that is, on the main facet of the simplex $n\bar{X}_n = t$ is $dS_{t,n}/S_{t,n}$, which is invariant under any group of transformations that leaves invariant the facet and which, expressed in terms of M_{n-1}, is the $n\bar{X}_n$-Thompson generator of order $n-1$, leading by projection to the generators of lower orders and the associated projectors and antiprojectors, and generating the family of simplicial distributions (and possibly distributions that are generalized functions), through mixing by mixers $h_n(t)$ of the generators, through projections, and through antiprojections. (Trivially, under conditions of convergence, $g_n(t) = h_n(t)/t^{n-1}$.) Among all mixers the chisquare with $2n$ d.o.f. mixer leads to the factorization of $f(M_k)$ as a product of negative exponential distributions, that is, the class of simplicial distributions contains a "completely factorable" family (which is due to the additive structure of the $n\bar{x}_n$ statistic). According to the viewpoint one may consider that the p.d.f. $f(M_k)$ generalizes the one dimensional $f(x_1)$, or generalizes the negative exponential model, or, even, the chisquare mixer. As n tends to infinity the $n\bar{X}$-Thompson distribution of order k, as well as all other of finite order, tends to a product of k negative exponentials. From a heuristic viewpoint this stems from the fact that the chisquare mixer leads to factorization for all n, and behaves asymptotically as a Dirac function, hence the latter has asymptotically the factoring property. Heuristically again, the asymptotic Dirac behavior of the chisquare mixer should be common to all admissible mixers in view of the increasing concentration of the volume of the simplex defined by $\bar{x}_n = \theta$ in the neighborhood of its main facet. The fact that one obtains a factorization (which itself maximizes entropy) into a product of those distributions (negative exponentials) that are maximum entropy under the restrictions $E[X_i] = \theta$ can be "explained" heuristically by a passage to the limit. As n increases, the family $\{g_n(T_n)\}$ is asymptotically equivalent to a product of negative exponentials. It is easy to prove that $G_1(x, n\bar{X}_n)$ converges almost surely to $F(x,\theta) = 1 - e^{-x/\theta}$. The uniform superficial measure on the main facet of the simplex is invariant under the multiplicative group $\{X_i^* = AX_i\}$, which transforms $n\bar{X}_n$ into $n\bar{X}_n^*$, and is reducible to the uniform superficial measure on the "unit facet" $n\bar{\varphi} = 1$. It can be expressed as a function of the maximal invariant $\{\varphi_i\} = \{x_i/n\bar{x}_n\}$; this induces $\{X_i/n\bar{X}_n\}$ as an ancillary statistic whose distribution is independent from $n\bar{X}_n$ (let us say, "orthogonal" to $n\bar{X}_n$), and can be interpreted as an "angle." One could equivalently prove from the completeness of $n\bar{X}_n$ that there exists a maximal ancillary statistic $\{\emptyset_i = \emptyset_i(X_i, n\bar{X}_n)\} = \{X_i/n\bar{X}_n\}$ that is independent from $n\bar{X}_n$ and whose c.d.f. $G_i(\varphi)$ provides the minimum variance unbiased estimator $G_i(x/n\bar{X}_n)$ of the c.d.f. $F_i(x)$ of X_i. The expression $\emptyset_i = X_i/n\bar{X}_n$ will be called the "Thompsonized form" of X_i, a notion which generalizes that of studentized variable. It can be interpreted as an "error," and one may view X_i as having "basic structure" $X_i = n\bar{X}_n \cdot \emptyset_i$. This basic structure reflects the "natural" change of variable, usually performed to study simplicial distributions. It is imposed by the choice of $n\bar{X}_n$ as an A.S. statistic, therefore, as a statistic defining a priviledged partition of the sample space. The simplicial family is that for which the conditional distribution of the "errors," as defined above, is maximum entropy. The process of obtaining the p.d.f. of any statistic $S(M_n) = W(n\bar{X}_n, \emptyset)$, when M_n is simplicial, involves only an integration with respect to $\emptyset$ (that is, solving the problem for the unit simplex)--integration which is valid for the whole class of simplicial distributions--and will depend, then, on the specific family $g_n(n\overline{X}_n)$ that is considered (one should note that this approach is not always the most practical).

A generalization of the exponential model and of the simplical class will be obtained, when one will assume that the conditional measure on the simplex is no longer uniform but is replaced by some positive measure μ (which we will assume to be absolutely continuous for the sake of convenience, defined up to a multiplicative constant, and properly restricted to meet all problems of convergence). We will say that this distribution is (trivially) μ maximum entropy, and we will generate by the process already described a class of "μ-simplicial" distributions and the μ-exponential model:

$$f(M_n) = \mu'(M_n)\ \exp(-n\bar{x}_n/\theta)/\theta^n \quad .$$

This model will be μ-maximum entropy and the "closest" to μ measure, under the constancy of the expected values of X_i's (one may refer to [5] and [12] for similar terminology).

The p.d.f. of $n\overline{X}_n$ is

$$h_n(t) = t^{n-1}q(t)g_n(t)/(n-1)! \ ,$$

where $q(t)$ is the ratio of the μ measure and the uniform measure of the main facet of the simplex. It may depend only on the $\varnothing_i$'s . The choice of an "almost" uniform measure on the simplicial facet will lead to distributions "almost" simplicial; such considerations are useful to study or "measure" departure from simpliciality. In specific fields of applications such measure as μ may be suggested by consideration of invariance, in the framwork of group theoretic considerations, or of "equivalence" of some sort in the framework of a generalized utility theory, or (and) may be regarded as a "rescaling" of the variables. If the measure involves unspecified parameters, the latter will be called "exogeneous" or "extrinsic," while the parameter for which $n\overline{X}_n$ is μ-absolutely sufficient (and Lebesgue sufficient) will be called "endogeneous" or "intrinsic." μ' may or may not be a product of n univariate densities and it may or may not have structure $\rho_n(H_n)$, where H_n is a statistic; if it does one has a reinterpretation of the corresponding subclass of μ-simplicial distributions in terms of a pair of absolutely sufficient statistics (an interesting example is the family constituted by a product of univariate gamma distributions).

In view of the asymptotic exponentionality of the uniform distribution on the main facet of the n simplex the negative exponential model may be reinterpreted as induced by a uniform distribution on a simplex in the proper Hilbert space. The geometrization of the random phenomena is then complete and exponentionality is reduced to simplicial structure. It should be noted that the $n\overline{X}_n$-Thompson generator of order one is a Beta-distribution and that the mixture of that sort of distribution by a mixer (the latter to be considered an a priori probability distribution in a Bayesian context or not), can be reinterpreted as the marginal distribution of some coordinate of a simplicial distribution. The mixture of $\lambda(1-x/\theta)^{\lambda-1}/\theta$ by $w(\theta)d\theta$ can be interpreted as a projection of $\lambda!\theta^{-h}w(\theta)$, where θ reads $(\lambda+1)\overline{X}_{\lambda+1}$.

Testing exponentionality.

Laurent [17], [26] has shown that the existence of the ancillary statistic $\{X_i/nX_n\}$ makes it very simple to obtain the distribution of the reduced or Thompsonized order statistic $\emptyset_{(i)} = X_{(i)}/nX_n$, as this distribution is simply, in reduced form, the conditional distribution of $X_{(i)}$, given the A.S. statistic $n\overline{X}_n$. The method presented in [17] is absolutely general and based on the fact that $P(X_{(i)} > x)$ admits

$$\pi(X_1,\ldots,X_n;x) = \sum_{p=n-i+1}^{n} (-1)^{p+n-i+1}\binom{p-1}{n-1}\binom{n}{p} I(X_1,\ldots,X_p;x),$$

where I denotes the indicator of the set $\bigcap_{i=1}^{p}\{X_i > x\}$, as an unbiased estimator, so that the distribution is obtained by Black-wellizing this unbiased estimator with respect to the sufficient statistic, followed by the proper change of variable, here, $\emptyset_{(i)} = X_{(i)}/nX_n$. The distribution is a linear function of the minimum variance unbiased estimators of the $[P(X > x)]^p$, $p = 1$ to n, which may be obtained by different techniques.

The existence of the ancillary statistic $\{\emptyset_i\}$ makes it possible, when the observations come in a "natural" order, to devise a parameter free test as to whether a distribution belongs to the simplicial class or not. The conditional distribution of X_j, given $X_1 + \ldots + X_j = j\overline{X}_j = u$, is independent from $j\overline{X}_j$ and also from X_{j+1}, hence from $(j+1)\overline{X}_{j+1}$ and X_{j+1};

therefore $Z_j = X_j/j\overline{X}_j$ is independent from $Z_{j+1} = X_{j+1}/(j+1)\overline{X}_{j+1}$. Their cumulative distributions are Thompson parameter free distributions $G_j(z)$, $G_{j+1}(z)$, and all $G_j(Z_j)$, $j = 2$ to n, are independent uniform rectangular $(0,1)$ distributions. To test whether a distribution is simplicial one may test whether $G_j = ((j-1)\overline{X}_{j-1}/j\overline{X}_j)^{j-1}$, $j = 2$ to n, are uniform rectangular $(0,1)$. In case it is taken for granted that M_n is a sample, then, this test can be used to test exponentionality.

In case the variables do not come in a natural order, one may use order statistics, and a similar reasoning, based on the exponentionality and the independence of the Y_i's, $Y_i = (n-i+1)(X_{(i)} - X_{(i-1)})$, will lead to testing whether the variables

$$G_j = [((j-1)\overline{X}_{j-1}+(n-j-1)X_{(j-1)})/(j\overline{X}_j + (n-j)X_{(j)})]^{j-1},\ j=2 \text{ to } n,$$

are independently rectangular uniform $(0,1)$.

As the class admits a completely factorized family, that can, therefore, be antiprojected, the existence of the ancillary statistic $\{X_i/n\overline{X}_n\}$ for all n, larger than one, allows parameter free statements ("predictions") on the behavior of k future observations $\xi_1,\ldots,\xi_k$ on the basis of n past random observations. One has:

$$\emptyset_i = \psi_i/(1+k\overline{\psi}),\quad \psi_i = \xi_i/k\overline{\xi}_k,$$

and the p.d.f. of the ψ_i's is easily obtained as:

$$f(\psi) = (n+k-1)!/(1+k\overline{\psi}_k)^{n+k}n!\ ;$$

a distribution that is a "dual" of the Thompson generator. This is a very general property which allows for the constructing of prediction regions and the like. One may note that the distribution of $n\overline{X}_n$ may be regarded as a generalization of the distribution of X_i, that is, $t^{n-1}g_n(t)/(n-1)!$ may be viewed as a generalization of $g_1(x)$, univariate projection of $g_n(t)$.

IV. The Rectangular Uniform Model. Cubic Distributions and Cubic Structure.

Similar work has been done, for the case of the uniform rectangular $(0,\theta)$ model, with absolutely sufficient statistic $X_{(n)}$, the largest coordinate of M_n. The results are rather trivial and the discontinuity in the distribution of M_n, given $X_{(n)}$, introduces slight modifications. One observes a uniform superficial density on the boundary of a half truncated cube. The distributions with structure $f(M) = g_n(X_{(n)})$, are called here "cubic distributions" and the ancillary statistic for them is $\{X_i/X_{(n)}\}$.

V. Other Models. Distributions With Absolutely Sufficient Statistic T_n . T_n Structure.

T_n-Class, Kernels, and Projectors. The results of the preceding paragraphs can be generalized to the case of any priviledged statistic T_n , taken a priori as "absolutely sufficient" (A.S.) --with respect to Lebesgue measure or some other positive pre-selected measure μ . Subject to the necessary conditions of convergence, such A.S. statistic is characterized, through the consideration of maximum entropy induced conditional distribution, of the vector observation M_n given $T_n = t$, by a T_n-Thompson generator of order n-1 , which in turn (restricting ourselves, only for convenience, to the case of absolutely continuous distributions), generates one of the classes with p.d.f.

$$f(M_n) = g_n(T_n) \ , \ f(M_n) = g_n(T_n)\mu' \ ,$$

or "T_n-distributions," which may or may not admit a completely factorized class. This generator induces by projection, or otherwise, generators of order k conditional distributions of M_k given $T_n = t$. The joint distribution of M_k and T_n provides a T_n-Thompson kernel of order k and a T_n-projector operator of order k , whose iversion, when possible, leads to the "antiprojection" of k variate distributions into n variate T_n distributions and the generalization of admissible univariate distributions to multivariate T_n distributions. A certain number of measure theoretic problems naturally arise; they will be put aside in the following paragraphs, where the approach is purely heuristic, as we are only interested today in the possible paradigmic value of general ideas. As many details were given in the exponential case, we will limit ourselves to a few remarks in order to avoid too much redundancy.

The concept of "absolute sufficiency" emerges as the property of a statistic which, in Fisherian semantics, "contains all information" given by the vector observation M_n on both the numerically unspecified parameters of the model and the analytic structure of the model itself, so that the statistic T_n is absolutely sufficient (A.S.) if the conditional distributions of M_n and any measurable function of M_n, given $T_n = t$, are parameter and distribution free. One will consider only, here, the case of scalar A.S. statistics, even though the vector case has been investigated.

The class of distributions with p.d.f. $f(M_n) = g_n(T_n)$ admits T_n as an A.S. statistic (Lebesgue absolute sufficiency or μ-absolute sufficiency, where μ is a positive measure; the class $f(M_n) = g_n(T_n,\theta)\mu'$ admits T_n as a "μ-A.S. statistic"). An A.S. statistic is sufficient. If $f(M_n)$ admits T_n as an A.S. statistic then M_n is trivially "locally" maximum entropy in that sense that for $T_n = t$, that is, on the A.S. partition defined by T_n, the density of M_n is uniform. Clearly, because of convergence requirements, not all statistics can be Lebesgue-A.S. but they may be μ-A.S. for a proper choice of μ. The condition that $g_n(T_n,\theta)$ be absolutely continuous (made here for convenience) would restrict even more the class of A.S. statistics.

T_n Structure. The superficial density on C_{t_o}, defined by $T_n = t_o$, that is induced by a uniform n-dimensional density in the neighborhood of C_{t_o} will be called maximum entropy for C_{t_o} with respect to that neighborhood. When C_{t_o} is defined as a member of the family $\{C_t\}$, defined by $\{T_n = t\}$, which defines such a neighborhood, the superficial distribution is called "T_n-maximum entropy induced." Granted all necessary conditions of convergence and smoothness such a superficial distribution is

$$f_{T_n}(M_n)d\sigma_{t,n} = d\sigma_{t,n}/R_n^*(t)\,|\text{grad } T_n| ,$$

where $R_n^*(t)$ denotes the measure of C_t (this is Lebesgue measure if T_n is A.S.; a μ induced measure, otherwise, T_n being then μ-A.S.). In case C_t is a smooth convex body with volume $R_n(t)$, the distribution is

$$f_{T_n}(M_n)d\sigma_{t,n} = d\sigma_{T,n}/|\text{grad } R_n(t)| .$$

If $|\text{grad } R_n(t)|$ is a constant, then, the distribution is also marginally maximum entropy on C_t (e.g., sphere, simplex). If T_n is A.S., then, the conditional distribution $f(M_n/T_n{=}t)d\sigma_{t,n}$ of M_n given $T_n = t$, expressed as a superficial density distribution on C_t, is the T_n-maximum entropy induced distribution $f_{T_n}(M_n)d\sigma_{t,n}$. The latter can be expressed as a function of n-1 coordinates of M_n (a "subsample" M_{n-1}), and is then called a T_n-Thompson generator of order n-1 . Conversely, starting with a statistic T_n that is, with a "T_n-structure," "local" maximum entropy conditions induce Thompson generators of order n-1 . The latters generate the class admitting T_n as a A.S. statistic; among them, possibly, the "T_n-Heaviside distribution"

$$g_n(T_n,\theta) = I(T_n,\theta)/R_n(\theta) ,$$

where $I(T_n,\theta)$ is the proper set indicator; and, possibly, the "T_n-Dirac distribution"

$$g_n(T_n,\theta) = \Delta(T_n-\theta)/R_n'(T_n) ,$$

where Δ denotes the Dirac function.
The distribution of T_n is

$$h_n(t) = R_n^*(t)g_n(t) .$$

These facts parallel a change of variables, where M_n coordinates are taken as a "radius" T_n , defining a priviledged partition of the sample space, and "angles" $\{\emptyset_i\}$, so that the superficial density on $T_n = t$ can be expressed as function of the "angles" alone.

The introduction of a μ measure, whether necessary or not to make T_n an A.S. statistic will depend, in an applied field, on considerations of group invariance, "equivalence," utility indifference or the like, and may be reinterpreted sometimes as a new, more "objective," scaling. In case the partition defined by μ is not the trivial maximal partition its intersection with the T_n induced partition will make it possible to reinterpret the pair T_n , μ as an A.S. pair of statistics.

Projective Statistic. We shall call $\{T_k\}$, k = 1 to n , a "projective sufficient (A.S.) family" with respect to T_n if the marginal distribution of M_k admits T_k as a sufficient (A.S.) statistic. If $\{T_k\}$ is an A.S. projective family, then the k indexed class $\{g_k(T_k)\}$ is closed under projection. A necessary and sufficient condition is that a T_n-Thompson generator of order k admits $\{T_k\}$ as an A.S. statistic. In the general case the projection on the k-dimensional space of a T_n-Thompson generator of order $n-1$, provides a T_n-Thompson generator of order k and consequently a T_n-projector of order k, that is, the conditional distribution $w(M_k,t)$ of M_k, given $t_n = t$ (conditional distribution of a "subsample" M_k), leads to the joint p.d.f.

$$w_{n,k}(M_k,t)R_n^*(t)g_n(t) = W_{n,k}(M_k,t)g_n(t)$$

of M_k and T_n, hence, the projection operator

$$\mathbb{P}_{n,k}(\cdot) = \int(\cdot)\, W_{n,k}(M_k,t)\,dt ,$$

which, if and when it can be inverted, leads to the antiprojection of $f_k(M_k)$ as

$$f_n(M_n) = \mathbb{P}_{n,k}^{-1}\, f_k(M_k) ,$$

and the T_n induced multivariate generalization of $f_1(x)$ as

$$f_n(M_n) = g_n(T_n) = \mathbb{P}_{n,1}^{-1}\, f_1(x) .$$

A Thompson generator of order k can be obtained by integrating out $n-k$ variables; sometimes it may be obtained directly from the consideration of the "angles" distribution; sometimes it is more simple to obtain the projection kernel first. In any case the problem depends only on the structure of the differential element dM_n. One may rewrite the differential element dM_n as $dM_k\, dM_{n-k}$ and express dM_{n-k} as a function of M_k, T_n, and "angles," to be integrated out. This is easily done when $\{T_i\}$ is a projective family, in which case dM_{n-k} can be expressed as a function of T_{n-k} and angles, and T_{n-k} as a function of T_k and T_n -- an example is the Kabe-Sverdrup's lemma that boils down to say that, in case $T_n = \sum_1^n X_i^2$, namely, in case polar coordinates are used, the projection of M_n on E_{n-k} is $T_{n-k} = T_n - T_k$ (in this case, the family $\{T_i\}$ is projective through physical, geometric projection).

The preceding notions can be generalized rather easily and in a similar vein to the case of a μ-A.S. statistic.

The case when the family $\{T_i\}$, $i \leq n$, is projective or contains a projective subsequence is of special interest. One can construct such subsequences. The differential element can be written as

$dM_n = dM_k \, dM_{n-k}$;

if T_{n-k} is any statistic such that $\int d\sigma_{T_{n-k}} / |\text{grad } T_{n-k}|$, taken $T_{n-k} = t$ converges as $R^*_{n-k}(t)$ and if T_{n-k} has structure $T_{n-k}(M_{n-k}) = \zeta[T_n(M_n) , T_k(M_k)]$, where ζ is differentiable, then the projective kernel is $W_{n,k} = R^*_{n-k}[\zeta(t_n,t_k)]\zeta'_{T_n}(t_n,t_k) = V(t_k,t_n)$; $f_k(M_k) = g_k(t_k)$ admits T_k as an A.S. , and the Thompson generator of order k is $V(t_k,t_n)/R^*_n(t_n)$.

Factorable Class. A sufficient condition to have a projective sequence $\{T_i\}$ is that the class $f(M_n) = g_n(T_n)$ contains a family $g^*_n(T_n) = \prod_1^n f^*_i(x_i)$, which be the product of the n univariate marginal distributions of the X_i's ("complete" factorization); we will say, then, that T_n is "measurewise completely factorable" (a special case is when M_n can be a random sample of n observations of a random variable X). Assuming all necessary properties of differentiability, and continuity, and positivity of the p.d.f. involved, one may show that the necessary and sufficient condition that T_n be completely factorable is that all derivatives T'_{n,x_i} of $T_n(x_1,\ldots,x_n)$, with respect to each x_i have structure $T'_{n,x_i} = u_i(x_i)/\ell_n(T_n)$, where $u(x_i)$ be a negative function of x_i alone and that there exist k and $a = T_n(a_1,\ldots,a_n)$ so that

$k \exp \int_a^{T_n(x_1,\ldots,x_n)} \ell_n(\xi)d\xi = \gamma_n(T_n(x_1,\ldots,x_n))$ be a density function or, equivalently, that there exist C and a_i such that $C \exp \int_{a_i}^{x_i} u_i(\xi)d\xi = q_i(x_i)$ be a density function; $k = \gamma_n(a)$, $C = (\gamma(a))^{1/n}$. Then $f^*_i(x_i) = q_i(x_i)$ and $g^*_n(T_n) = \gamma_n(T_n)$.

One should note that the p.d.f. $f^*_i(x_i) = q_i(x_i)$ is that p.d.f. with maximum entropy under the constraint that the expected value of $U_i(X_i)$ (where $U_i(x) = \int_{a_i}^{x_i} u_i(\xi)d\xi$) be a constant; that is $\int f^*_i(x_i)dx_i$ is the "closest" to Lebesgue measure under such constraints.

When T_n is completely factorable the T_n-maximum entropy superficial distribution on $T_n = t$ is

$$\ell_n(t)d\sigma_{t,n}/R^*(t)|\text{grad } u| = dM_{n-1}, \ \ell_n(t)/R^*_n(t)u_n(x_n) .$$

T_{n-k} can be expressed as a function of T_k and T_n, $T_{n-k} = \zeta(T_k, T_n)$; hence, the projector:

$$\mathbb{P}_{n,k}(\cdot) = \int R^*_{n-k}[\zeta(t,\tau_k)]\zeta'_{\tau_n}(t,\tau_k)(\cdot)dt ,$$

which defines a symbolic calculus relating $f_k(M_k) = g_k(\tau_k)$ and $f_n(M_n) = g_n(\tau_n)$. The antiprojector $\mathbb{P}^{-1}_{n,k}$ may realize the T_n generalization of univariate to multivariate distributions; the $f^*_i(x_i)$ will, of course, have always a trivial generalization: their products.

Equivalently the mixing of the T_n-Thompson generator of order k by $h_n(t)dt$, as long as $h_n(T_n)/R^*_{T_n}(T_n)$ is an n-dimensional p.d.f., will generate at least a subclass $g_k(T_k)$ admitting T_k as an A.S. statistic.

Similarly there is a symbolic calculus relating the distribution of T_k and T_n, defined by

$$h_k(t_k) = R^*_k(t_k)\int R^*_{n-k}[\zeta(t_k,t_n)]\zeta'(t_k,t_n)h_n(t_n)dt_n/R^*_n(t_n) .$$

When one can mix the T_n-Thompson generators of order k by the Dirac mixer $\Delta(T_n - \theta)$, after T_n has been properly normalized by a function of n, one obtains

$$f_{T_n}(M_k;\theta) = R^*_{T_{n-k}}[\zeta(T_k,\theta)]\zeta'(T_k,\theta)/R^*_{T_n}(\theta) ,$$

which is called a T_n-Thompson distribution of order k and admits, of course, T_k as an A.S. statistic. In case T_n is symmetric in $X_1,\ldots,X_n$, that is, invariant in all permutations of $X_1,\ldots,X_n$ --the case where the class admits M_n as a sample-- the projective sequence $\{T_n\}$ is defined for all n, and n appears explicitly in the formula of $f_{T_n}(x_1,\ldots,x_k;\theta)$; then, as n tends to infinity, one may expect, under conditions that are to be investigated, that $f_{T_n}(x_1,\ldots,x_k;\theta)$ tends to the product $\prod_1^k f^*_i(x_i)$. A heuristic justification of this conjecture rests on the fact that there exists a mixer $h_n(t)$ that leads to that factorization for every n, hence for large n, and behaves asymptotically as a Dirac function; under this conjecture the maximum entropy properties of M_n under constraint $T_n = t$ is kept at the limit under constraint $E[u(x)] = \theta$; and the univariate model that results from factorization appears as the limit of the distribution of M_n on the "structure" $T_n = t$, or as the distribution of a coordinate of M in a subspace of a proper Hilbert space--thus giving a purely geometrical interpretation of the phenomena. The case when the class contains a completely factored family that is not a sample distribution family and the sequence $\{T_n\}$ is defined for all n is worth investigation (a certain amount of work has been done in different cases when the family $\{T_n\}$ defines different families of ellipsoids, leading to different definitions of "ellipsoidal" distributions).

Characteristic Functions. There is no room to discuss here the characteristic function aspect of these problems. This aspect is very important for the class of spherical distributions, due essentially to the fact that spherical distributions' characteristic functions are also spherical as functions of their arguments, and remain formally invariant under projection; and to a lesser, but still great, extent to the fact that the completely factorized family, the product of univariate identical normal distributions, is invariant under Fourier's transform. Such properties do not hold true for other A.S. statistics than the radius vector, even though some aspects of invariance under projection are kept, and even though an alternative definition of characteristic function as an ad hoc multivariate transform may exhibit such invariant properties. It is, however, clear that, under projectivity of $\{T_i\}$, the characteristic function of M_n is a univariate transform of $g_n(T_n)$; the latter is, therefore, a generalization of, or an analog to, the Hankel transform of the spherical case. In case the family admits a T_n-Dirac distribution, then the latter's characteristic function provides the kernel of the univariate transform just mentioned. Criteria of projectivity can be established on the basis of characteristic function characterization but they are not simple enough to be very useful for practical applications.

Ancillary Statistics, Errors and Prediction. An A.S. statistic T_n will be called "complete" if, for at least one family of the class admitting T_n as an A.S. statistic, there exists no nontrivial estimator of zero (if T_n is completely factorable and M_n is a sample from a continuous p.d.f., then, T_n is complete). If T_n is complete, then, the conditional c.d.f. $G_i(x;t)$ of X_i, given $T_n = t$, is the minimum variance unbiased estimator of the marginal c.d.f. $F_i(x;\theta)$. If G is continuous in x_1 (if it is not, a similar argument can be used with slight modifications in the conclusion), then, $G_i(X_i,t)$ is uniform rectangular $(0,1)$, hence independent from θ , and, because of completeness in one case, independent from T_n . Consequently $G(X_i;T_n)$ is independent from T_n . There exists, therefore, a set of univariate maximal ancillary statistics $\{\emptyset_1(X_i;T_n)\}$, which play the part of "angles" whose joint distribution is given by the T_n-maximum entropy induced superficial distribution on $T_n = t$, and define an ancillary partition of the sample space. To define an absolutely sufficient complete statistic is to decide that the distribution of the vector observation be invariant in a "rotation" $\emptyset$ leaving T_n invariant; it is to conceive X_i as "structured" according to $X_i = X_i(T_n,\emptyset_i)$, a regression of X_i on T_n with a reduced or Thompsonized "error" or "residual" $\emptyset_i$. In applications such conceptualizations will stem more often than not from group theoretic considerations. In some cases it will be either because a certain invariant variety will be used as the basis of an absolutely sufficient partition or because a certain maximal invariant will be chosen as ancillary statistic; then the family itself, which is induced by the T_n-maximum entropy generator, will be invariant under the corresponding groups of transformations acting on the variables and the parameters.

These ancillary statistics, at least in the case where the T_n family admits M_n as a sample, provide the tools to make probability statements on the behavior of future observations on the basis of random "past" observations. In the case of one future observation, $\emptyset_{n+1} = \varphi_{n+1}(X_{n+1};T_{n+1})$ is also $\emptyset_{n+1} = \psi_{n+1}(X_{n+1},T_n)$. In the case of k future observations $M_k(X_{n+1},\ldots,X_{n+k})$ $\emptyset_{n+i} = \varphi_{n+i}(X_{n+i};T_{n+k}) = \psi_{n+i}(X_{n+i},\ldots,X_{n+k};T_n)$, $i = 1$ to k , and the conditional distribution of the "subsample" M_k given T_{n+k} , which, in reduced form, is the distribution of the $\emptyset$'s , provides the distributions of the ψ's , from which prediction regions can be obtained (for applications to bombing problems involving the nontrivial bivariate and multivariate normal family see [19], [20]).

Testing Hypotheses. Many different tests of hypotheses as to whether a vector observation M_n follow a T_n distribution, when the parameter θ involved in the model is unspecified, can be developed on the basis of the ancillary statistics induced by T_n. Among those the following test can be used when T_n is projective, completely factorable, and complete.

Let $G_i(x;t)$ be the conditional c.d.f. of X_i given $T_i = t$, where $T_i = T_i(X_1,\ldots,X_i)$ is A.S. for $M_i(X_1,\ldots,X_i)$, (assume continuity of G_i with respect to x, for the sake of shortening the argument); then, the Y_i's, $i = 2$ to n, defined by $Y_i = G_i(X_i;T_i)$ are ancillary statistics that are independently rectangular, uniformly distributed on $(0,1)$, and this property characterizes the family $g_n(T_n)$. The change of variables from X to Y, T_n, parallels the classical change of variables which is generally used in the proof of the Dirichlet integral (and, therefore, for the study of the negative exponential and the gamma model). Consequently, the testing of the uniform rectangularity of the Y_i's is equivalent to that of the absolute sufficiency of T_n. Testing, however, will rest on the arbitrary choice of the observation X_1. But tests against a specified alternative hypothesis, such as H_1 that T_n is A.S. with respect to a measure μ, involve all the coordinates of M_n. Assuming μ with density μ' for the sake of convenience, the testing statistic will be

$$L = \mu'(M_n)R^*_n(T_n)/R^*_{\mu,n}(T_n) ,$$

where $R^*_{\mu,n}(t)$ is the measure of $C_t = \{T_n = t\}$, with respect to measure μ.

Similarly, testing the absolute sufficiency of T_n can be done by applying the Smirnov test to $G_i(x;t)$, if $G_i(x;t)$ is the same for all i's--which is the case for the spherical and the simplicial structures--but the $G(X_i,t)$ are not independent. In case independence is taken for granted, testing the spherical and the simplicial structures is testing normality and exponentiality and consequently this paragraph provides such tests as by-products.

VI. Conclusions.

The concept of structure or the equivalent concept of A.S. statistic (an entropy theoretic notion) can be a unifying concept in building, classifying, and interpreting many univariate and multivariate models and distributions, that may, thus, find a new rationale through considerations of uniformity and homogeneity. Such considerations, expressions of a certain principle of "sufficient reason," lead to a "geometrization" or many so-called random phenomena as they relate them to a "structure," and are consistent with or motivated by utility theoretic or group theoretic interpretations. Generalizations and extensions from the basic generators and kernels induced by the A.S. statistic can be done by projections, anti-projections, mixtures, changes of scale, etc., and allow for a distinction between the endogeneous and the exogeneous parameters of a model. As many physical systems are practically closed and evaluating to a state of maximum entropy under the constraints defining their structures, distributions admitting A.S. statistic should be often relevant to physics, but other systems, such as social systems, should be investigated as well. In many cases a system that, *per se*, and isolated, does not exhibit maximum homogeneity with respect to Lebesgue measure may be "uniform" with respect to another measure μ, or will appear as maximum entropy induced when embedded in a higher dimensional space, of which it is a manifold, the latter being defined by a system of constraints that constitutes what we have called a "structure"; the structure explaining away and geometrically absorbing the original lack of uniformity. One may find the concept of "structure" a more convenient semantic tool of description than the notion of "mechanism" of random phenomena. One may note that some structures are privileged in that the state of maximum entropy on the structure is the same as that which is conditionally induced by the embedding of the structure in a family; those are the structures for which $\text{grad } R^*(T_n)$ is constant--examples: simplicial and spherical structures.

VII. Final Remarks on Structures.

A *structure* is a partition of the sample space into equivalence classes on the basis of an equivalence relation induced by group, utility, information theoretic considerations. It is associated with a "natural" conditional probability measure common to all equivalence classes, which, from several viewpoints (group invariance, maximal entropy, etc.), generalizes Lebesgue measure, may be considered an *a priori* measure, and defines the integral geometry associated with the structure, that is, an integral geometry on the equivalence classes. Such *a priori* measure through asymptotic considerations induces probability models that allow for geometrical and information theoretical interpretations in Hilbert space; it can be mixed by mixers to generate a class having the structure as its absolutely sufficient partition. Given a structure the class of distributions that admits that structure as its likelihood partition admits it as an absolutely sufficient partition. The common properties of the class reflect the common integral geometry of their structure. Assumptions of independence or others define subsets of such a class and many families of distributions can be usefully related and characterized as specific subsets of a class with a given structure. Many theorems valid for a specific family can be easily generalized to all the families with the same structure. In many applications the model used is often overspecified when represented by a family of distributions while what is actually needed is the consideration of a class with a given structure (example: normality is often assumed when sphericity is meant because of an implicit assumption made on the independence of the observations). Under projection, antiprojection, and mixing manipulations "natural" generalizations of probability distributions can be made and "natural" paths between them be traced. Often a convenient instrument of proof, the structure is, at worst, a semantic relabeling of known material and, at best, a paradigm to geometrize randomness and interpret the notion or equal probability and equivalent events, which mathematicians-philosophers such as Fréchet have characterized as "at bottom, whether explicit or implicit, the basic notions on which the foundations of Probability Theory are resting."

Selected Bibliography

[1] Aczel, J. (1966). Lectures on Functional Equations and Their Applications. New York.

[2] Anderson, T. W. (1958). Introduction to Multivariate Statistical Analysis. New York.

[3] Borel, E. (1914). Introduction géométriques à quelques Théories physiques. Paris.

[4] Calcul des Probabilités (1951). Congrès International de Philosophie des Sciences (1949). Paris.

[5] Campbell, L. L. (1970). Equivalence of Gauss's Principle and Minimum Information Estimation of Probabilities. Ann. Math. Statist. 41, 1011.

[6] David, H. A. (1970). Order Statistics. New York.

[7] Deltheil, R. (1926). Traité du Calcul des Probabilités et ses Applications. Probabilités Géométriques. Paris.

[8] Erdelyi, A. and al. (1953). Higher Transcendental Functions. New York.

[9] Fraser, D. A. S. (1968). The Structure of Inference. New York.

[10] Good, I. J. (1963). Maximum Entropy for Hypothesis Formulation. Ann. Math. Statist. 34, 911.

[11] Jaynes, E. T. (1957). Information Theory and Statistical Mechanics. Phys. Rev. 106, 620.

[12] Kabe, D. G. (1964). Some Contributions to Multivariate Statistical Analysis. Ph.D. Thesis. Wayne State University Detroit.

[13] Kabe, D. G. (1965). Generalization of Sverdrup's Lemma and Its Applications to Multivariate Distribution Theory. Ann. Math. Statist. 36, 671.

[14] Kabe, D. G. (1968). Some Distribution Problems of Order Statistics From Exponential and Power Function Distributions. Canad. Math. Bull. 4.2.

[15] Karlin, S. (1968). A First Course in Stochastic Processes. New York.

[16] Kullback, S. (1959). Information Theory and Statistics. New York.

[17] Laurent, A. G. (1963). Conditional Distribution of the Reduced i-th Order Statistics of the Exponential Model. Ann. Math. Statist. 34, 652.

[18] Laurent, A. G. (1955). Distribution d'échantillons quand la population de référence est Laplace-Gaussienne de paramètres inconnus. Journ. Soc. Statist. Paris. 10, 262.

[19] Laurent, A. G. (1957). Bombing Problems, A Statistical Approach. Operations Research. 5, 75.

[20] Laurent, A. G. (1962). Bombing Problems, A Statistical Approach, II. Operations Research. 10, 380.

[21] Laurent, A. G. (1956). Generalizations of Thompson's Distributions. Ann. Math. Statist. 27, 1184.

[22] Laurent, A. G. (1960). Generalizations of Thompson's Distributions II. Ann. Math. Statist. 31, 526.

[23] Laurent, A. G. (1960). Generalizations of Thompson's Distributions III. Ann. Math. Statist. 32, 238.

[24] Laurent, A. G. (1968). Bombing Problems III. Contributed Paper. O.R.S.A. San Francisco.

[25] Laurent, A. G. (1959). Minimum Variance Unbiased Estimators; Generalizations of Thompson's Distributions; Random Bases. W.S.U. Research Memorandum.

[26] Laurent, A. G. (1965). Probability Distributions, Functional Moments, Empty Cell Test. Classical and Contagious Discrete Distribution, International Symposium. Montreal 1963. Calcutta.

[27] Laurent, A. G. (1972). Distributions on the Sphere and Other Structures and Related Distributions. Manuscript.

[28] Levy, P. (1951). Problèmes Concrets d'Analyse Fonctionnelle. Paris.

[29] Lieberman, G. J. and Resnikoff, G. J. (1955). Sampling Plans by Inspection by Variables. Journ. Amer. Statist. Assoc. 270, 457.

[30] Likes, J. (1968). Difference of Two Ordered Observations Divided by the Standard Deviation in Case of Negatively Exponential Distribution. Metrika. 12, 161.

[31] Lord, R. D. (1954). The Use of Hankel Transforms in Statistics I. Biometrika. 41, 44. II. Biometrika. 41, 344.

[32] Maxwell, J. C. (1860). Illustration of the Dynamical Theory of Gases. Taylor's Philosoph. Magaz. 19, 19.

[33] Piaget, J. (1968). Le Structuralisme. Paris.

[34] Renyi, A. (1962, 1966). Wahrscheinlichketsrechnung. Berlin.

[35] Statistical Research Group, Columbia University (1948). Sampling Inspection, Principles, Procedures and Tables. New York.

[36] Tate, R. F. (1959). Unbiased Estimation: Function of Location and Scale Parameters. Ann. Math. Statist. 30, 341.

[37] Thomas, D. H. (1970). Some Contributions to Radial Probability Distributions, Statistics, and the Operational Calculi. Ph.D. Thesis. Wayne State University, Detroit.

[38] Thompson, W. R. (1935). On a Criterion for the Rejection of Observations and the Distribution of the Ratio of Deviation to Sample Standard Deviation. Ann. Math. Statist. 6, 214.

[39] Whittle, P. (1971). Optimization Under Constraints. London.

[40] Wiener, N. (1948)(1961). Cybernetics. Boston.

[41] Wilks, S. (1962). Mathematical Statistics. New York.

Andre G. Laurent
Department of Mathematics
Wayne State University
Detroit 2, Michigan
U.S.A.

A FEW REMARKS ON THE EXACT DISTRIBUTIONS OF CERTAIN MULTIVARIATE STATISTICS - II

A. M. Mathai

Summary

This article deals with a method of obtaining exact distributions of statistics having certain structures. Consider the statistic T having the structure $T = [X_1^{a_1} \dots X_k^{a_k}]/[X_{k+1}^{a_{k+1}} \dots X_m^{a_m}]$ where $a_1, \dots, a_m$ are rational numbers, $X_1, \dots, X_m$ are either all Gamma variates or all Beta variates or some of them Gamma and some Beta variates, with arbitrary parameters. It is easy to see that a large number of test criteria, under the null hypothesis, for testing hypotheses on multinormal populations and several other such statistics are structurally of the type T mentioned above. Whenever the s-th moment exists it is easily seen that the density of T is an H-function. In order to have practical utility one needs a computable representation of this H-function. After pointing out different types of problems which come under an H-function this article deals with a computable representation of a general H-function, thus giving the exact distributions, in computable forms, of a wide variety of test statistics. This series representation given in this article is for the general case when the poles of the integrand are not restricted to be simple. Thus the results given in this article on an H-function are not available in the literature. The technique of Calculus of residues and properties of Psi and Zeta functions are used in deriving the results.

1. Introduction.

The H-function is the most generalized Special Function studied so far. All other Special Functions such as Meijer's G-function, Bessel-, Legendre-, Whittaker, Struve-, Mittag-Leffler- and Hypergeometric functions are all special cases of an H-function. An H-function is defined as a Mellin-Barnes type integral as follows.

$$H(z) = H_{p,q}^{m,n}\left[z\,\middle|\,\begin{matrix}(a_1,\alpha_1),(a_2,\alpha_2),\dots,(a_p,\alpha_p)\\(b_1,\beta_1),(b_2,\beta_2),\dots,(b_q,\beta_q)\end{matrix}\right] = \frac{1}{2\pi i}\int_L h(s)z^{-s}ds \tag{1.1}$$

where

$$(1.2) \qquad h(s) = \frac{\prod_{j=1}^{m} \Gamma(b_j + \beta_j s) \prod_{j=1}^{n} \Gamma(1 - a_j - \alpha_j s)}{\prod_{j=m+1}^{q} \Gamma(1 - b_j - \beta_j s) \prod_{j=n+1}^{p} \Gamma(a_j + \alpha_j s)}$$

$i = (-1)^{1/2}$; m,n,p,q are non-negative integers such that, $0 \le n \le p$, $1 \le m \le q$; $\alpha_j (j=1,\ldots,p)$, $\beta_j (j=1,\ldots,q)$ are positive numbers; $a_j (j=1,\ldots,p)$, $b_j (j=1,\ldots,q)$ are complex numbers such that,

$$(1.3) \qquad \alpha_j(b_h + v) \neq \beta_h(a_j - 1 - \lambda), \quad \text{for} \quad v,\lambda = 0,1,\ldots$$
$$h = 1,2,\ldots,m; \; j=1,\ldots,n$$

and L is a contour separating the points,

$$(1.4) \qquad -s = (b_j + v)/\beta_j \;, \; (j=1,\ldots,m; \; v=0,1\ldots)$$

and

$$(1.5) \qquad -s = (a_j - 1 - v)/\alpha_j \;, \; (j=1,\ldots,n; \; v=0,1,\ldots) \quad .$$

The contour L exists on account of (1.3).

Let,

$$(1.6) \qquad \mu = \sum_{j=1}^{q} \beta_j - \sum_{j=1}^{p} \alpha_j \quad \text{and} \quad \beta = \prod_{j=1}^{p} \alpha_j^{\alpha_j} \prod_{j=1}^{q} \beta_j^{-\beta_j}$$

The function $H(z)$ exists for all $z \neq 0$ if $\mu > 0$ and for $0 < |z| < \beta^{-1}$ if $\mu = 0$.

A detailed study of the H-function was done by Braaksma (1963). One form or the other of the function was studied by Pincherle (1888), see Erdélyi (1953, p. 49), Mellin (1910) and Barnes (1908). Fox (1961) studied the H-function as a symmetrical Fourier kernel. Functions close to an H-function occur in the study of the solutions of certain functional equations by Bochner (1958) and Chandrasekharan and Narasimhan (1962). Most of the study is centered around the properties of the function rather than its practical utility. Its application in solving some problems in statistical theory of distributions is pointed out in Mathai (1970). In the following sections some statistical problems are pointed out where an H-function occurs.

2. Generalized Gamma Variates.

Consider a real stochastic variable X_j with the density function $f_j(x_j)$ given by,

$$(2.1) \qquad f_j(x_j) = \frac{\beta_j \, a_j^{\alpha_j/\beta_j}}{\Gamma(\alpha_j/\beta_j)} x_j^{\alpha_j - 1} e^{-a_j x_j^{\beta_j}}, x_j>0,\ a_j>0,\ \alpha_j>0,\ \beta_j>0$$

and $f_j(x_j) = 0$ elsewhere. X_j is called a generalized Gamma variate and a generalized Gamma distribution is very important in Reliability Analysis and in Life Testing Models. The vast amount of literature in these fields may be seen from the bibliographies Mendenhall (1958) and Govindarajulu (1964). Distributions such as Gamma, chi-square, Exponential, Raleigh, Erlangian, Waiting time, Folded Normal, are all particular cases of a Generalized Gamma distribution. Now let

$$(2.2) \qquad Y = (X_1 \ldots X_m)/(X_{m+1} \ldots X_k)$$

where $X_1,\ldots,X_k$ are independently distributed according to (2.1) for $j=1,\ldots,k$. The h-th moment of Y is given by,

$$(2.3) \qquad E(Y^h) = C \prod_{j=1}^{m} \{\Gamma(\frac{\alpha_j}{\beta_j} + \frac{h}{\beta_j})/a_j^{h/\beta_j}\} \prod_{j=m+1}^{k} \{\Gamma(\frac{\alpha_j}{\beta_j} - \frac{h}{\beta_j})/a_j^{-h/\beta_j}\}$$

where

$$(2.4) \qquad C = [\prod_{j=1}^{k} \Gamma(\frac{\alpha_j}{\beta_j})]^{-1}$$

and E denotes 'mathematical expectation'. When (2.3) exists the density function of Y, denoted by $g(y)$, is given by the inverse Mellin transform. That is,

$$(2.5) \qquad g(y) = \frac{1}{2\pi i} \int_{c-i\infty}^{c+i\infty} [E(Y^h)]\, y^{-h-1} dh$$

and from (1.1) it is easily seen that $g(y)$ is an H-function.

3. Beta Variates.

A real stochastic variable X_j with the density function,

$$(3.1) \qquad h_j(x_j) = \frac{\Gamma(\alpha_j+\beta_j)}{\Gamma(\alpha_j)\,\Gamma(\beta_j)}\, x_j^{\alpha_j-1}(1-x_j)^{\beta_j-1}, \quad 0<x_j<1,\ \alpha_j>0,\ \beta_j>0$$

and $h_j(x_j) = 0$ elsewhere is called a Beta variate. Let,

$$(3.2) \qquad Y = X_1^{\delta_1} X_2^{\delta_2} \cdots X_k^{\delta_k}$$

where X_j , $j=1,\ldots,k$ are independent Beta variates with the density functions (3.1) for $j=1,2,\ldots,k$ and δ_j , $j=1,\ldots,k$ are some constants. The h-th moment of Y is given by,

$$(3.3) \qquad E(Y^h) = \prod_{j=1}^{k} E[(X_j^{\delta_j})^h]$$

$$= C \prod_{j=1}^{k} \frac{\Gamma(\alpha_j+\delta_j h)}{\Gamma(\alpha_j+\beta_j+\delta_j h)}$$

where

$$(3.4) \qquad C = \prod_{j=1}^{k} \frac{\Gamma(\alpha_j+\beta_j)}{\Gamma(\alpha_j)}$$

If (3.3) exists then evidently the density function of Y is an H-function. Product of independent Beta variates occur in a number of statistical problems such as the distribution of the likelihood ratio criteria for testing multivariate statistical hypotheses, Anderson (1958).

4. A Series Representation.

In a problem such as the ones pointed out in sections 2 and 3, one needs the H-function to be put into a computable form. For testing statistical hypotheses one needs the percentage points which can be obtained if the exact density function of the test criterion is representable in a computable form. The Mellin-Barnes type integral representation as given in (1.1) will not be of much help. Braaksma (1963) gives a series representation of $H(z)$ as follows. When,

$$(4.1) \qquad \beta_h(b_j+\lambda) \neq \beta_j(b_h+v) \text{ for } j\neq h;\ j,h=1,\ldots,m;\ \lambda,v=0,1,\ldots$$

$$(4.2)\qquad H(z)=\sum_{h=1}^{m}\sum_{v=0}^{\infty}\frac{\prod_{j=1}^{m}{}'\,\Gamma(b_j-\beta_j(b_h+v)/\beta_h)\prod_{j=1}^{n}\Gamma(1-a_j+\alpha_j(b_h+v)/\beta_h)}{\prod_{j=m+1}^{q}\Gamma(1-b_j+\beta_j(b_h+\ )/\beta_h)\prod_{j=n+1}^{p}\Gamma(a_j-\alpha_j(b_h+\ )/\beta_h)}$$

$$\times\ \frac{(-1)^v\, z^{(b_h+v)/\beta_h}}{v\,!\,\beta_h}$$

where $\prod_{j=1}^{m}{}'$ means the product of the factors with $j=1,\ldots,m$ save $j=h$. The condition (4.1) says that the poles are restricted to be simple. But from (2.5) and (3.3) it is evident that in practical situations usually the poles are of higher order and (4.1) occurs only in very special cases. So the aim of this article is to obtain a general series expansion for $H(z)$ when the poles are not restricted to be simple. This will be achieved with the help of the following lemmas and by using theorem 1 of Braaksma (1963) which in effect says that $H(z)$ is available as the sum of the residues of $h(s)z^{-s}$ at the poles of $\prod_{j=1}^{m}\Gamma(b_j+\beta_j s)$ where $h(s)$ is given in (1.2). Before giving the lemmas a few convenient notations will be developed here.

Consider $\Gamma(b_j+\beta_j s)$. The poles are given by the equation,

$$(4.3)\qquad -s = (b_j + v)/\beta_j\ ,\ v=0,1,\ldots$$

Similarly the poles of $\Gamma(b_h + \beta_h s)$, $h \neq j$, are obtained from,

$$(4.4)\qquad -s = (b_h + \lambda)/\beta_h\ ,\ \lambda = 0,1,\ldots.$$

There may exist a pair of values (v_1, λ_1) such that

$$(4.5)\qquad (b_j + v_1)/\beta_j = (b_h + \lambda_1)/\beta_h\ .$$

Then evidently, the point $-s = (b_j + v_1)/\beta_j$ is a pole of order two if the point does not coincide with the pole of any other Gamma of $h(s)$. If the point coincides with the poles coming from $r-1$ other Gammas of the set $\Gamma(b_j + \beta_j s)$, $j=1,\ldots,m$ then the point gives a pole of order r . For convenience, we will assume throughout this article that no singularity of any Gamma of the denominator of $h(s)$ coincides with any singularity of any Gamma of the set

$\Gamma(b_j + \beta_j s)$, $j=1,\ldots,m$. However, the technique developed in this article is applicable to other situations as well after making some minor adjustments. In order to distinguish poles of all orders we will consider the following equations. For a fixed j consider the equations,

$$(4.6)\qquad \frac{b_1 + v^{(j1)}_{j_1\ldots j_m}}{\beta_1} = \frac{b_2 + v^{(j2)}_{j_1\ldots j_m}}{\beta_2} = \ldots = \frac{b_m + v^{(jm)}_{j_1\ldots j_m}}{\beta_m}$$

The following convention is used in writing down (4.6). For a fixed j , $j_r = 0$ or 1 for $r=1,2,\ldots,m$. If $j_r = 0$ then $(b_r + v^{(jr)}_{j_1\ldots j_m})/\beta_r$ is to be excluded from the equations in (4.6). $v^{(jj)}_{j_1\ldots j_m}$ is a value of v in (4.3). Evidently the possible values are $0,1,2,\ldots$. $v^{(jr)}_{j_1\ldots j_m}$ denotes the number corresponding to $v^{(jj)}_{j_1\ldots j_m}$ when the equation,

$$(4.7)\qquad \frac{b_j + v^{(jj)}_{j_1\ldots j_m}}{\beta_j} = \frac{b_r + v^{(jr)}_{j_1\ldots j_m}}{\beta_r}$$

is satisfied by some values of $v^{(jj)}_{j_1\ldots j_m}$ and $v^{(jr)}_{j_1\ldots j_m}$. Thus $v^{(jr)}_{j_1\ldots j_m}$ may or may not exist. Under these notations,

$$(4.8)\qquad 0 \le j_1 + j_2 + \ldots + j_m \le m$$

for every fixed j and $j_1 + \ldots + j_m$ denotes the order of the pole at $-s = (b_j + v^{(jj)}_{j_1\ldots j_m})/\beta_j$. If $j_1 + \ldots + j_m = r$ then in (4.6) there will be r elements if $(b_k + v^{(jk)}_{j_1\ldots j_m})/\beta_k$, $k=1,2,\ldots,m$ are called the elements in (4.6). In the above notation a pole of order r will be considered r times. In order to avoid duplication we will always assume that,

$$(4.9)\qquad j_1 = j_2 = \ldots = j_{j-1} = 0$$

while considering the points corresponding to (4.3). If $j_r = 0$ for $r=1,\ldots,m$, $r\neq j$ then $-s = (b_j + v^{(jj)}_{j_1\ldots j_m})/\beta_j$ gives simple poles. If $j_1 + \ldots + j_m = 0$ then the corresponding point is not a pole to be considered. Let,

(4.10) $$S^{(jj)}_{j_1\cdots j_m} = \{v^{(jj)}_{j_1\cdots j_m}\} .$$

That is, $S^{(jj)}_{j_1\cdots j_m}$ is the set of all values $v^{(jj)}_{j_1\cdots j_m}$ takes for given $j_1, j_2,\ldots,j_m$. With this notation the H-function can be written as follows.

(4.11) $$H(z) = \Sigma^m_{j=1} \Sigma_{S^{(jj)}_{j_1\cdots j_m}} R_j$$

where R_j is the residue of $h(s)z^{-s}$ at the pole $-s = (b_j + v^{(jj)}_{j_1\cdots j_m})/\beta_j$ and $\Sigma_{S^{(jj)}_{j_1\cdots j_m}}$ denotes the summation over all sets $S^{(jj)}_{j_1\cdots j_m}$. The quantity R_j will be evaluated with the help of the following lemmas.

<u>Lemma 1</u>. If $\Delta(s)$ denotes a product of Gamma functions with a pole of order k at $s = a$ then the residue R of $\Delta(s)z^{-s}$ at $s = a$ is given by the following expression.

(4.12) $$R = \frac{1}{(k-1)!}\frac{\partial^{k-1}}{\partial s^{k-1}} (s-a)^k\Delta(s)z^{-s} , \text{ at } s = a$$

$$= \frac{z^{-a}}{(k-1)!} \Sigma^{k-1}_{r=0}\binom{k-1}{r}(-\log z)^{k-1-r}[\Sigma^{r-1}_{r_1=0}\binom{r-1}{r_1}A_o^{(r-1-r_1)}$$

$$\times \Sigma^{r_1-1}_{r_2=0}\binom{r_1-1}{r_2} A_o^{(r_1-1-r_2)} \ldots] B_o$$

where,

(4.13) $$B_o = (s-a)^k\Delta(s) , \text{ at } s = a$$

and

(4.14) $$A_o^{(t)} = \frac{\partial^{t+1}}{\partial s^{t+1}} \log[(s-a)^k\Delta(s)], \text{ at } s = a , t \geq 0 .$$

The proof follows from the following observations.

(4.15) $$\frac{\partial^{k-1}}{\partial s^{k-1}}[(s-a)^k\Delta(s)z^{-s}] = z^{-s}\{\frac{\partial}{\partial s} + (-\log z)\}^{k-1}[(s-a)^k\Delta(s)]$$

$$= z^{-s} \Sigma^{k-1}_{r=0}\binom{k-1}{r}(-\log z)^{k-1-r}\frac{\partial^r}{\partial s^r}[(s-a)^k\Delta(s)] .$$

But,

$$\frac{\partial^r}{\partial s^r} B = \frac{\partial^{r-1}}{\partial s^{r-1}} [AB] = \Sigma_{r_1=0}^{r-1} \binom{r-1}{r_1} A^{(r-1-r_1)} \frac{\partial^{r_1}}{\partial s^{r_1}} B , \tag{4.16}$$

where,

$$B = (s-a)^k \Delta(s) \quad \text{and} \quad A = \frac{\partial}{\partial s} \log B . \tag{4.17}$$

Now the proof follows from (4.15) and (4.16).

Lemma 2.

$$\frac{\partial^r}{\partial s^r} \{\log \prod_{j=1}^{k} \Gamma(d_j + \delta_j s)\}$$

$$= \Sigma_{j=1}^{k} \delta_j \Psi(d_j + \delta_j s) , \text{ for } r = 1 \tag{4.18}$$

$$= (-1)^r (r-1)! \Sigma_{j=1}^{k} \delta_j^r \rho(r, d_j + \delta_j s) , \text{ for } r \geq 2 \tag{4.19}$$

where Psi function and the generalized zeta function are defined as follows.

$$\Psi(z) = \frac{d}{dz} \log \Gamma(z) = -\gamma + (z-1) \Sigma_{n=0}^{\infty} \frac{1}{(n+1)(z+n)} \tag{4.20}$$

where γ is the Euler's constant; $\gamma = 0.577 \ldots$.

$$\rho(s,v) = \Sigma_{n=0}^{\infty} \frac{1}{(n+v)^s} , \; R(s) > 1 , \; v \neq 0,-1,-2,\ldots. \tag{4.21}$$

where $R(.)$ denotes the real part of $(.)$. Now with the help of lemmas 1 and 2, R_j of (4.11) can be easily obtained.

Theorem 1. $H(z)$ is given in (4.11) where,

$$R_j = \frac{(j_1+\ldots+j_m) Z^{(b_j+v_{j_1\ldots j_m}^{(jj)})/\beta_j}}{(j_1+\ldots+j_m)!} \Sigma_{r=0}^{j_1+\ldots+j_m-1} \binom{j_1+\ldots+j_m-1}{r} \tag{4.22}$$

$$\times (-\log z)^{j_1+\ldots+j_m-1-r} [\Sigma_{r_1=0}^{r-1} \binom{r-1}{r_1} C_j^{(r-1-r_1)}$$

$$\times \Sigma_{r_2=0}^{r_1-1} \binom{r_1-1}{r_2} C_j^{(r_1-1-r_2)} \ldots.] D_j ,$$

(4.23) $$D_j = \{\prod_{h=1}^{m} \Gamma[b_h-\beta_h(b_j+v^{(jj)}_{j_1\cdots j_m})/\beta_j+j_h(v^{(jh)}_{j_1\cdots j_m}+1)]$$

$$\times \prod_{h=1}^{n} \Gamma[1-a_h+\alpha_h(b_j+v^{(jj)}_{j_1\cdots j_m})/\beta_j]\}/\{\prod_{h=1}^{m}[\beta_h^{jh}(-1)^{j_h(v^{(jh)}_{j_1\cdots j_m})}$$

$$\times (v^{(jh)}_{j_1\cdots j_m}!)^{jh}] \prod_{h=m+1}^{q} \Gamma[1-b_h+\beta_h(b_j+v^{(jj)}_{j_1\cdots j_m})/\beta_j]$$

$$\times \prod_{h=n+1}^{p} \Gamma[a_h-\alpha_h(b_j+v^{(jj)}_{j_1\cdots j_m})/\beta_j]\};$$

(4.24) $$C_j^{(0)} = \Sigma_{h=1}^{m} \beta_h \Psi[b_h-\beta_h(b_j+v^{(jj)}_{j_1\cdots j_m})/\beta_j+j_h(v^{(jh)}_{j_1\cdots j_m}+1)]$$

$$+ \Sigma_{h=1}^{m} \beta_h\ j_h[\frac{1}{1} + \frac{1}{2} + \ldots + \frac{1}{v^{(jh)}_{j_1\cdots j_m}}]$$

$$- \Sigma_{h=1}^{n} \alpha_h\ \Psi[1-a_h + \alpha_h(b_j + v^{(jj)}_{j_1\cdots j_m})/\beta_j]$$

$$+ \Sigma_{h=m+1}^{q} \beta_h\ \Psi[1-b_h + \beta_h(b_j + v^{(jj)}_{j_1\cdots j_m})/\beta_j]$$

$$- \Sigma_{h=n+1}^{p} \alpha_h\ \Psi[a_h - \alpha_h(b_j + v^{(jj)}_{j_1\cdots j_m})/\beta_j];$$

(4.25) $$C_j^{(t)}\ ,\ t \geq 1\ ,\ = (-1)^{t+1}\ t!\ \{\Sigma_{h=1}^{m} \beta_h^{t+1}\ \rho[t+1,\ b_h -$$

$$- \beta_h(b_j + v^{(jj)}_{j_1\cdots j_m})/\beta_j + j_h(v^{(jj)}_{j_1\cdots j_m}+1)]$$

$$+ \Sigma_{h=1}^{m} \beta_h^{t+1}\ j_h\ [\frac{1}{(-1)^{t+1}} + \frac{1}{(-2)^{t+1}} +\ldots+ \frac{1}{(-v^{(jh)}_{j_1\cdots j_m})^{t+1}}]$$

$$+ \Sigma_{h=1}^{n}(-\alpha_h)^{t+1}\ \rho[t+1,\ 1-a_h + \alpha_h(b_j + v^{(jj)}_{j_1\cdots j_m})/\beta_j]$$

$$- \Sigma_{h=m+1}^{q}(-\beta_h)^{t+1}\ \rho[t+1,\ 1-b_h + \beta_h(b_j + v^{(jj)}_{j_1\cdots j_m})/\beta_j]$$

$$- \Sigma_{h=n+1}^{p}(\alpha_h)^{t+1}\ \rho[t+1,\ a_h-\alpha_h(b_j + v^{(jj)}_{j_1\cdots j_m})/\beta_j]\}$$

The $\Psi(.)$ and $\rho(.,.)$ are defined in (4.20) and (4.21) respectively. The proof follows from lemmas 1 and 2 in the light of the following observations. The pole at $-s = (b_j+v^{(jj)}_{j_1\ldots j_m})/\beta_j$ is of order $j_1+\ldots+j_m$. Hence the residue at this point is given by

$$(4.26)\qquad R_j = \frac{(j_1+\ldots+j_m)}{(j_1+\ldots+j_m)!}\frac{\partial^{j_1+\ldots+j_m-1}}{\partial s^{j_1+\ldots+j_m-1}}\left(\frac{\beta_j s+b_j+v^{(jj)}_{j_1\ldots j_m}}{\beta_j}\right)^{j_1+\ldots+j_m}$$

$$\times\ h(s)\ z^{-s}$$

$$= \frac{(j_1+\ldots+j_m)}{(j_1+\ldots+j_m)!}\frac{\partial^{j_1+\ldots+j_m-1}}{\partial s^{j_1+\ldots+j_m-1}}\left\{\left[\prod_{h=1}^{m}\left(\frac{\beta_h s+b_h+v^{(jh)}_{j_1\ldots j_m}}{\beta_h}\right)^{j_h}\right]h(s)z^{-s}\right\}$$

at $-s = (b_j + v^{(jj)}_{j_1\ldots j_m})/\beta_j$, where $h(s)$ is given in (1.2) and when $j_1+\ldots+j_m = 0$, $R_j=0$. Now (4.26) can be simplified by using the following procedure. For example, consider,

$$(4.27)\qquad \left(s + \frac{b_h+v^{(jh)}_{j_1\ldots j_m}}{\beta_h}\right)^{j_h}\Gamma(b_h + \beta_h s)$$

$$= [(b_h + \beta_h s + v^{(jh)}_{j_1\ldots j_m})^{j_h}(b_h + \beta_h s + v^{(jh)}_{j_1\ldots j_m}-1)^{j_h}\ldots$$

$$\ldots(b_h+\beta_h s)^{j_h}\Gamma(b_h + \beta_h s)]/[\beta_h^{j_h}(b_h + \beta_h s + v^{(jh)}_{j_1\ldots j_m}-1)^{j_h}$$

$$\ldots(b_h + \beta_h s)^{j_h}]$$

$$= \Gamma[b_h + \beta_h s + j_h(v^{(jh)}_{j_1\ldots j_m}+1)]/[\beta_h^{j_h}(b_h+\beta_h s+v^{(jh)}_{j_1\ldots j_m}-1)^{j_h}$$

$$\ldots(b_h + \beta_h s)^{j_h}],\ \text{since}\ j_h = 0\ \text{or}\ 1\ .$$

Now (4.27) at $-s = (b_j + v^{(jj)}_{j_1\ldots j_m})/\beta_j$ yields,

$$\Gamma[b_h-\beta_h(b_j+v^{(jj)}_{j_1\ldots j_m})/\beta_j+j_h(v^{(jh)}_{j_1\ldots j_m}+1)]/[\beta_h^{j_h}(-1)^{j_h(v^{(jh)}_{j_1\ldots j_m})}$$

$$\times(v^{(jh)}_{j_1\ldots j_m}!)^{j_h}]\ .$$

Now the application of lemmas 1 and 2 proves the theorem. For the sake of illustration a particular case will be evaluated here in detail.

5. An Illustration. Consider the following particular case of $H(z)$.

(5.1) $$H(z) = H_{0,2}^{2,0} [z|_{(2,2),(3,4)}] = \frac{1}{2\pi i} \int_L \Gamma(2+2s)\Gamma(3+4s)z^{-s}ds .$$

In order to expand (5.1) we examine the solutions of the equations corresponding to (4.6), namely,

(5.2) $$\frac{2+v}{2} = \frac{3+\lambda}{4} .$$

Evidently, $(v,\lambda) = (0,1),(1,3),(2,5),\ldots.$
According to the notation used,

(5.3) $$S_{11}^{(11)} = \{v_{11}^{(11)}\} = \{0,1,2,\ldots\} ,$$

which gives poles of order 2. There is no point corresponding to $v_{10}^{(11)}$ and

(5.4) $$S_{01}^{(22)} = \{v_{01}^{(22)}\} = \{0,2,4,\ldots\} ,$$

which corresponds to poles of order unity coming from the second Gamma. Now $H(z)$ can be written down by using theorem 1 as follows.

(5.5) $$H(z) = \Sigma_{S_{11}^{(11)}} z^{(b_1+v_{11}^{(11)})/\beta_1}[-\log z + c_1^{(o)}] D_1$$

$$+ \Sigma_{S_{01}^{(22)}} [z^{(b_2+v_{01}^{(22)})/\beta_2}] D_2$$

(5.6) $$= \Sigma_{v=0}^{\infty} z^{(2+v)/2}(-\log z + c_1^{(o)})D_1 + \Sigma_{v=0}^{\infty} z^{(3+2v)/4} D_2 ,$$

where

(5.7) $$D_1 = [\beta_1\beta_2(-1)^{v_{11}^{(11)}+v_{11}^{(12)}} (v_{11}^{(11)}!)(v_{11}^{(12)}!)]^{-1}$$

$$= \frac{(-1)^{v+1}}{8v!(2v+1)!} ;$$

$$(5.8) \qquad c_1^{(o)} = (\beta_1+\beta_2)\Psi(1) + \beta_1(\frac{1}{1} + \frac{1}{2} + \ldots + \frac{1}{v_{11}^{(11)}}) + \ldots$$

$$+ \beta_2(\frac{1}{1} + \frac{1}{2} + \ldots + \frac{1}{v_{11}^{(12)}})$$

$$= 2\Psi(v+1) + 4\Psi(2v+2) \quad .$$

The simplification in (5.8) is effected by using the result

$$(5.9) \qquad \Psi(n+1) = 1 + \frac{1}{2} + \ldots + \frac{1}{n} + \Psi(1) \ .$$

$$D_2 = \Gamma[b_1-\beta_1(b_2+v_{01}^{(22)})/\beta_2]/[\beta_2(-1)^{v_{01}^{(22)}} (v_{01}^{(22)}!)]$$

$$= \Gamma(\frac{1}{2} - v)/[4(2v)!] \ .$$

6. Discussion.

The series expansion given in theorem 1 is a very general expansion under the general conditions of existence of the H-function and when the poles of the integrand are not restricted to be simple. Evidently theorem 1 generalizes (4.2). In theorem 1 $H(z)$ is put into a computable form. That is, for given numerical parameter values and z, numerical values of $H(z)$ can be computed. While applying $H(z)$ to statistical distributional problems involving real stochastic variables usually one needs,

$$(6.1) \qquad F(x) = \int_0^x H(t)dt \ ,$$

in order to compute percentage points. In these problems, term by term integration of $H(z)$ in theorem 1 will give $F(x)$ into a convenient form for computation. In the H-function, if $\alpha_j (j=1,\ldots,p)$ and $\beta_j (j=1,\ldots,q)$ are all rational numbers then evidently,

$$(6.2) \qquad \alpha_j = m_j \alpha \ , \ j=1,\ldots,p \ \text{ and } \ \beta_j = m_j' \alpha \ , \ j=1,\ldots,q$$

where m_j and m_j' are positive integers and α is some positive number. Now replacing (αs) by s and then expanding the Gammas by using Gauss-Legendre multiplication formula, namely,

$$(6.3) \qquad \Gamma(nz) = (2\pi)^{(1-n)/2} n^{nz-\frac{1}{2}} \prod_{j=0}^{n-1} \Gamma(z + \frac{j}{n}) \ , \ n=1,2,\ldots$$

the H-function can be reduced to a Meijer's G-function. Applications of a G-function in statistical problems are mentioned in Mathai (1970a). The Gamma, psi and zeta functions appearing in H(z) of theorem 1 can be simplified by using the properties of these functions which are given in Erdélyi (1953).

References

1. Anderson, T. W. (1958). An Introduction to Multivariate Statistical Analysis, Wiley, New York.

2. Barnes, E. W. (1908). A new development of the theory of the hypergeometric functions. Proc. London Math. Soc. (2), 6, 141.

3. Bochner, S. C. (1958). On Riemann's functional equation with multiple gamma factors. Ann. of Math. 67, 29.

4. Braaksma, B. L. J. (1963). Asymptotic expansions and analytic continuation for Bornes-integrals. Compositio Mathematica. 15, 239.

5. Chandrasekharan, K. and Narasimhan, R. (1962). Functional equations with multiple gamma factors and the average order of arithmetical functions. Ann. of Math. 76, 93.

6. Erdélyi, A. et. al. (1953). Higher Transcendental Functions. Vol. I. McGraw-Hill, New York.

7. Fox, C. (1961). G and H-functions as symmetrical Fourier kernels. Trans. Amer. Math. Soc. 98, 395.

8. Govindarajulu, Z. (1964). A supplement to Mendenhall's bibliography on life testing and related topics. J. Amer. Statist. Assoc. 59, 1231.

9. Mathai, A. M. (1970). Applications of Generalized Special Functions in Statistical Theory of Distributions. Indian Statistical Institute (RTS), Calcutta, India.

10. Mathai, A. M. (1970a). Statistical theory of distributions and Meijer's G-function. Metron. 28, 122.

11. Meijer, C. S. (1946). On the G-Function I-VIII, Proc. Kon. Neder. Acad. Van. Wet. 49, 227, 334, 457, 632, 765, 936, 1063, 1165.

12. Mellin, H. J. (1910). Math. Ann. 68, 305.

13. Mendenhall, W. (1958). A bibliography on life testing and related topics. Biometrika. 45, 521.

A. M. Mathai
Department of Mathematics
McGill University
Montreal, Canada

TESTING AND ESTIMATION FOR STRUCTURES WHICH ARE CIRCULARLY SYMMETRIC IN BLOCKS[1]

Ingram Olkin

1. Introduction

There has been considerable study of models in which the observations arise in a structured form. These structures may yield covariance matrices which exhibit special patterns or symmetries. An early example of patterned covariance structure is the intraclass correlation model (in which all the variances and all the covariances are homogeneous) considered by Wilks (1946). Another example is the spherical model in which all the variances are equal and the covariances are zero. A wide class of structured models, called radex models, was introduced by Guttman (1954). In these models, test scores are generated from components which may be viewed as having a special geometrical structure, and hence the more recent name simplex models. Although Guttman (1954, 1957) provided examples of data which approximated simplex structure, there was little work in developing estimators of the parameters or in designing tests of hypotheses. The paper of Wilks (1946) was concerned with inference for the particular intraclass correlation model. This model has now been studied in some detail in various contexts. Similarly, tests for sphericity also have been studied in detail (for references, see Gleser, 1966).

The first general study of some simplex models was that of Mukherjee (1966), in which solutions of the maximum likelihood equations are discussed. In a subsequent paper, Mukherjee (1970) was able to obtain explicit maximum likelihood estimators for a certain class of simplex models.

Another general paper which has relevance to some of the Guttman's simplex models is that of Anderson (1970). The method of analysis considered by both Anderson (1970) and Mukherjee (1970) is based on covariance matrices Σ which may have a representation of the form

$$\Sigma = \Theta_1 A_1 + \dots + \Theta_m A_m \,, \tag{1.1}$$

where the matrices A_j are known and the Θ's are unknown. Many patterned covariance matrices may be expressed in the form of (1.1).

---

[1] Supported in part by the National Science Foundation Grant 32326X at Stanford University and by Educational Testing Service.

Jöreskog (1970) provides a general discussion of the solution of the maximum likelihood equations for simplex models.

Although general methods were considered in these papers, greater depth can be achieved in the study of some particular patterns. One such pattern, called the circumplex by Guttman, was considered by Olkin and Press (1969), in which an extensive study was made of a hierarchy of models. In addition to studying the covariance structure, patterns among the means are also considered. Olkin and Press provide not only the likelihood ratio tests of various hypotheses, but also different approximations to the null and non-null distributions. It is of interest to note that the genesis of their study is a physical model in which observations are made at the vertices of a regular polygon. Because of stationarity [see Olkin and Press (1969)], a circularly symmetric model is generated, which is identical to that of the circumplex.

In the present paper we extend the circularly symmetric model to the case where the symmetries are exhibited in blocks, and show how maximum likelihood estimators (MLE) and likelihood ratio tests (LRT) can be obtained.

2. Preliminaries on the Circularly Symmetric Model

Before providing an extension of the model, we first review the circularly symmetric model, and show how the reduction to a canonical form enables us to obtain easily the MLE and LRT. In both Guttman's circumplex model and in that considered by Olkin and Press (1969), we have random variables $x_1,\ldots,x_k$ with means $\mu_1,\ldots,\mu_k$ and covariance matrix Σ_c, which is a circular symmetric matrix. A circular symmetric matrix A_c is given by

$$(2.1)\qquad A_c = \begin{bmatrix} a_1 & a_2 & \cdots & a_r \\ a_r & a_1 & \cdots & a_{r-1} \\ & & \cdots & \\ & & \cdots & \\ a_2 & a_3 & \cdots & a_1 \end{bmatrix}, \text{ where } a_j = a_{r-j+2},\ j=2,\ldots,r.$$

Every symmetric matrix has a representation

$$(2.2)\qquad A_c = \Gamma' D_\tau \Gamma \quad ,$$

where $\tau_1,\ldots,\tau_r$ are the eigenvalues of A_c, $D_\tau = \mathrm{diag}(\tau_1,\ldots,\tau_r)$, and $\Gamma = (\gamma_{jk})$ is orthogonal.

A key point in the development is the fact that if A_c is circularly symmetric, then

(2.3) $\tau_j = \tau_{r-j+2}$, $j = 2,\ldots,r$;

furthermore, the elements γ_{jk} are given by

(2.4) $\gamma_{jk} = r^{-1/2}\{\sin 2\pi r^{-1}(j-1)(k-1) + \cos 2\pi r^{-1}(j-1)(k-1)\}$,

which are independent of the elements of A_c .

Another way to express (2.1) [see Wise (1955)] is

(2.5) $A_c = a_1W_0 + a_2W_1 + \ldots + a_rW_{r-1}$,

where

(2.6) $W_0 = I$, $W_j = \begin{pmatrix} 0 & I_{r-j} \\ I_j & 0 \end{pmatrix}$, $j = 1,\ldots,r-1$.

Remark

Since $a_j = a_{r-j+2}$, we may combine terms having the same coefficients to yield terms $W_j + W_{r-j+2}$. But $W_{r-j+2} = W_j'$, so that $W_j + W_{r-j+2}$ is symmetric. It is easily verified that $W_j = W_1^j$, so that all matrices $(W_j + W_j')$ $j = 0,\ldots,r-1$ are commutative, and hence may be diagonalized by the same orthogonal matrix. This fact will be used later.

Suppose we have a sample of size N , $(x_{1\alpha},\ldots,x_{p\alpha})$, $\alpha = 1,\ldots,N$, from a normal population with mean vector μ and covariance matrix Σ . By sufficiency we may consider the mean vector $\bar{x}$, which has a $N(\mu,\Sigma/N)$ distribution, and the cross product matrix S , which has a Wishart distribution, $\mathcal{W}(\Sigma;p,n)$, $n = N-1$, with density function

$$p(S) = c(p,n)|S|^{(n-p-1)/2}|\Sigma|^{-(n/2)} \exp[-\tfrac{1}{2}\operatorname{tr}\Sigma^{-1}S] ,$$

where

$$c(p,n) = 2^{-pn/2}\pi^{-p(p-1)/4}\ [\prod_1^p \Gamma(\tfrac{1}{2}(n-i+1))]^{-1} .$$

Now transform $\bar{x}$ and S by

(2.7) $y = \sqrt{N}\,\bar{x}\Gamma'$, $V = \Gamma S\Gamma'$,

so that

(2.8) $y \sim N(\nu,\bar{\Sigma})$, $V \sim \mathcal{W}(\bar{\Sigma};p,n)$,

where $\nu = \sqrt{N}\,\mu\Gamma$ and $\bar{\Sigma} = \Gamma\Sigma\Gamma'$. If Σ is circular, then $\bar{\Sigma}$ is diagonal, and estimators of the parameters are readily available.

In the following extension, we make use of these ideas to afford an analogous simplification for the extended problem.

3. Block Circularity

The extended model may be generated in various ways. In terms of the physical model mentioned in Section 1, we again have a point source located at the geocenter of a regular polygon of p sides, from which a signal is transmitted. Identical signal receivers are positioned at the p vertices. However, now the signal received at the i-th vertex is characterized by k components, and is denoted by $x_i = (x_{i1},\dots,x_{ik})$. The main assumption is that the covariance matrices depend only on the number of vertices separating the two receivers, so that

$$\text{Cov}(x_i, x_{i+m}) = \Sigma_m = \Sigma_{p-m} \quad , \quad m = 0,\dots,p \quad , \tag{3.1}$$

where each Σ_m is a $k \times k$ matrix. Thus, for example, if $p=4$ and 5 , we obtain

$$\begin{bmatrix} \Sigma_0 & \Sigma_1 & \Sigma_2 & \Sigma_1 \\ \Sigma_1 & \Sigma_0 & \Sigma_1 & \Sigma_2 \\ \Sigma_2 & \Sigma_1 & \Sigma_0 & \Sigma_1 \\ \Sigma_1 & \Sigma_2 & \Sigma_1 & \Sigma_0 \end{bmatrix} , \quad \begin{bmatrix} \Sigma_0 & \Sigma_1 & \Sigma_2 & \Sigma_2 & \Sigma_1 \\ \Sigma_1 & \Sigma_0 & \Sigma_1 & \Sigma_2 & \Sigma_2 \\ \Sigma_2 & \Sigma_1 & \Sigma_0 & \Sigma_1 & \Sigma_2 \\ \Sigma_2 & \Sigma_2 & \Sigma_1 & \Sigma_0 & \Sigma_1 \\ \Sigma_1 & \Sigma_2 & \Sigma_2 & \Sigma_1 & \Sigma_0 \end{bmatrix} . \tag{3.2}$$

In terms of Guttman's circumplex model, vectors of scores are generated from a structured model as follows. For simplicity, consider the special case in which there are five tests t_1, t_2, t_3, t_4, t_5 made up from 5 components c_1, c_2, c_3, c_4, c_5 , where each c_i is a k-dimensional vector:

$$\begin{aligned} t_1 &= c_1 + c_2 + c_3 , \\ t_2 &= \phantom{c_1 + {}} c_2 + c_3 + c_4 , \\ t_3 &= \phantom{c_1 + c_2 + {}} c_3 + c_4 + c_5 , \\ t_4 &= c_1 \phantom{{} + c_2 + c_3} + c_4 + c_5 , \\ t_5 &= c_1 + c_2 \phantom{{} + c_3 + c_4} + c_5 . \end{aligned}$$

If c_1, c_2, c_3, c_4, c_5 are on a circle [Guttman (1954) provides a rationale for this], so that

$$\text{Cov}(c_i, c_{i+k}) = \Delta_k \quad , \quad \text{for} \quad i = 1,2,\dots,5 \quad ,$$

with $\Delta_1 = \Delta_4$, $\Delta_2 = \Delta_3$ [because the "distance" from c_1 to c_2 is that of c_1 to c_5 , and the "distance" from c_1 to c_3 is that of c_1 to c_4], then we obtain (3.2) for $p = 5$ as the covariance matrix.

4. Reduction to Canonical Form

The critical question at this point is whether a reduction to a canonical form is possible for the block circular case. Although not obvious, by using Kronecker products it becomes straightforward to see that such a reduction is possible, indeed.

First we note several facts concerning the Kronecker product

$$A \otimes B \equiv (a_{ij}B) \quad .$$

If $A: m \times n$, $B: p \times q$, then $A \otimes B$ is an $mp \times qn$ matrix. We would like to generate block circular matrices as in (3.2). To do this, the matrix we use is (2.5) and form

$$\Sigma = (W_0 \otimes \Sigma_0) + (W_1 \otimes \Sigma_1) + \ldots + (W_{p-1} \otimes \Sigma_{p-1}) \ , \tag{4.1}$$

where the matrices W_j are defined in (2.5), and $\Sigma_j = \Sigma_{p-j}$, $j = 1,\ldots,p-1$. For example, when $p = 4$, we obtain

$$\Sigma = \begin{pmatrix} \Sigma_0 & 0 & 0 & 0 \\ 0 & \Sigma_0 & 0 & 0 \\ 0 & 0 & \Sigma_0 & 0 \\ 0 & 0 & 0 & \Sigma_0 \end{pmatrix} + \begin{pmatrix} 0 & \Sigma_1 & 0 & 0 \\ 0 & 0 & \Sigma_1 & 0 \\ 0 & 0 & 0 & \Sigma_1 \\ \Sigma_1 & 0 & 0 & 0 \end{pmatrix} + \begin{pmatrix} 0 & 0 & \Sigma_2 & 0 \\ 0 & 0 & 0 & \Sigma_2 \\ \Sigma_2 & 0 & 0 & 0 \\ 0 & \Sigma_2 & 0 & 0 \end{pmatrix} + \begin{pmatrix} 0 & 0 & 0 & \Sigma_1 \\ \Sigma_1 & 0 & 0 & 0 \\ 0 & \Sigma_1 & 0 & 0 \\ 0 & 0 & \Sigma_1 & 0 \end{pmatrix}$$

$$= \begin{bmatrix} \Sigma_0 & \Sigma_1 & \Sigma_2 & \Sigma_1 \\ \Sigma_1 & \Sigma_0 & \Sigma_1 & \Sigma_2 \\ \Sigma_2 & \Sigma_1 & \Sigma_0 & \Sigma_1 \\ \Sigma_1 & \Sigma_2 & \Sigma_1 & \Sigma_0 \end{bmatrix}$$

Next we need several well-known facts concerning Kronecker products:

$$(A_1 \otimes B_1)(A_2 \otimes B_2) = A_1A_2 \otimes B_1B_2 \ , \tag{4.2}$$

$$A \otimes (B + C) = (A \otimes B) + (A \otimes C) \ , \tag{4.3}$$

$$(A + B) \otimes C = (A \otimes C) + (B \otimes C) \ . \tag{4.4}$$

Applying these facts to (4.1), we obtain

$$\begin{aligned} &(\Gamma \otimes I)\Sigma(\Gamma' \otimes I) \\ &\quad = (\Gamma W_0 \Gamma' \otimes \Sigma_0) + (\Gamma W_1 \Gamma' \otimes \Sigma_1) + \ldots + (\Gamma W_{p-1} \Gamma' \otimes \Sigma_{p-1}) \ . \end{aligned} \tag{4.5}$$

Recall that $W_0 = I$, $W_j = W_1^j$, $j = 1,\ldots,p-1$. Consequently, if

Γ diagonalizes $W_1 + W_{p-1} = W_1 + W_1'$, i.e., $\Gamma(W_1 + W_1')\Gamma' = \mathrm{diag}(\epsilon_1,\ldots,\epsilon_k)$, then $\Gamma(W_j + W_j')\Gamma' = D_\epsilon^j$. But the matrix Γ defined by (2.3) is exactly that orthogonal matrix which diagonalizes $W_j + W_j'$, and the diagonal elements ϵ_j are the p roots of unity. Thus

$$(\Gamma \otimes I)\Sigma(\Gamma' \otimes I) = \mathrm{Diag}(\psi_1,\psi_2,\ldots,\psi_p) \equiv D_\psi \quad , \tag{4.6}$$

where the matrices ψ_j are positive definite and satisfy

$$\psi_j = \psi_{p-j+2} \quad , \quad j = 2,\ldots,p \ . \tag{4.7}$$

As in the case when the blocks are single elements, we may now use (2.6) and (2.7) as our starting point, noting that we have pk variates instead of p .

Remark

If we wish to recapture estimates of Σ_j , we may do so from (4.6), namely,

$$\Sigma = (\Gamma' \otimes I)\ D_\psi\ (\Gamma \otimes I) \ . \tag{4.8}$$

Indeed (4.8) yields simple linear equations of the form

$$\Sigma_\alpha = a_{\alpha 1}\psi_1 + \ldots + a_{\alpha p}\psi_p \quad , \quad \alpha = 1,\ldots,p \quad , \tag{4.9}$$

where the coefficients a_{ij} are functions of the γ_{ij} .

5. Hypotheses for Symmetric Structures in the Covariance Matrix and the Likelihood Functions

The following hypotheses represent block versions of (1) sphericity, (2) intraclass correlation, (3) circular symmetry, and (4) a general matrix:

$$H_1 : \Sigma = \mathrm{Diag}(\Sigma_0,\ldots,\Sigma_0) \quad , \tag{5.1}$$

$$H_2 : \Sigma = \begin{bmatrix} \Sigma_0 & \Sigma_1 & \cdots & \Sigma_1 \\ \Sigma_1 & \Sigma_0 & \cdots & \Sigma_1 \\ & & \cdots & \\ \Sigma_1 & \Sigma_1 & & \Sigma_0 \end{bmatrix} \quad ,$$

$$H_3 : \Sigma = \Sigma_c \quad ,$$

$$H_4 : \Sigma > 0 \quad .$$

In terms of the canonical representation, we may now test

hypotheses such as

(a) sphericity versus intraclass correlation,

(b) sphericity versus circular symmetry,

(c) intraclass correlation versus circular symmetry,

(d) circular symmetry versus general structure.

Because of the canonical form (4.6), the parameter space for each of the hypotheses $H_1 - H_4$ becomes

(5.2)

$$\omega_1 = \{\psi: \psi_1 = \ldots = \psi_p > 0 \ , \text{ given } \psi_j = \psi_{p-j+2} \ , \ j=2,\ldots,p\} \ ,$$

$$\omega_2 = \{\psi: \psi_1 > 0 \ , \ \psi_2 = \ldots = \psi_p > 0 \ , \text{ given } \psi_j = \psi_{p-j+2} \ , \ j=2,\ldots,p\} \ ,$$

$$\omega_3 = \{\psi: \psi_1 > 0 \ , \ \psi_2 = \psi_{p-j+2} > 0 \ , \ j=2,\ldots,p\} \ ,$$

$$\omega_4 = \{\Sigma: \Sigma > 0\} \ .$$

Following the procedure of Olkin and Press (1969), we may obtain the maxima of the likelihood functions $L(y,V)$ over the regions ω_i and ω_j, and thereby generate the likelihood ratio tests (LRT) for hypothesis H_i versus H_j by

$$\lambda_{ij} = \frac{\sup_{\omega_i} L(y,V)}{\sup_{\omega_j} L(y,V)} \tag{5.3}$$

Because the circular symmetric model is equivalent to the condition $\psi_2 = \psi_p$, $\psi_3 = \psi_{p-1}$, etc., it is clear that we will want to pool the covariance matrices V_{22} with V_{pp}, V_{33} with $V_{p-1,p-1}$, etc. in estimating the common ψ_2, ψ_3, etc. Thus, it will simplify our notation if we write

$$(V_1,\ldots,V_{m+1}) = (V_{11},V_{22} + V_{pp},\ldots,V_{mm} + V_{m+2,m+2},V_{m+1,m+1}), \tag{5.4}$$

when $p = 2m$ is even, and

$$(V_1,\ldots,V_{m+1}) = (V_{11},V_{22} + V_{pp},\ldots,V_{m+1,m+1} + V_{m+2,m+2}) \ , \tag{5.5}$$

when $p = 2m + 1$ is odd. For later use we define

$$V_j = V_{p-j+2} \ , \quad j = 2,\ldots,p \ . \tag{5.6}$$

Since the mean vector y and the V_j's are independently distributed, when Σ is circular, we have as a canonical model:

<u>Mean Vector</u>

$$y_1 \sim N(\nu_1,\psi_1), \ y_j \sim N(\nu_j,\psi_j), \ \psi_j = \psi_{p-j+2} \ , \ j = 2,\ldots,p \ . \tag{5.7}$$

Covariance Matrices

$$(5.8)\qquad V_1 \sim \mathcal{W}(\psi_1;k,n),\quad V_j \sim \mathcal{W}(\psi_j;k,2n),\quad j = 2,\dots,m\ ,$$

$$V_{m+1} \sim \mathcal{W}(\psi_{m+1};k,n),\quad \text{when}\quad p = 2m\ .$$

$$(5.9)\qquad V_1 \sim \mathcal{W}(\psi_1;k,n),\quad V_j \sim \mathcal{W}(\psi_j;k,2n),\quad j = 2,\dots,m+1$$

when $p = 2m + 1$.

The maxima of the likelihoods may now be obtained in a straightforward manner, and to a certain extent, the results parallel those in Olkin and Press (1969). The results are based on the assumption that the mean vectors $\nu_1,\dots,\nu_p$ are unknown. A slight modification in the development yields analogous results when the means are known.

We now list the maxima of the likelihood function for the various models using (5.7) - (5.9) as our starting point. In each case the maximum involves a common term

$$(5.10)\qquad c(V) = c(pk,n)(2\pi)^{-pk/2}|V|^{(n-pk-1)/2}e^{-pkN/2}\ .$$

Spherical Model

$$(5.11)\qquad \sup_{\omega_1} L(y,V) = \frac{c(V)\ (pN)^{pkN/2}}{|\sum_1^p V_{ii}|^{pN/2}}\ .$$

Intraclass Correlation Model

$$(5.12)\qquad \sup_{\omega_2} L(y,V) = \frac{c(V)\,[N^p(p-1)^{p-1}]^{kN/2}}{|V_{11}|^{N/2}|\sum_2^p V_{ii}|^{(p-1)N/2}}\ .$$

Circular Model

$$(5.13)\qquad \sup_{\omega_3} L(y,V) = \frac{c(V)\ N^{pkN/2}}{|V|^{N/2}}\ .$$

General Case

$$(5.14)\qquad \sup_{\omega_4} L(y,V) = \frac{c(V)\ N^{pkN/2}}{|V|^{N/2}}\ .$$

For each of the hypotheses considered when the hypothesis is true, the LRT is distributed as a product of independent beta variates, so that the procedure of Box (1949) may be used to obtain an approximation for the null distribution. We here present an approximation to $O(N^{-2})$. Because some of the hypotheses are closely allied to testing for the equality of covariance matrices, the Bartlett modification may be preferable to the LRT [see Anderson (1958) for details concerning this test]. Also, because some of the hypotheses are nested, we may use the procedure of Gleser and Olkin (1972) to provide an easier evaluation of the needed constants.

6. Likelihood Ratio Tests and Their Approximate Null Distribution for Testing Symmetric Structures

Using the results (5.11) - (5.13), we may readily form the LRT for various hypotheses. In some instances the results for $p = 2m$ or $p = 2m + 1$ differ; whenever possible, we combine our results for even and odd p by using the parameter m.

6.1 Test for Sphericity Versus Circularity

From (5.11) and (5.13), the LRT, λ_{13}, is given by

$$\lambda_{13}^{2/N} = \frac{p^{pk} \prod_{1}^{p} |V_j|}{2^{2k(p-m-1)} \left| \sum_{1}^{p} V_{ii} \right|^{p}} \quad . \tag{6.1}$$

The modified Bartlett statistic, L_{13}, is a simple function of λ_{13}, namely, $L_{13} = \lambda_{13}^{n/N}$. Using the result of Anderson (1958, p. 254) we obtain the approximate null distribution:

$$P\{-\rho \log L_{13} \le z\} \cong P\{\chi_f^2 \le z\} + 0(n^{-2}) \quad , \tag{6.2}$$

where

$$f = \frac{1}{2} mk(k + 1) \quad ,$$

$$\rho = 1 - \frac{1}{n} \frac{[p(3m + 3 - p) - 2](2k^2 + 3k - 1)}{12mp(k + 1)} \quad .$$

6.2 Test for Intraclass Correlation Model Versus Circularity

From (5.12) and (5.13), the LRT, λ_{23}, is given by

$$\lambda_{23}^{2/N} = \frac{(p-1)^{(p-1)k} \prod_{2}^{p} |V_j|}{2^{2k(p-m-1)} \left| \sum_{2}^{p} V_{ii} \right|^{p-1}} \tag{6.3}$$

As in Section 6.1, the modified Bartlett statistic, L_{23}, is given by $L_{23} = \lambda_{23}^{n/N}$. Similarly,

$$P\{-\rho \log L_{23} \le z\} \cong P\{\chi_f^2 \le z\} + 0(n^{-2}) \quad , \tag{6.4}$$

where

$$f = \frac{1}{2} mk(k+1) \quad ,$$

$$\rho = 1 - \frac{1}{n} \frac{(3mp - p^2 - 3m + p - 3)(2k^2 + 3k - 1)}{12(p-1)m(k+1)} \quad .$$

6.3 Tests for Circular Versus General Structure

From (5.13) and (5.14), the LRT is given by

(6.5) $$\lambda_{34}^{2/N} = \frac{2^{2k(p-m-1)} |V|}{\prod_1^p |V_j|} \quad .$$

In order to show that this statistic is distributed as a product of independent beta variates, note that

$$\frac{|V|}{\prod_1^p |V_j|} = \frac{|V|}{\prod_1^p |V_{ii}|} \frac{|V_{22}|\,|V_{pp}|}{|V_{22} + V_{pp}|} \cdots \frac{|V_{mm}|\,|V_{m+2,m+2}|}{|V_{mm} + V_{m+2,m+2}|} \quad ,$$

if $p = 2m$, and

$$\frac{|V|}{\prod_1^p |V_j|} = \frac{|V|}{\prod_1^p |V_{ii}|} \frac{|V_{22}|\,|V_{pp}|}{|V_{22} + V_{pp}|} \cdots \frac{|V_{m+1,m+1}|\,|V_{m+2,n+2}|}{|V_{m+1,m+1} + V_{m+2,m+2}|} \quad ,$$

if $p = 2m + 1$. Under the null hypothesis, these statistics are independently distributed. Furthermore, each term is known to be distributed as a product of independent beta variables, Anderson (1958, Chapters 9, 10). Consequently, we may use the following result of Gleser and Olkin (1972): If $Z = \prod_1^G Z_i$, and appropriate regularity conditions prevail, then

$$P\{-2 \log Z \le z\} \cong P\{\chi_f^2 < \rho z\} \quad ,$$

where $f = \sum_1^G f_i$, $\rho = \sum_1^G f_i \rho_i / f$, and the f_i and ρ_i are obtained by applying the Box procedure to each Z_i .

In the present case, we may let $Z_1 = |V| / \prod_1^p |V_{ii}|$. This is the test statistic for testing for independence in a covariance matrix [see Anderson (1958, p. 233)]. Here $f_1 = \frac{1}{2} k^2 p(p-1)$, $\rho_1 = 1 - [2k(p+1) + 9]/6N$.

The remaining statistics Z_j are of the form $|V_{11}|\,|V_{22}|/|V_{11} + V_{22}|$, which is the test statistic for testing for the equality of two covariance matrices [see Anderson (1958, p. 255)]. For each such test, we obtain

$$f_j = \frac{1}{2} k(k+1) \quad , \quad \rho_j = 1 - \frac{2k^2 + 3k - 1}{12N(k+1)} \quad .$$

Note that $p - m - 1$ is equal to $m - 1$ when $p = 2m$, and is equal to m when $p = 2m + 1$. Thus, in either case of p even or odd, we obtain

$$f = \frac{1}{2} k^2 p(p-1) + \frac{1}{2} (p - m - 1) k(k + 1) \quad , \tag{6.6}$$

$$\rho = 1 - \frac{k}{24fN}\{2kp(p-1)(2kp+2k+q) + (p-m-1)(2k^2+3k-1)\} \quad .$$

The final approximation to the null distribution is then given by

$$p\{-\rho \log \lambda_{34} \leqq z\} = P\{\chi_f^2 \leq z\} + 0(N^{-2}) \quad ,$$

where f and ρ are given by (6.6).

7. Tests for Means Given That the Covariance Matrix is Circular

When the population covariance matrix has no special structure, and we wish to test that the mean vector is zero, the appropriate test is Hotelling's T^2. However, when we know that there is a circular structure, we can take advantage of this information in constructing a test.

From the canonical form (5.7) - (5.9), we see that under $H: \nu = 0$, we should estimate ψ_1 by $V_{11} + y_1'y_1$, and we should estimate ψ_j by

$$V_{jj} + V_{p-j+2,p-j+2} + y_j'y_j + y_{p-j+2}'y_{p+j+2} \quad ,$$

$j = 2,\ldots,p$. Thus the LRT statistic is given by

$$\lambda^{2/N} = \prod_1^p \frac{|V_j|}{|V_j + W_j|} \quad , \tag{7.1}$$

where when $p = 2m + 1$

$$Z_1 = y_1'y_1 \ , \ Z_j = y_j'y_j + y_{p-j+2}'y_{p-j+2} \ , \ j = 2,\ldots,m + 1 \quad , \tag{7.2}$$

and when $p = 2m$,

$$Z_1 = y_1'y_1 \ , \ Z_j = y_j'y_j + y_{p-j+2}' \ , \ Z_{m+1} = y_{m+1}'y_{m+1} \quad ; \tag{7.3}$$

$j = 2,\ldots,m$. Each component of the product in (7.1) is distributed as a U-statistic [see Anderson (1958, p. 193)]. Thus, we may again use the Lemma of Gleser and Olkin (1972) to obtain an approximation to the null distribution of the LRT. To do this we need to know the degrees of freedom f, and the value of ρ. For ratios $|V_j|/|V_j + Z_j|$, which do not involve pooling, e.g., $j = 1$, we have

$$f = k \quad , \quad \rho = 1 - \frac{k^2}{(k + 1)n} \quad ;$$

for terms which involve pooling, e.g., $j = 2,\ldots,m$, we have

$$f = 2k \quad , \quad \rho = 1 - \frac{k - 1}{2n} \quad .$$

Consequently, the overall value of f is

$$f = pk \quad , \quad \rho = 1 - \frac{[k^3 + (k^2 - 1)(p - m - 1)]}{2f(k + 1)n} \quad .$$

Remark

The methods outlined lend themselves to the development of other tests concerning the means. For example, Olkin and Press (1969) consider the test that the mean vectors are equal when there is circular symmetry. This model may be extended to test that the mean vectors are equal in blocks, when the covariance matrix is circularly symmetric in blocks. Similarly, we may simultaneously test for the equality of the mean vectors and circular symmetry. The key point in the development of such tests is to start with the canonical form (5.7) - (5.9), from which the LRT may be readily obtained.

References

1. Anderson, T. W. (1958). An Introduction to Multivariate Statistical Analysis. Wiley, New York.

2. Anderson, T. W. (1970). Estimation of covariance matrices which are linear combinations or whose inverses are linear combinations of given matrices, pp. 1-24 in Essays in Probability and Statistics, University of North Carolina Press, Chapel Hill.

3. Box, G. E. P. (1949). A general distribution theory for a class of likelihood criteria. Biometrika 36, 317.

4. Gleser, L. (1966). A note on the sphericity test. Ann. Math. Statist. 37, 464.

5. Gleser, L. J. and Olkin, I. (1972). A note on Box's general method of approximation for the null distributions of likelihood criteria. Submitted for publication.

6. Guttman, L. (1954). A new approach to factor analysis: the radex, pp. 258-348, 430-433 in Mathematical Thinking in the Social Sciences (ed. by P. F. Lazarsfeld), The Free Press, Glencoe, Ill.

7. Guttman, L. (1957). Empirical verification of the radex structure of mental abilities and personality traits. Educ. Psychol. Meas. 17, 391.

8. Jöreskog, K. G. (1970). Estimation and testing of simplex models. British J. Math. Statist. Psychol. 23, 121.

9. Mukherjee, B. N. (1966). Derivation of likelihood ratio tests for the Guttman quasi-simplex covariance structures. Psychometrika 31, 97.

10. Mukherjee, B. N. (1970). Likelihood ratio tests of statistical hypotheses associated with patterned covariance matrices in psychology. British J. Math. Statist. Psychol. 23, 89.

11. Olkin, I. and Press, S. J. (1969). Testing and estimation for a circular stationary model. Ann. Math. Statist. 40, 1358.

12. Wilks, S. S. (1946). Sample criteria for testing equality of means, equality of variances, and equality of covariances in a normal multivariate distribution. Ann. Math. Statist. 17, 257.

13. Wise, J. (1955). The autocorrelation functions and the spectral density function. Biometrika 42, 151.

Ingraham Olkin
Department of Statistics
Stanford University
Stanford, California

UNIVARIATE DATA FOR MULTI-VARIABLE SITUATIONS: ESTIMATING VARIANCE COMPONENTS

S. R. Searle

Summary. The concepts of variance components models are outlined, along with estimation procedures associated with them and the difficulties involved in the unbalanced data that are so often available for such models. Several specific unsolved problems arising from these models are also described.

1. Introduction. Variance components models come within the wide purview of multivariate analysis envisaged by Dempster [1971] although they do so in a rather special sense: they are truly multivariate models but with the peculiarity that available data are only univariate. Hence this paper's title.

The classical form of a linear model is

$$\underline{y} = \underline{X}\underline{\beta} + \underline{e} \tag{1}$$

where $\underline{y}$ is a vector of observations on a random variable Y, $\underline{\beta}$ is a vector of parameters to be estimated, $\underline{X}$ is a matrix of known values and $\underline{e}$ is a vector of residual error terms. The elements of $\underline{\beta}$ are called constants or fixed effects, and are regression slopes, main effects, or interactions, depending on the context. In this way the formulation in (1) embraces all analysis of variance models, regression models and mixtures of the two, namely analysis of covariance models. In all these cases the elements of $\underline{\beta}$ are never envisaged as random variables. They are parameters to be estimated, generically referred to as fixed effects.

In contrast, there are models in which some of the elements of $\underline{\beta}$ are random variables. Usually they are the effects corresponding, in traditional analysis of variance situations, to the levels of one or more factors (or interactions). When the factors are such that the levels in the data can be considered as a random sample from a population of levels, the corresponding effects are random variables and are called random effects. In this case it is not the effects themselves, as elements of $\underline{\beta}$, that are the parameters of interest, but their variances. Hence the name variance components models or, more usually, random effects models.

Current interest in random effects models is quickening, with numerous papers about them published in recent years, including two in the inaugural volume of the Journal of Multivariate Analysis. Many of these deal with the more difficult aspects of estimation that arise from unequal-subclass-number data: for example, the multiplicity of estimation procedures, the intractability of criteria for judging between them and the prescience of negative estimates of parameters (namely variances) that are by definition positive.

2. Examples and models. Variance components have had a long use in genetics. Suppose a male animal has many progeny, a Holstein bull for example, which, through the use of artificial insemination, has sired many cows. If x_{ij} is the milk yield of the jth daughter of the ith sire a suitable model is

$$x_{ij} = \mu + \alpha_i + e_{ij} \quad \text{for} \quad i = 1,2,\ldots,a \qquad j = 1,2,\ldots,n_i \ . \tag{2}$$

The parameter μ is a general mean, α_i is the effect due to the ith sire (a random effect from a population of α's that has zero mean and variance σ_α^2) and the e_{ij}'s are the usual random error terms, uncorrelated with the α's, having zero mean and variance σ_e^2. Thus $\sigma_x^2 = \sigma_\alpha^2 + \sigma_e^2$ and the problem is to estimate σ_α^2 and σ_e^2. The sire's effect α_i on his progeny's milk yield represents a random half of the sire's genetic make-up, so that $\sigma_\alpha^2 = \frac{1}{4}\sigma_G^2$ where σ_G^2 is the (additive) genetic variance of milk yield. The ratio of this to σ_x^2, namely $h = \sigma_G^2/\sigma_x^2 = 4\sigma_\alpha^2/(\sigma_\alpha^2 + \sigma_e^2)$ which, known as heritability, is of great use in animal breeding programs. For example, expected increases in yield arising from selecting a high-yielding fraction of animals to be parents of the next generation are proportional to h.

Another example is crossing 2 varieties of corn, using pollen from replicate males (tassels) of one variety on replicate females (silks) of the other variety, the sample of males and females used in each case being considered random samples from the varieties concerned. If x_{ijk} is the yield of the kth plant resulting from crossing the ith male with the jth female the model

$$x_{ijk} = \mu + m_i + f_j + (mf)_{ij} + e_{ijk} \tag{3}$$

is appropriate. Here m_i, f_j and $(mf)_{ij}$ are uncorrelated random variables with zero means and homoscedastic with variances

σ_m^2, σ_f^2 and σ_{mf}^2 respectively. These variables are also uncorrelated with the error terms e_{ijk} which have zero mean and variance σ_e^2.

There are also models involving both random and fixed effects. A non-biological example given in Thompson [1963] is that of analyzing the muzzle velocity x_{ij} of the ith shell fired from a gun using the jth of several measuring instruments. Here we have

$$x_{ij} = \mu + s_i + m_j + e_{ij} \tag{4}$$

where s_i is the effect due to the ith shell and m_j is the bias in the jth measuring instrument. Since the shells used are a random sample of shells, the s_i are random effects, whereas the m_j are fixed effects.

A model such as (4) is usually called a mixed model, involving as it does both fixed and random effects. But since in all models μ is a fixed effect and the error terms are random, all models can be considered as mixed. To distinguish the 2 kinds of effects a generalization of the fixed effects model (1) is

$$\underline{y} = \underline{X\beta} + \underline{Zu} + \underline{e} \tag{5}$$

for the mixed effects model. Here $\underline{\beta}$ is the vector of fixed effects (including coefficients of covariates if present), $\underline{u}$ is the vector of random effects and $\underline{e}$ is the vector of error terms. $\underline{X}$ and $\underline{Z}$ are matrices of known values corresponding to the incidence of the fixed and random effects respectively in $\underline{y}$. Properties generally attributed to the random variables in $\underline{u}$ and $\underline{e}$ are

$$E(\underline{u}) = \underline{0}\ ,\ E(\underline{e}) = \underline{0}\ ,\ E(\underline{ue}') = \underline{0}\ ,$$

$$\text{var}(\underline{u}) = E(\underline{uu}') = \underline{D} \quad \text{and} \quad \text{var}(\underline{e}) = E(\underline{ee}') = \underline{R}\ .$$

Hence

$$E(\underline{y}) = \underline{X\beta} \tag{6}$$

and

$$\text{var}(\underline{y}) = \underline{ZDZ}' + \underline{R} \equiv \underline{V}\ , \text{ say .} \tag{7}$$

The problem is to estimate not only $\underline{\beta}$ but also $\underline{D}$ and $\underline{R}$. Usually $\underline{R}$ is taken as $\sigma_e^2\underline{I}$ and $\underline{D}$ is taken as a diagonal matrix. For example, in reformulating (2) in the form of (5), the vector $\underline{u}$ would be $\underline{u}' = \underline{\alpha}' = (\alpha_1\ \alpha_2\ \ldots\ \alpha_a)$ so that $\underline{D}$ would be $\underline{D} = \sigma_\alpha^2\underline{I}_a$. Similarly for (3), with a males, $i = 1,2,\ldots,a$ and b females, $j = 1,2,\ldots,b$, $\underline{D}$ would be

$$\underline{D} = \begin{bmatrix} \sigma_m^2 \underline{I}_a & \underline{0} & \underline{0} \\ \underline{0} & \sigma_f^2 \underline{I}_b & \underline{0} \\ \underline{0} & \underline{0} & \sigma_{mf}^2 \underline{I}_{ab} \end{bmatrix} . \tag{8}$$

In general, we can specify $\underline{u}$ as

$$\underline{u}' = [\underline{u}_1' \ \underline{u}_2' \ \cdots \ \underline{u}_\theta' \ \cdots \ \underline{u}_K'] \tag{9}$$

and $\underline{Z}$ correspondingly as

$$\underline{Z} = [\underline{Z}_1 \ \underline{Z}_2 \ \cdots \ \underline{Z}_\theta \ \cdots \ \underline{Z}_K] \tag{10}$$

where $\underline{u}_\theta'$ is the vector of N_θ effects for the θth factor (main effects or interaction factor), for $\theta = 1,2,\ldots,K$. Customarily the elements of $\underline{u}_\theta$ are assumed to have zero mean, be uncorrelated and have uniform variance σ_θ^2, and to be uncorrelated with elements of all other $\underline{u}$'s ; i.e., $E(\underline{u}_\theta) = \underline{0}$, $\text{var}(\underline{u}_\theta) = \sigma_\theta^2 \underline{I}_{N_\theta}$, $E(\underline{u}_\theta \underline{u}_\varphi') = \underline{0}$ for $\theta \neq \varphi$ and $E(\underline{u}_\theta \underline{e}') = \underline{0}$. Hence $\underline{D}$ is a diagonal matrix of the matrices $\sigma_\theta^2 \underline{I}_{N_\theta}$ and so can be written as

$$\underline{D} = \sum_{\theta=1}^{K}{}^{+} \sigma_\theta^2 \underline{I}_{N_\theta} \tag{11}$$

where Σ^+ denotes the operation of a direct (Kronecker) sum of matrices. An example is (8). Then $\underline{V}$ in (7) becomes, using $\underline{R} = \sigma_e^2 \underline{I}$,

$$\underline{V} = \sum_{\theta=1}^{K} \sigma_\theta^2 \underline{Z}_\theta \underline{Z}_\theta' + \sigma_e^2 \underline{I} \tag{12}$$

which, by defining $\underline{Z}_{K+1} \equiv \underline{I}$ and $\sigma_{K+1}^2 \equiv \sigma_e^2$ can be further generalized as

$$\underline{V} = \sum_{\theta=1}^{K+1} \sigma_\theta^2 \underline{Z}_\theta \underline{Z}_\theta' . \tag{13}$$

The problem is to estimate the variance components σ_e^2 and σ_θ^2 for $\theta = 1,2,\ldots,K$.

Additional generality could be given to the model by assuming $\text{var}(\underline{u}_\theta)$ to be $\underline{\Sigma}_{\theta\theta}$ say, rather than $\sigma_\theta^2 \underline{I}_{N_\theta}$, and more still by assuming $\text{cov}(\underline{u}_\theta \underline{u}_\varphi')$ to be $\underline{\Sigma}_{\theta\varphi}$ say, rather than zero. Then $\underline{D}$ would be $\underline{D} = \{\underline{\Sigma}_{\theta\varphi}\}$ for $\theta, \varphi = 1,2,\ldots,K$. However, there are difficulties enough in estimating $\underline{D}$ when it has the simple form shown in (11)--i.e., estimating the σ^2's--so that (11) is the form usually assumed.

3. Available data. Situations for which random effects models are appropriate often yield, especially in biology and economics, what can be called "messy data." The data are often voluminous in extent and frequently include large samples of the random effects. However, these same data often stem from survey-like situations and seldom do they have a uniform number of observations in each submost subclass. Not only may the numbers vary greatly but many subclasses may be empty, having no observations at all--in some cases as many as 30% or more of the subclasses being empty. At first thought the prospects of having efficient estimation procedures for such data are gloomy, and indeed they are. But the pressing need by biologists, economists and others for variance components estimates that they can use reliably in their work is such that development of efficient estimators for their kinds of data is worth pursuing.

Dichotomizing data according to the number of observations in the subclasses, namely data having equal subclass numbers (which we call balanced data) or data having unequal subclass numbers (which we call unbalanced data) is pertinent to variance component estimation because in the one case estimation is easy and well documented and in the other it is difficult with unsolved problems. The easy case is balanced data; the difficult case is unbalanced data. We deal first, and briefly, with the easy case.

4. Estimation: balanced data. Consideration of variance components models in balanced data led to estimation methods based on the mean squares of classical analyses of variance. Expected values of these mean squares are linear functions of the variance components and equating them to observed values gives linear equations in the components, the solutions of which are taken as estimators. Suppose $\underline{m}$ and $\underline{\sigma}^2$ are vectors of analysis of variance mean squares and variance components respectively for some set of data. Writing the expected value of $\underline{m}$ as $\underline{P}\underline{\sigma}^2$ we have

$$E(\underline{m}) = \underline{P}\underline{\sigma}^2 \tag{14}$$

and the equations for deriving $\hat{\underline{\sigma}}^2$ as an estimator of $\underline{\sigma}^2$ are

$$\underline{m} = \underline{P}\hat{\underline{\sigma}}^2 \; . \tag{15}$$

For random models the elements of $\underline{m}$ are all the mean squares of the appropriate analysis of variance and in mixed models they are the mean squares whose expectations contain no fixed effects. In both cases (for balanced data) $\underline{P}$ is nonsingular and the estimators are $\underline{P}^{-1}\hat{\sigma}^2$.

Unbiasedness is a well-evident property of estimators obtained from (14) and (15), e.g., Winsor and Clarke [1940]; but establishment of other properties has been relatively recent; e.g., minimum variance quadratic unbiasedness and, under normality assumptions, minimum variance unbiasedness, Graybill and Hultquist [1961]. Normality assumptions for the random effects also lead, as usual, to the analysis of variance sums of squares having χ^2-distributions (multiplied by constants) with the result that confidence intervals and test statistics for hypotheses about certain linear combinations of the components can be derived. However, the linear combinations of χ^2-variables that constitute the estimators have coefficients that involve the unknown components and so the distributions of the estimators are unknown. For example, for the model (2) with $n_i = n$ for all i

$$\hat{\sigma}^2 = \frac{n\sigma_\alpha^2 + \sigma_e^2}{n(a-1)}\chi^2_{a-1} - \frac{\sigma_e^2}{an(n-1)}\chi^2_{a(n-1)} \qquad (16)$$

where the χ^2_r symbol here denotes a variable distributed as χ^2 on r degrees of freedom. Furthermore, since the estimators involve differences between such variables, as in (16), their distributions involve sums of confluent hypergeometric functions as in Robinson [1965] and Wang [1967]. However, one characteristic of the estimators that can be derived is their variances, because the χ^2-variables of which they are linear combinations are independent. Unbiased estimators of these variances are also available, Ahrens [1965] (see also Searle [1971a]).

Maximum likelihood estimation using normality assumptions leads _pro forma_ to almost the same estimators as given by the analysis of variance method summarized in (14) and (15). However, the estimators can be negative so they cannot truly be maximum likelihood estimators since these would stem from maximizing the likelihood over the parameter space which, for variance components, is non-negative. Herbach [1959], Thompson [1962] and Thompson and Moore [1963] discuss this problem.

Estimates obtained from (15) are, on some occasions, negative. This is clearly embarrassing because the corresponding parameters, being variances, are essentially positive. Many awkward moments arise between a consulting statisticain and his client when explanation of this peculiarity is called for and found wanting. Various unsatisfactory alternatives are listed in Searle [1971a] but the need for developing essentially positive estimators remains.

5. Estimation: unbalanced data. The innocent looking difference between balanced and unbalanced data has widespread ramifications in the task of estimating variance components. It leads to a variety of methods of estimation, properties of which are mostly unknown (save for unbiasedness which is almost universally achieved). This, in combination with the largely empirical nature of the criteria used for deriving the estimation methods, makes it almost impossible to pass judgement on the different estimators. Furthermore, the algebra involved is horrendous ("algebraic heroics" are Hartley's [1967] words) and computing procedures are correspondingly difficult even for large computers; e.g., inverting matrices of order 1000 and greater. Nevertheless, the practical need for efficient estimators of variance components from unbalanced data is sufficiently compelling to pursue the problems involved.

5a. Basic development. The development of estimators has basically been that of a variety of quadratic forms in the observation vector $\underline{y}$ having expected values that are linear combinations of the variance components. Thus if $\underline{y}'\underline{\underline{Q}}\underline{y}$ is one such quadratic form we know from (6) and (7), without any assumptions about the form of the distribution of $\underline{y}$, that

$$E(\underline{y}'\underline{\underline{Q}}\underline{y}) = tr(\underline{\underline{Q}}\underline{\underline{V}}) + \underline{\beta}'\underline{\underline{X}}'\underline{\underline{Q}}\underline{\underline{X}}\underline{\beta} \ . \tag{17}$$

$\underline{\underline{Q}}$ is therefore chosen so that

$$\underline{\underline{Q}}\underline{\underline{X}} = \underline{0} \ . \tag{18}$$

Then, analogous to (14) and (15), if $\underline{q}(\underline{y})$ is a vector of such quadratic forms with

$$E[\underline{q}(\underline{y})] = \underline{\underline{P}}\underline{\sigma}^2 \ , \tag{19}$$

the equations for deriving $\hat{\underline{\sigma}}^2$ as an estimator of $\hat{\underline{\sigma}}^2$ are

$$\underline{\underline{P}}\underline{\sigma}^2 = \underline{q}(\underline{y}) \ . \tag{20}$$

Now with balanced data there is an 'obvious' set of mean squares (quadratic forms) to use in $\underline{m}$ in the estimation procedure of (14) and (15), and the resulting estimators have been shown to have certain desirable properties in addition to unbiasedness. In contrast, with unbalanced data there is no single set of quadratic forms satisfying (18) that are 'obvious' for use in (19) and (20). A variety of suggestions have been made, some of them involving more equations in (19) than there are variance components. In this case equations (20) are over-identified, but provided $\underline{P}$ of (19) has full column rank, a 'least squares' solution can be obtained as

$$\hat{\underline{\sigma}}^2 = (\underline{P}'\underline{P})^{-1}\underline{P}'\underline{q}(\underline{y}) \ . \tag{21}$$

5b. Adaptations of analysis of variance. Until quite recently the quadratic forms $\underline{y}'\underline{Q}\underline{y}$ suggested for use in $\underline{q}(\underline{y})$ of (19) and (20) have been chosen by analogy with classical analysis of variance procedures. Three such analogies given in Henderson [1953] have received widespread use and attention. The first uses unbalanced data analogies of sums of squares used in analyses of variance of balanced data. For example, for the model (3) with $i = 1,2,\ldots,a$ and $j = 1,2,\ldots,b$ and $k = 1,2,\ldots,n_{ij}$, the interaction sum of squares for balanced data ($n_{ij} = n$ for all i and j) is

$$n\sum_{i=1}^{a}\sum_{j=1}^{b}(\bar{x}_{ij.}-\bar{x}_{i..}-\bar{x}_{.j.}-\bar{x}_{...})^2 \equiv n\sum_{i=1}^{a}\sum_{j=1}^{b}\bar{x}_{ij.}^2-bn\sum_{i=1}^{a}\bar{x}_{i..}^2-an\sum_{j=1}^{b}\bar{x}_{.j.}^2 + abn\bar{x}_{...}^2 \ , \tag{22}$$

using familiar bar and subscript dot notation for means. Analogous to the right-hand side of this identity Henderson's [1953] first method suggests using for unbalanced data

$$\sum_{i=1}^{a}\sum_{j=1}^{b}n_{ij}\bar{x}_{ij.}^2 - \sum_{i=1}^{a}n_{i.}\bar{x}_{i..}^2 - \sum_{j=1}^{b}n_{.j}\bar{x}_{.j.}^2 + n_{..}\bar{x}_{...}^2 \ . \tag{23}$$

Although not a sum of squares (it is not a positive definite quadratic form), this and its counterparts for the error and 2 main effects sums of squares for (3) do provide 4 elements for $\underline{q}(\underline{y})$ in (19) and so yield estimators from (20).

The second Henderson method, described by Searle [1968] in matrix terminology, is intended for mixed models like (5). Balanced data present no difficulties in estimating variance components in mixed models because the analysis of variance sums of squares for the random effects factors have expected values free of the fixed effects. But with unbalanced data there is need for eliminating these fixed effects. An apparently easy procedure is to first estimate the fixed effects as

$$\tilde{\underline{\beta}} = \underline{L}\underline{y} \tag{24}$$

say, and then estimate the variance components from $\underline{y}$ corrected for $\tilde{\underline{\beta}}$ in the form

$$\underline{z} = \underline{y} - \underline{X}\tilde{\underline{\beta}} , \tag{25}$$

for which the model is, from (5),

$$\underline{z} = (\underline{X} - \underline{X}\underline{L}\underline{X})\underline{\beta} + (\underline{Z} - \underline{X}\underline{L}\underline{Z})\underline{u} + (\underline{I} - \underline{X}\underline{L})\underline{e} . \tag{26}$$

Henderson's Method 2 chooses $\underline{L}$ to reduce this to a random model that is, apart from the error terms, directly suited to his first method. However, the choice of $\underline{L}$ is not unique and it necessitates preclusion of models for $\underline{y}$ that contain interactions between fixed and main effects, Searle [1968]. These are rather severe limitations.

A recent use of (24) and (25) is made by Wallace and Hussain [1969, sec. 5.4] for eliminating covariates from a mixed model. In using $(\underline{X}'\underline{X})^{-1}$ for $\underline{L}$ they certainly eliminate $\underline{\beta}$ from (26), but their then estimating variance components from familiar analysis of variance mean squares of the z's (their data is balanced) predicates the assumption that the model for $\underline{z}$ is $\mu^*\underline{1} + \underline{Z}\underline{u} + \underline{e}$, for some μ^* . This is incorrect, as is evident from (26).

The third Henderson [1953] method uses the reduction in sums of squares calculated when fitting constants. Suppose $R(\underline{\beta}_1, \underline{\beta}_2)$ and $R(\underline{\beta}_1)$ are the reductions in sum of squares for fitting

$$y = \underline{X}_1\underline{\beta}_1 + \underline{X}_2\underline{\beta}_2 + \underline{e} , \tag{27}$$

and $\underline{y} = \underline{X}_1\underline{\beta}_1 + \underline{\epsilon}$ respectively. Then the expectation under the model (27) of

$$R(\underline{\beta}_2|\underline{\beta}_1) = R(\underline{\beta}_1,\underline{\beta}_2) - R(\underline{\beta}_1)$$

is

$$E\ R(\underline{\beta}_2|\underline{\beta}_1) = tr\{\underline{X}_2'[\underline{I}-\underline{X}_1(\underline{X}_1'\underline{X}_1)^{-}\underline{X}_1']\underline{X}_2\ E(\underline{\beta}_2\underline{\beta}_2')\} + \sigma_e^2[r(\underline{X}_1\underline{X}_2)-r(\underline{X}_1)]. \tag{28}$$

The importance of this result is that $E\ R(\underline{\beta}_2|\underline{\beta}_1)$ contains no terms in $\underline{\beta}_1$. Therefore for the mixed model $\underline{y} = \underline{X}\underline{\beta} + \underline{Z}\underline{u} + \underline{e}$ of (5), reductions in sums of squares of the form $R(\underline{B}_2|\underline{\beta}_1)$ can be derived having expected values free of the fixed effects so long as $\underline{\beta}_1$ always contains $\underline{\beta}$. For example using (9), (10) and (11) with $K = 2$ the model is

$$\underline{y} = \underline{X}\underline{\beta} + \underline{X}_1\underline{u}_1 + \underline{Z}_2\underline{u}_2 + \underline{e} \qquad (29)$$

and expected values of $R(\underline{u}_1|\underline{\beta},\underline{u}_2)$ and $R(\underline{u}_2|\beta,\underline{u}_1)$ will by (28) be linear functions of σ_1^2, σ_e^2 and σ_2^2, σ_e^2 respectively, with $E[\underline{y}'\underline{y} - R(\underline{\beta},\underline{u}_1,\underline{u}_2)]$ being a multiple of σ_e^2 in the usual way. Note that $E[R(\underline{u}_1,\underline{u}_2|\underline{\beta})]$ will be a linear function of σ_1^2, σ_2^2 and σ_e^2 so that there are 4 elements of $\underline{q}(\underline{y})$ for (19) and (20) with only 3 variance components to estimate. This is the problem of over-identifiability referred to earlier. It is also discussed in Searle [1971a,b] for the model (3) and in Mount and Searle [1972] for a covariance model.

5c. Maximum likelihood. On the basis of normality assumptions Hartley and Rao [1967] consider maximum likelihood estimation for mixed models. This involves equations that are extremely complicated in the estimators: using $\underline{V}$ of (12) they are

$$\underline{X}'\underline{\tilde{V}}^{-1}\underline{X}\underline{\tilde{\beta}} = \underline{X}'\underline{\tilde{V}}^{-1}\underline{y} \quad ,$$

$$(\underline{y} - \underline{X}\underline{\tilde{\beta}})'\underline{\tilde{V}}^{-1}(\underline{y} - \underline{X}\underline{\tilde{\beta}}) = N \ ,$$

and

$$\mathrm{tr}(\underline{\tilde{V}}^{-1}\underline{Z}_\theta\underline{Z}_\theta') = (\underline{y} - \underline{X}\underline{\tilde{\beta}})'\underline{\tilde{V}}^{-1}\underline{Z}_\theta\underline{Z}_\theta'\underline{\tilde{V}}^{-1}(\underline{y} - \underline{X}\underline{\tilde{\beta}}) \ . \qquad (30)$$

Numerical solution by the method of steepest ascent is suggested. However, the computing procedures are neither easy, nor yet widely available. Nor are they attractive for the typically large-sized data set of variance components models, because $\underline{V}^{-1}$ has order equal to the number of observations. And with unbalanced data $\underline{V}$ is in no sense a patterned matrix and permits of no easy analytical inverse.

Although explicit maximum likelihood estimators cannot be obtained their large sample variances can. In fact their large sample variance-covariance matrix is

$$[\mathrm{var}(\underline{\tilde{\sigma}}^2)]^{-1} = \left\{\frac{1}{2}\,\mathrm{tr}\left(\underline{V}^{-1}\frac{\partial\underline{V}}{\partial\sigma_\theta^2}\,\underline{V}^{-1}\frac{\partial\underline{V}}{\partial\sigma_\varphi^2}\right)\right\} \quad \text{for} \quad \theta,\ \varphi = 1,2,\ldots,K+1. \qquad (31)$$

This is derived in Searle [1970] where explicit elements of the matrix on the right-hand side are obtained for the 2-way nested classification, those for the 3-way nested classification being given in Rudan and Searle [1971]. Unfortunately all attempts at obtaining $\underline{V}^{-1}$ analytically for the 2-way crossed classification have failed, see Rudan and Searle [1971a]. Deriving this inverse for use in (31) remains an unsolved problem.

The computing difficulties associated with numerical inversion of matrices of large order that arise with $\underline{V}^{-1}$ in the maximum likelihood method would also occur in trying to use numerical methods for obtaining sampling variances from (31). Similar difficulties can arise in (28) where $(\underline{X}_1' \underline{X}_1)^-$ can be large; however, its order is only the number of effects in $\underline{\beta}_1$ of (27), which is usually far less than the number of observations, the order of $\underline{V}$. Furthermore, in at least one case of widespread application the 2-way crossed classification, explicit computing procedures for (28) are available, e.g., Searle [1971b, pp. 483-4].

<u>5d. Minimization criteria</u>. Several new quadratic forms for use in $\underline{q}(\underline{y})$ of (19) and (20) have recently been suggested, derived by setting up estimation criteria that seem appropriate. Rao [1971a], in suggesting that earlier methods other than maximum likelihood are "<u>ad hoc</u>" has developed, Rao [1970, 1971a], what he calls the MINQUE method, a method of <u>min</u>imum <u>n</u>orm <u>q</u>uadratic <u>u</u>nbiased <u>e</u>stimation. This entails establishing $\underline{y}'\underline{Q}\underline{y}$ as an estimator of $\Sigma\, p_i\sigma_i^2$ by choosing $\underline{Q}$ so that, as in (18), $\underline{Q}\underline{X} = \underline{0}$ and, in terms of (13), $\mathrm{tr}(\underline{Q} \sum_{\theta=1}^{K+1} \underline{Z}_\theta \underline{Z}_\theta')^2$ is a minimum. Writing

$$\underline{W} = \sum_{\theta=1}^{K+1} \underline{W}_\theta \quad \text{for} \quad \underline{W}_\theta = \underline{Z}_\theta \underline{Z}_\theta'$$

this means minimizing $\mathrm{tr}(\underline{Q}\underline{W})^2$. Rao [1971a] gives a variety of theorems useful to this kind of minimization problem, the results for this particular case being that $\hat{\underline{\sigma}}^2$ is derived from

$$\underline{S}\, \hat{\underline{\sigma}}^2 = \underline{q}(\underline{y})$$

where

$$\underline{S} = \{\mathrm{tr}[\underline{R}\underline{W}_\theta \underline{R}\, \underline{W}_{\theta'}]\} \quad \text{for} \quad \theta, \theta' = 1,2,\ldots,K+1$$

$$\underline{q}(\underline{y}) = \{\underline{y}'\underline{R}\underline{W}_\theta \underline{R}\underline{y}\} \qquad \text{for} \quad \theta = 1,\ldots,K+1$$

and

$$\underline{R} = \underline{W}^{-1}[\underline{I} - \underline{X}(\underline{X}'\underline{W}^{-1}\underline{X})^-\underline{X}'\underline{W}^{-1}] .$$

In Rao [1971b] this development is extended to minimum variance estimation, MIVQUE, and minimum mean square estimation, MIMSQUE.

Whilst these solutions to the porblem are to be applauded, they appear to have two strikes against them insofar as practical usage is concerned. First, $\underline{W}^{-1}$ is a matrix of order equal to the number of observations, as is $\underline{V}^{-1}$; and second, to quote Rao [1971b],

> "In all the formulas for estimating $\Sigma p_i\sigma_i^2$ the true σ_i^2 appear. In practice we use a priori values or a given set of values at which a minimum is sought."

In addition to these difficulties they also have the deficiencies of other estimators, namely that they are not necessarily non-negative and their distributions are unknown.

Estimators similar to those of Rao have also been suggested by LaMotte [1970] who calls his procedure QUESOM, quadratic unbiased estimation orthogonal to the mean. And Townsend and Searle [1971] have developed explicit expressions for the BQUE's of σ_α^2 and σ_e^2 in the random model $y_{ij} = \alpha_i + e_{ij}$ for unbalanced data, a BQUE being a best (in the minimum variance sense) quadratic unbiased estimator.

6. Some specific problems. The particulars are now given of some unfinished problems, both small and large.

6a. Variances of binomial probabilities. Sometimes the probability parameter p of the binomial distribution can be considered a random variable; e.g., the conception rate for each bull of a population of bulls available for use in artificial breeding; or hatchability rate of a hen's eggs in poultry. The analysis of data on such variables is often an analysis of variance of the appropriate (0,1) variable representing success and failure. This analysis is tantamount to a weighted analysis of variance of the $\hat{p}$'s corresponding to the p's. A simple relationship exists between the variance components of the (0,1) variables and those of the population of p's, even for unbalanced data. However, as indicated in Gates and Searle [1971], unweighted analyses of variance calculations can also yield unbiased estimators of the variance components of the p's. Although the two methods are the same for balanced data they are not for unbalanced data and in this case investigation is needed into their relative efficiency. Assuming a beta distribution for the p's may also yield distribution properties for the estimators, at least for balanced data.

6b. Models with covariates. The coefficients of covariates in a covariance model are nothing more than fixed effects and can be handled in accord with (27) and (28); i.e., so long as $\underline{X}_1$ of (27) always includes the covariates, (28) will yield variance components estimators unencumbered by the covariates. For example, consider the model

$$y_{ij} = \mu + \sum_{t=1}^{c} \beta_t x_{tij} + u_{1i} + e_{ij}$$

which can be written as

$$\underline{y} = \mu\underline{i} + \underline{X}_1\underline{\beta}_1 + \underline{Z}_1\underline{u}_1 + e \quad .$$

Then for n_i observations in the ith level of the random effects $u_{11},\ldots,u_{1a}$ we have $\underline{Z}_1 = \sum_{i=1}^{a}{}^{+} \underline{1}_{n_i}$ where $\underline{1}_{n_i}$ is a vector of n_i unities. The variance component to be estimated, in addition to σ_e^2, is σ_1^2 corresponding to the random effects of $\underline{u}_1$. Since the model is that of the 1-way classification with covariance, the estimator of $\hat{\sigma}_e^2$ is

$$\hat{\sigma}_e^2 = \left[\sum_{i=1}^{a}\sum_{j=1}^{n_i}(y_{ij} - \bar{y}_{i.})^2 - \underline{w}'\underline{W}_1^{-1}\underline{w}\right] /(N - a - c)$$

where $\underline{w}$ is the vector of within-group sums of products of the covariates with the y's and $\underline{W}_1$ is the matrix of within-group sums of squares and products of the covariates. The estimator of σ_1^2 is obtained from using (28), with its $\underline{X}_1$ and $\underline{\beta}_1'$ now being $[\underline{i}\ \underline{X}_1]$ and $[\mu'\ \underline{\beta}_1']$ and its $\underline{Z}_2$ and $\underline{u}_2$ now being $\underline{Z}_1$ and $\underline{u}_1$. The result is

$$\hat{\sigma}_1^2 = \frac{R(\underline{u}_1|\mu,\underline{\beta}_1) - (a - 1)\hat{\sigma}_e^2}{N - \sum_{i=1}^{a} n_i^2/N - tr(\underline{T}^{-1}\underline{U}_1\underline{U}_1')}$$

where

$$\underline{T} = \left\{\sum_{i=1}^{a}\sum_{j=1}^{n_i}(x_{tij} - \bar{x}_{t..})(x_{t'ij} - \bar{x}_{t'..})\right\} \quad \text{for } t,t' = 1,\ldots,c$$

and

$$\underline{U}_1\underline{U}_1' = \left\{\sum_{i=1}^{a} n_i^2(\bar{x}_{ti.} - \bar{x}_{t..})(\bar{x}_{t'i.} - \bar{x}_{t'..})\right\} \quad \text{for } t,t' = 1,\ldots,c \ .$$

Mount and Searle [1972] derive these results. They also obtain results for the 2-way cross-classification with covariates, one observation per cell and both factors of the classification being random effects factors; i.e., the model (30) with $X = [\underline{i}\ \underline{X}_1]$ and $\underline{\beta} = [\mu\ \underline{\beta}_1]$ as above. Further extensions of this application of (28) to covariate models are needed.

6c. Components of covariance. The simplest form of dispersion matrix for the random effects of a model is $\sum_{\theta=1}^{K} \sigma_\theta^2 \underline{I}_{N_\theta}$ shown in (11); although simple it involves difficult estimation problems, as we have seen. A more general form, $\{\underline{\Sigma}_{\theta\varphi}\}$ for $\theta,\varphi = 1,\ldots,K$ is discussed following (13). Rao [1971a] calls this a covariance components model. But it is not, for it is nothing more than a variance components model with covariances among the random effects. A true components of covariance model is one for 2 (or more) observable variables having a covariance between them; it is this covariance whose components are of interest. This is the components of covariance model that biologists have used for many years; e.g., it is the basis of procedures for estimating genetic correlations given in Hazel [1943]. Suppose, for example, we have observations on the staple length and crimp of the fleeces shorn from ewes sired by a number of rams. If x_{ij} and y_{ij} are the 2 observations from the jth ewe sired by the ith ram, suitable random effects models might be

$$x_{ij} = \mu + \alpha_i + e_{ij}$$

and (32)

$$y_{ij} = \mu' + \alpha_i' + e_{ij}'$$

having variance components σ_α^2, σ_e^2 and $\sigma_{\alpha'}^2$, $\sigma_{e'}^2$ respectively. The components of covariance are the covariances $\sigma_{\alpha\alpha'}$ and $\sigma_{ee'}$ between α_i and α_i' and between e_{ij} and e_{ij}', with

$$\sigma_{xy} = \sigma_{\alpha\alpha'} + \sigma_{ee'} \quad . \tag{33}$$

Estimation of components of covariance of the nature described for (33) is no more difficult than is estimation of components of variance. On all occasions, components of covariance estimators will be the same linear combinations of the same bilinear forms in $\underline{x}$ and $\underline{y}$ as variance components estimators are of quadratic forms in $\underline{x}$ and in $\underline{y}$. However, the need for bilinear forms can be bypassed by using quadratic forms in $\underline{x} + \underline{y}$ and relying on the identities

$$\sigma_{xy} \equiv \tfrac{1}{2}(\sigma_{x+y}^2 - \sigma_x^2 - \sigma_y^2)$$

and

$$\underline{x}'\underline{Q}\underline{y} \equiv \tfrac{1}{2}[(\underline{x} + \underline{y})'\underline{Q}(\underline{x} + \underline{y}) - \underline{x}'\underline{Q}\underline{x} - \underline{y}'\underline{Q}\underline{y}] \ . \qquad (34)$$

Thus for any data set in which the vectors of components of variance of x and y , $\underline{\sigma}_x^2$ and $\underline{\sigma}_y^2$ respectively, are estimated in accord with (20) by

$$\hat{\underline{\sigma}}_x^2 = \underline{P}^{-1}\underline{q}(\underline{x}) \quad \text{and} \quad \hat{\underline{\sigma}}_y^2 = \underline{P}^{-1}\underline{q}(\underline{y}) \ ,$$

the vector of components of covariance will be estimated by

$$\begin{aligned} \hat{\sigma}_{xy} &= \tfrac{1}{2}\underline{P}^{-1}[\underline{q}(\underline{x} + \underline{y}) - \underline{q}(\underline{x}) - \underline{q}(\underline{y})] \\ &= \tfrac{1}{2}(\hat{\underline{\sigma}}_{x+y}^2 - \hat{\underline{\sigma}}_x^2 - \hat{\underline{\sigma}}_y^2) \ . \end{aligned} \qquad (35)$$

Hence components of covariance can be estimated directly from the estimated components of variance of the two variables concerned and of their sum.

Investigation of properties of estimated components of covariance is also needed. Their variances, for example, on the basis of normality can be derived using

$$\text{cov}(\underline{x}'\underline{Q}\underline{x}, \underline{x}'\underline{Q}\underline{y}) = 2\ \text{tr}(\underline{Q}\underline{C})^2 + 4\ \underline{\mu}_x'\underline{Q}\underline{C}\underline{Q}\underline{\mu}_y \qquad (36)$$

where $\underline{C}$ is the matrix of covariances between $\underline{x}$ and $\underline{y}$ and $\underline{\mu}_x = E(\underline{x})$ and $\underline{\mu}_y = E(\underline{y})$. Similarly

$$\text{cov}(\underline{x}'\underline{Q}\underline{x}, \underline{x}'\underline{Q}\underline{y}) = \text{tr}(\underline{Q}\underline{V}_x\underline{Q}\underline{C}) + \text{tr}(\underline{Q}\underline{V}_x)^2 + 2\underline{\mu}_x'\underline{Q}\underline{C}\underline{Q}\underline{\mu}_x + 2\underline{\mu}_x'\underline{Q}\underline{V}_x\underline{Q}\underline{\mu}_y , \qquad (37)$$

where $\underline{V}_x$ and $\underline{V}_y$ are the variance - covariance matrices of $\underline{x}$ and $\underline{y}$ respectively; and

$$\text{var}(\underline{x}'\underline{Q}\underline{y}) = \text{tr}(\underline{Q}\underline{C})^2 + \text{tr}(\underline{Q}\underline{V}_y\underline{Q}\underline{V}_x) + 2\underline{\mu}_x'\underline{Q}\underline{C}\underline{Q}\underline{\mu}_y + \underline{\mu}_x'\underline{Q}\underline{V}_y\underline{Q}\underline{\mu}_x + \underline{\mu}_y'\underline{Q}\underline{V}_x\underline{Q}\underline{\mu}_y . \qquad (38)$$

These expressions come from the general form of the covariance between any two bilinear forms in normal variables, e.g., Searle [1971b, p. 66].

6d. Criteria for estimation. The various estimation procedures originating from analysis of variance calculations undoubtedly arose as a matter of convenience and because they seemed "obvious." The only known property of the resulting estimators is unbiasedness (and the χ^2-nature of $\hat{\sigma}_e^2$ under normality). This property is retained in the MINQUE, MIVQUE and BQUE procedures, and others are added.

Although these procedures stem from desirable criteria they do not overcome the problem of yielding negative estimators which are such an embarrassment. Furthermore, the property of unbiasedness itself merits questioning in the case of variance components estimators. This is so because with unbalanced data from random models the concept of repetitions of similarly structured data and associated repetitions of estimators is often not appropriate -- more data, maybe, but not necessarily with the same pattern of unbalancedness. Replications of data cannot therefore be thought of as mere resamplings of the data already available. This situation appears to demand that consideration of expected values over repeated samplings should take into account the varying numbers of observations that arise from sample to sample. Also to be taken into account is the fact that random model data are often available in such large quantities that additional data may involve other populations. Investigation of these and other ideas is needed to develop estimators that have properties more in keeping with them than do those currently available. Some form of model unbiasedness is one possibility that has been suggested, Searle [1968]; or estimators for which the probability of small deviations from true value is maximized rather than minimizing the probability of large deviations. Even though, as Eisenhart [1968] points out, this was the idea that Gauss rejected in favour of his minimum-mean-squared-error approach, it may not be inappropriate for variance components situations where data sets are often large and non-replicable in the usual sense.

A standard procedure for comparing different estimators is by means of their sampling variances. With this in view, expressions for the sampling variances of variance components estimators under normality assumptions have been derived in several places; e.g., Searle [1956, 1958, 1961, 1970], Rohde and Tallis [1969], Searle and Rudan [1971b]. Further work is needed to derive from Rohde and Tallis explicit expressions for particular cases; and a great deal of work is needed in comparative studies, using these results.

6e. <u>Computing difficulties</u>. Reference has already been made to some of the computing difficulties involved in calculating estimates from some of the estimators discussed. These largely revolve around the difficulty of inverting matrices of very large order, such as calculating $\underline{V}^{-1}$. In addition to these computing difficulties the resulting estimates are such that their distributional properties are mostly unknown, or known only in terms of the unknown variance components parameters. Furthermore these properties themselves can involve computing headaches. Two questions therefore arise: (i) Can we develop variance components estimators based on much simpler calculations than are needed now--such as using rank order statistics, perhaps? (ii) Can we numerically investigate the properties of present estimators in a manner which will yield definitive information about their behaviour for variations in the unknown variance components and variations in the numbers of observations in the subclasses? Investigation of these two questions surely seems worthwhile.

6f. Defining unbalancedness. Referring to the values that the numbers of observations take on in a data set as an n-pattern, one of the preceding questions is to what extent does the n-pattern affect the properties of an estimator? Unfortunately the algebra of most properties usually involves the n-pattern in such a complicated way that the effect of different n-patterns cannot be studied analytically. For example, assuming normality in the model (2), the sampling variance of the analysis of variance estimator of σ_α^2 is

$$v(\hat{\sigma}_\alpha^2) = \frac{2\sigma_e^4 N^2(N-1)(a-1)}{(N^2-S_2)^2(N-a)} + \frac{4\sigma_e^2\sigma_\alpha^2 N}{N^2-S_2} + \frac{2\sigma_\alpha^4(N^2S_2+S_2^2-2NS_3)}{(N^2-S_2)^2} \tag{39}$$

where

$$N = \sum_{i=1}^{a} n_i \ , \quad S_2 = \sum_{i=1}^{a} n_i^2 \quad \text{and} \quad S_3 = \sum_{i=1}^{a} n_i^3 \ .$$

The involvement of the n_i's in this expression is clearly such that investigating its behaviour for changes in the n_i's is out of the question. It seems, therefore, that analytical comparisons of estimators are likely to be quite intractable and recourse must be made to numerical studies.

Even if the behaviour of expressions like (39) in terms of the n_i's could be delineated it would be advantageous to be able to summarize an n-pattern in terms of some characteristic, such as a measure of unbalancedness. The behaviour of (39) could then be described in terms of this measure. The possibility of doing this may be remote, however, because preliminary indications are that even in the simplest of cases the effect of the n-pattern on properties of estimators is apparently itself a function of the variance components being estimated. The effects of unbalancedness therefore appear to differ according to the values of the true variance components. This suggests that unbalancedness may have to be defined in terms of the components being estimated, an unsatisfactory conclusion from the point of view of considering the effect of unbalancedness on estimation.

Numeric studies for comparing estimators also involve a problem concerning n-patterns. It is that of planning a set of n-patterns to be used, in conjunction with sets of parameter values. Deciding on the latter usually poses no great difficulty because only a small number of variance components are involved. But deciding on a set of n-patterns in a situation of having infinitely many choices; for example, in a 2-way crossed classification of rows and columns, the planning of a set of n-patterns demands answering such questions as how many rows, how many columns, how many empty cells, and how many observations in the cells that are not empty? The sky's the limit, a fact which makes it exceedingly difficult to plan a set of n-patterns that are sufficiently disparate to encompass an interesting range but which also differ in some logical manner in such a way that effects on the properties of the estimators might be apparent. One possibility is to draw samples of n_{ij}'s from some distribution, provided a useful, realistic and tractable distribution can be postulated. Even then, this course of action would provide little information on just exactly how it is that the characteristic of unbalancedness affects estimation procedures. Investigation on this problem is therefore badly needed.

References

1. Ahrens, H. [1965]. Standardfehler geschätzter Varianzkomponenten eines unbalanzierten Versuchplanes in r-stufiger hierarchischer Klassifikation. Monatsb. Deutsch. Akad. Wiss., Berlin, 7, 89.

2. Dempster, A. P. [1971]. An overview of multivariate data analysis. J. Multivariate Analysis, 1, 316.

3. Eisenhart, C. [1968]. Discussion of Searle [1968]. Biometrics, 24, 784.

4. Gates, C. E. and Searle, S. R. [1971]. Estimating variance components of binomial frequencies. Paper BU-368-M in the Biometrics Unit, Cornell University, Ithaca, N. Y.

5. Graybill, F. A. and Hultquist, R. A. [1961]. Theorems concerning Eisenhart's Model II. Ann. Math. Statist., 32, 261.

6. Hazel, L. N. [1943]. The genetic basis for constructing selection indexes. Genetics, 28, 476.

7. Hartley, H. O. [1967]. Expectation, variances and covariances of ANOVA mean squares by 'synthesis.' Biometrics, 23, 105. Correction 23, 853.

8. Hartley, H. O. and Rao, J. N. K. [1967]. Maximum likelihood estimation for the mixed analysis of variance model. Biometrika, 54, 93.

9. Henderson, C. R. [1953]. Estimation of variance and covariance components. Biometrics, 9, 226.

10. Herbach, L. H. [1959]. Properties of Model II type analysis of variance tests. Ann. Math. Statist., 30, 939.

11. LaMotte, L. R. [1970]. A class of estimators of variance components. Technical Report No. 10, Dept. of Statistics, University of Kentucky, Lexington.

12. Mount, T. D. and Searle, S. R. [1972]. Estimating variance components in covariance models. Paper BU-403-M in the Biometrics Unit, Cornell University, Ithaca, N. Y. Submitted to Econometrica.

13. Rao, C. R. [1970]. Estimation of heteroscedastic variances in linear models. J. Amer. Statist. Assoc., 65, 161.

14. Rao, C. R. [1971a]. Estimation of variance and covariance components - MINQUE theory. J. Multivariate Analysis, 1, 257.

15. Rao, C. R. [1971b]. Minimum variance quadratic unbiased estimation of variance components. J. Multivariate Analysis, 1, 445.

16. Robinson, J. [1965]. The distribution of a general quadratic form in normal variables. Aust. J. Statist. 7, 110.

17. Rohde, C. A. and Tallis, G. M. [1969]. Exact first- and second-order moments of estimates of components of covariance. Biometrika, 56, 517.

18. Rudan, J. W. and Searle, S. R. [1971a]. Attempts at inverting the variance-covariance matrix of the 2-way crossed classification, unbalanced data, random model. Paper BU-353-M in the Biometrics Unit, Cornell University, Ithaca, N. Y.

19. Rudan, J. W. and Searle, S. R. [1971b]. Large sample variances of maximum likelihood estimators of variance components in

the 3-way nested classification, random model, with unbalanced data. Biometrics, 27, 1087.

20. Searle, S. R. [1956]. Matrix methods in variance and covariance components analysis. Ann. Math. Statist., 27, 737.

21. Searle, S. R. [1958]. Sampling variances of estimates of components of variance. Ann. Math. Statist., 29, 167.

22. Searle, S. R. [1961]. Variance components in the unbalanced 2-way nested classification. Ann. Math. Statist., 32, 1161.

23. Searle, S. R. [1968]. Another look at Henderson's methods of estimating variance components. Biometrics, 24, 749.

24. Searle, S. R. [1970]. Large sample variances of maximum likelihood estimators of variance components. Biometrics, 26, 505.

25. Searle, S. R. [1971a]. Topics in variance components estimation. Biometrics, 27, 1.

26. Searle, S. R. [1971b]. Linear Models. Wiley, New York.

27. Thompson, W. A., Jr. [1962]. The problem of negative estimates of variance components. Ann. Math. Statist., 33, 273.

28. Thompson, W. A., Jr. [1963]. Precision of simultaneous measurement procedures. J. Am. Statist. Assoc., 58, 474.

29. Thompson, W. A., Jr. and Moore, J. R. [1963]. Nonnegative estimates of variance components. Technometrics, 5, 441.

30. Townsend, E. C. and Searle, S. R. [1972]. Best quadratic unbiased estimation of variance components from unbalanced data in the 1-way classification. Biometrics, 27, 643.

31. Wallace, T. D. and Hussain, A. [1969]. The use of error components models in combining time series with cross section data. Econometrica, 37, 55.

32. Wang, Y. Y. [1967]. A comparison of several variance component estimators. Biometrika, 54, 301.

33. Winsor, C. P. and Clarke, G. L. [1940]. A statistical study of variation in the catch of plankton nets. Sears Foundation J. Marine Res., 3, 1.

S. R. Searle
Biometrics Unit
Cornell University
Ithaca, New York
U.S.A.

A SEQUENTIAL SOLUTION OF WILCOXON TYPES FOR A SLIPPAGE PROBLEM

A. K. Sen and M. S. Srivastava*

In this paper a sequential solution of Wilcoxon types is proposed for the so-called slippage problem under some mild regularity conditions on the cdf $F(x - \theta_i)$ of the populations π_i, $i = 0,1,\ldots,k$, $k \geq 2$, when both the error probabilities are controlled at some specified level and the amount of the slippage of π_i from π_0, i.e., $\theta_i - \theta_0$ is d, where $d > 0$ and is specified. An upper bound for the expected sample size is given from which the asymptotic efficiency follows trivially.

1. Introduction. Let $F(x - \theta_i)$ denote the cdf of the population π_i, $i = 0,1,\ldots,k$, $k \geq 2$. We will call π_0 the control population and $\pi_i (i = 1,2,\ldots,k)$, the ith treatment population. We say $H = 0$ when $\theta_\ell = \theta_0$, $\ell = 1,2,\ldots,k$, and $H = j$ when $\theta_\ell = \theta_0$ for $\ell \neq j$ and $\theta_j = \theta_0 + d$, where $d > 0$ and is specified. The decision D_i is preferred if $H = i$, $i = 0,1,\ldots,k$. The problem is to find a procedure for choosing one of the $k + 1$ decisions $D_0, D_1,\ldots,D_k$, so that

$$\text{(1.1)} \qquad \begin{aligned} &\liminf_{d\to 0} P_0(D_0) \geq 1 - \alpha, \quad 0 < \alpha < 1 \\ &\liminf_{d\to 0} P_j(D_j) \geq 1 - \beta, \quad 0 < \beta < 1, \quad j = 1,2,\ldots,k, \end{aligned}$$

where $P_j(D_i)$ denote the probability of choosing D_i when $H = j$. By symmetry, $P_j(D_j)$ is independent of j, $j = 1,2,\ldots,k$. The form of F is not known but we shall assume that $F \in \mathcal{F}$ where

$$\mathcal{F} = \Big\{ \text{class of absolutely continuous } F(\text{with } F' = f \text{ a.e.}) \text{ such that}$$

$$\text{(C1)} \qquad \lim_{d\to 0} d^{-1} \int_{\infty}^{\infty} [F(x+d) - F(x)]dF(x) = \int_{-\infty}^{\infty} f^2(x)dx < \infty$$

---

*Research supported partially by National Research Council of Canada.

Since F is not considered known, no fixed sample size procedure will work. We, therefore, propose a sequential stopping rule in which $N \equiv N(d)$ is a random variable and $N(d) \to \infty$ a.s. as $d \to 0$. We propose the following Wilcoxon type procedures.

(a) All the observations are ranked and the procedure is based on the sum of the ranks of each population.

(b) Two-sample Wilcoxon type in which the observations of π_i is combined with π_0 and ranked. The procedure is based on the sum of the ranks of π_i among (π_0, π_i).

(c) Two-sample Wilcoxon type Hodges-Lehmann estimator.

(d) One-sample Wilcoxon type Hodges-Lehmann estimator. For this, we assume that F is symmetric.

We show that (1.1) is met for all the above four procedures. We also show that

$$E^{\frac{1}{2}}(N) \leq s^{0\frac{1}{2}} + 2n_0^{\frac{1}{2}} + 1 , \tag{1.2}$$

where $E(N)$ denotes the expected sample size. s^0 is the sample size for a corresponding fixed sample procedure (when F is known) and n_0 is a small positive integer dependent only on k and α. Asymptotic efficiency in the sense of Chow and Robbins [3] follows trivially from (1.2).

It should be mentioned that this problem has been considered several times in the statistical literature (see e.g. [8] and [10] to [15]) but this seems to be the first nonparametric solution.

2. Notation and Preliminaries. Let X_{ij} denote the jth observation from population π_i $(i = 0,1,\ldots,k;\ j = 1,\ldots,n)$. We assume that $\{X_{is}\}$ is a sequence of mutually independent random variables with

$$P(X_{ij} \leq x) = F(x-\theta_i) , \quad (i=0,1,\ldots,k;\ j=1,\ldots,n) \tag{2.1}$$

where F satisfies condition (C1). As, it will be apparent later, we may, without any loss of generality assume that

$$\theta_0 = 0 . \tag{2.2}$$

Let

$$C = \int_{-\infty}^{\infty} f^2(x)\,dx , \quad \text{and} \quad \gamma^2 = 1/12 . \tag{2.3}$$

Let the rank of X_{ij} be denoted by $R_{ij}^{(m)}$ where $m = (k+1)n$. That is,

$$R_{ij}^{(m)} = \text{Number of } X_{rs}\text{'s in all } (k+1) \text{ populations which are } \leq X_{ij}$$

$$(2.4) \qquad = \Sigma_{s=1}^{n} \Sigma_{r=0}^{k} I(X_{ij} \geq X_{rs})$$

where $I(A)$ denotes the indicator function of the set A. Let

$$(2.5) \qquad Z_{in} = (1/nm)\left[\sum_{j=1}^{n} R_{ij}^{(m)} - \sum_{j=1}^{n} R_{0j}^{(m)}\right] \equiv T_{in} - T_{0n}, \quad i=1,2,\ldots,k$$

and

$$(2.6) \qquad \underline{Z}_n' = (Z_{1n},\ldots,Z_{kn}) .$$

Then the procedure (a) is based on the statistic $\underline{Z}_n'$. The prodedure (b) is based on the statistic $\underline{U}_n' = (U_{1n},\ldots,U_{kn}) - \frac{1}{2}$ where

$$(2.7) \qquad U_{in} = \frac{1}{n^2} \Sigma_{s=1}^{n} \Sigma_{j=1}^{n} I(X_{ij} \geq X_{0s}), \quad i=1,2,\ldots,k .$$

Note that when the observations of the i-th and 0-th populations are ranked together, the sum of the ranks of the i-th population is given by $n^2 U_{in} + n(n+1)/2$.

The two-sample Hodges-Lehmann estimate of $\underline{\mu} = (\mu_1,\ldots,\mu_k)$, $\mu_i = \theta_i - \theta_0$, based on $\underline{U}_n'$ is given by

$$(2.8) \qquad \underline{V}_n' = (V_{1n},\ldots,V_{kn})$$

where V_{in} is the median of n^2 differences $X_{ij} - X_{0\ell}$, $j,\ell = 1,2,\ldots,n$.

And the one-sample Hodges-Lehmann estimate of $\underline{\mu}$ is given by (when F is symmetric)

$$(2.9) \qquad \underline{W}' = (W_1, \ldots, W_k)$$

where

$$(2.10) \qquad W_i = \tilde{\theta}_i - \tilde{\theta}_0 ,$$

and $\tilde{\theta}_i$ is the median of $n(n+1)/2$ averages $\frac{1}{2}(X_{ij} + X_{ik})$ with

$j \leq k$, $j,k = 1,2,\ldots,n$.

The procedure (c) and (d) are based on $\underline{V}$ and $\underline{W}$ respectively.

Let

(2.11) $\varepsilon_0 = (0,\ldots,0)$; $\varepsilon_i = (0,\ldots,0, d, 0,\ldots,0)$,

d at the ith place and zero elsewhere, and;

(2.12) $n = 0(d^{-2})$.

Then it follows from Hoeffeding's [6] result on U-statistics (see Lehmann [9]) that $\underline{Z}_n$, $\underline{U}_n$, CV_n and CW_n all have the same limiting multivariate normal distribution. Under $H = i$, it is given by

(2.13) $N(\varepsilon_i C, 2\gamma^2 n^{-1}\Sigma)$ where $\Sigma = (r_{ij})$ with $r_{ii} = 1$ and $r_{ij} = \frac{1}{2}$, $i \neq j$.

It should be mentioned that condition (2.12) is not required for the asymptotic normality of V_n and W_n .

3. Fixed-Sample Procedure (F known). Let $Y_1,\ldots,Y_k$ be $N(0,\Sigma)$ where 2Σ is as given in (2.13). Let the constants $m_{k,\alpha}$, $m_{k-1,b}$, and z_a be defined respectively by

$$P\left\{\sup_{1\leq i\leq k} Y_i > m_{k,\alpha}\right\} = \alpha ,$$

(3.1) $$P\left\{\sup_{1\leq i\leq k-1} Y_i > m_{k-1,b}\right\} = b , \text{ and}$$

$$P\{Y_i > z_a\} = a .$$

Define constants a , b and δ as follows

$$m_{k,\alpha} = z_a \, \delta(1-\delta)^{-1}$$

(3.2) $$m_{k-1,b} = z_a \, (1-\delta)^{-1}$$

$$a + b = \beta .$$

Tables of $m_{k,\alpha}$ are given by Gupta [5] and Bechhofer [2]. Suppose n^0 is defined by

(3.3) $$n^0 = \text{smallest integer} \geq (\gamma m_{k-1,b}/dC)^2 = \left[\gamma z_a (1-\delta)^{-1}/dC\right]^2 = \left| \gamma m_{k,\alpha}/\delta dC \right|^2 .$$

Then when F is known, a fixed-sample of size n^0 can be taken and either of the four procedures can be adopted for choosing one of the $(k+1)$ decisions:

(a) Choose decision D_0 if $\sup_{1 \leq i \leq k} Z_{in} \leq Cd\delta$; otherwise choose decision D_i if Z_{in} is the largest of the Z_{jn}'s $(1 \leq j \leq k)$.

The procedures in (b), (c) and (d) are the same as in (a) except that Z_{jn}'s are replaced respectively by U_{in}, CV_{in} and CW_{in}.

We will now show that (1.1) is met for procedure in (a); the proofs for other procedures are similar. From (3.3) it is clear that

$$n^0 = 0(d^{-2}) .$$

Hence the results of (2.13) will be applicable. First we note that

(3.4)
$$\begin{aligned}
\liminf_{d\to 0} P_0(D_0) &= \liminf_{d\to 0} P_0\left[\sup_{1\leq i\leq k} Z_{in^0} \leq Cd\delta\right] \\
&= \liminf_{d\to 0} P_0\left[(n^0/\gamma^2)^{\frac{1}{2}} \sup_{1\leq i\leq k} Z_{in^0} \leq (n^0/\gamma^2)^{\frac{1}{2}} Cd\delta\right] \\
&\geq \liminf_{d\to 0} P_0\left[(n^0/\gamma^2)^{\frac{1}{2}} \sup_{1\leq i\leq k} Z_{in^0} \leq m_{k,\alpha}\right] \\
&= 1 - \alpha , \quad \text{from (3.1) and (3.2)} .
\end{aligned}$$

Next

(3.5)
$$\begin{aligned}
\liminf_{d\to 0} P_1(D_1) &= \liminf_{d\to 0}\left[1 - \Sigma_{i=2}^{k} P_1(D_i) - P_1(D_0)\right] \\
&\geq 1 - \limsup_{d\to 0} P_1\left[\sup_{2\leq i\leq k}\left(Z_{in^0} - Z_{1n^0}\right) > 0\right] \\
&\quad - \limsup_{d\to 0} P_1\left[Z_{1n^0} < Cd\delta\right]
\end{aligned}$$

$$= 1 - \lim\sup_{d\to 0} P_1\left[\sup_{2\le i\le k}\left(Z_{in^0} - Z_{1n^0} + Cd\right) > Cd\right]$$

$$- \lim\sup_{d\to 0} P_1\left[\left(Z_{1n^0} - Cd\right) < Cd(\delta-1)\right]$$

$$= 1 - \lim\sup_{d\to 0} P_1\left[(n^0/\gamma^2)^{\frac{1}{2}}\sup_{2\le i\le k}\left(Z_{in^0} - Z_{1n^0} + Cd\right) > (n^0/\gamma^2)^{\frac{1}{2}}Cd\right]$$

$$- \lim\sup_{d\to 0} P_1\left[(n^0/\gamma^2)^{\frac{1}{2}}\left(Z_{1n^0} - Cd\right) > (n^0/\gamma^2)^{\frac{1}{2}}Cd(1-\delta)\right]$$

$$= 1 - b - a = 1 - \beta \;.$$

An asymptotically equivalent fixed sample procedure can be obtained by taking a sample of size

(3.6) $$s^0 = \text{smallest integer} \ge (\tfrac{1}{3})\left[m_{k-1,b}/(G(d) - \tfrac{1}{2})\right]^2$$

where G is the cdf of $\frac{1}{2}\left(X_{ij} - X_{is}\right)$, $j \neq s$. Notice that

$$\lim_{d\to 0} d^{-1}\left[G(d) - \tfrac{1}{2}\right] = 2\int_{-\infty}^{\infty} f^2(x)\,dx = 2C \;.$$

4. <u>The Sequential Procedure</u>. In the preceding section we assumed that $C = \int_{-\infty}^{\infty} f^2(x)\,dx$ was known. But we usually do not know f and it is just this lack of knowledge that stimulates the use of ranks. When C is unknown (which is usually the case), no fixed sample size procedure can meet our requirements. We propose sequential procedures in this section.

Let

$$U_n = 2\left(\sum_{s=0}^{k}\sum_{i=1}^{n}\sum_{j=i+1}^{n} I\left(-2d\le X_{s,j}-X_{s,i}\le 2d\right)\right) \Big/ dm(n-1) \;.$$

Then $\{U_n\}$ forms a reverse martingale and hence as $n\to\infty$, $d\to 0$

(4.1) $$U_n \to 4C \quad \text{a.s.}$$

Hence we define our sequential stopping rule as follows

N = smallest integer $n \geq n_0$ such that

$$(k+1)^{-1} \sum_{s=0}^{k} \sum_{i=1}^{n} \sum_{j=i+1}^{n} I\left\{-2d \leq X_{s,i} - X_{s,j} \leq 2d\right\} \tag{4.2}$$

$$\geq m_{k-1,b}(n-1)(n/3)^{\frac{1}{2}} - n$$

where n_0 is so chosen as to make the right side of (4.2) positive.

When sampling is stopped at $N = n$, then any of the four procedures (a)-(d) of Wilcoxon type given for fixed-sample case can be adopted here to choose among the (k+1) decisions $(D_0, D_1, \ldots, D_k)$. We describe only procedure (a) here. The decision D_0 is taken if at $N = n$, $\sup_{1 \leq i \leq k} Z_{in} \leq \gamma m_{k,\alpha}(1/n)^{\frac{1}{2}}$. Otherwise the decision D_i is chosen if $\overline{Z}_{in}$ is the largest of the Z_{jn}'s, $j = 1, 2, \ldots, k$.

In the next section we show that (1.1) is met for procedure (a); the proof for others is analogous.

5. Proof of (1.1) for the procedure in (a).

Let

$$Y_n = \frac{n(n-1)(G(d) - \frac{1}{2})}{\left[(k+1)^{-1} \sum_{s=0}^{k} \sum_{i=1}^{n} \sum_{j=i+1}^{n} I\left\{-2d \leq X_{s,i} - X_{s,j} \leq 2d\right\}\right] + n}$$

$g(n) = n^{\frac{1}{2}}$ and $t = m_{k-1,b}/(G(d) - \frac{1}{2})$. Then $Y_n > 0$ a.s. and $\lim_{n\to\infty} Y_n = 1$ a.s. from (4.1). Also $g(n) > 0$, $\lim_{n\to\infty} g(n) = \infty$, $\lim_{n\to\infty} (g(n)/g(n-1)) = 1$. Then for each $t > 0$, N of (4.2) can be defined as

$$N \equiv N(t) \text{ the smallest } n \geq n_0 \text{ such that} \tag{5.1}$$
$$Y_n \leq g(n)/t .$$

By Lemma 1 of Chow and Robbins [3], it follows that N is well defined and nondecreasing as a function of t and that

$$\lim_{t\to\infty} N = \infty \text{ a.s.}, \quad \lim_{t\to\infty} E(N) = \infty \tag{5.2}$$

$$\lim_{t\to\infty} g(N)/t = 1 \quad \text{a.s.}$$

From (5.2) and the results of Section 3 above it follows that we need only verify Anscombe's [1] condition (C2) to establish (1.1). That is, we need to verify that for any $i = 0,1,\ldots,k$ given any small positive ε and η, there is a large v and small positive c such that for any $n > v$,

$$P\left\{|T_{in'} - T_{in}| < \varepsilon w_n \text{, simultaneously for all integers } n' \text{ such that } |n' - n| < cn\right\} > 1 - \eta \tag{5.3}$$

where $w_n^2 = 2\gamma^2/n$.

The rest of this section is devoted to proving (5.3).

It is clearly sufficient to prove

$$P\left\{\sup_{n\le n'\le n(1+c)} |T_{0n'} - T_{0n}| \ge \varepsilon w_n\right\} \le \eta/2 .$$

Let $\mu_{0r} = E\left\{I\left(X_{0j} \ge X_{rs}\right)\right\}$ and $b_{0rjs} = I\left(X_{0j} \ge X_{rs}\right) - \mu_{0r}$.

Then

$$|T_{0n'} - T_{0n}| \le (k+1)^{-1}(n^{-2} - n'^{-2})\left|\sum_{r=1}^{k}\sum_{j=1}^{n}\sum_{s=1}^{n} b_{0rjs}\right|$$

$$+ (k+1)^{-1} n^{-2}\left\{\left|\sum_{r=1}^{k}\sum_{j=1}^{n}\sum_{s=n+1}^{n'} b_{0rjs}\right|\right.$$

$$\left. + \left|\sum_{r=1}^{k}\sum_{j=n+1}^{n'}\sum_{s=1}^{n} b_{0rjs}\right| + \left|\sum_{r=1}^{k}\sum_{j=n+1}^{n'}\sum_{s=n+1}^{n'} b_{0rjs}\right|\right\}$$

$$= A_1 + A_{2n'} + A_{3n'} + A_{4n'} \qquad \text{(say)} .$$

By applying Hoeffding's [7] inequalities on U-statistics it is simple to show that by suitable choice of v and c

$$P\left\{\sup_{n\le n'\le n(1+c)} A_1 \ge \varepsilon w_n/4\right\} \le \eta/8$$

for $n > v$. In order to consider the next term, i.e. $A_{2n'}$, let

$$F_{r0j} = F_r\left(X_{0j}\right) = P\left(X_{rs} \le X_{0j} \mid X_{0j}\right) .$$

Then

$$P\left\{\sup_{n\le n'\le n(1+c)} A_{2n'} \ge \varepsilon w_n/4\right\}$$

$$\le \sum_{r=1}^{k} \left\{P\left(\sup_{n\le n'\le n(1+c)} \left| n^{-2} \sum_{s=n+1}^{n'} \sum_{j=1}^{n} \left(I\left(X_{0j} \ge X_{rs}\right) - F_{r0j}\right)\right| \ge \varepsilon w_n/8\right) + P\left(n^{-2}\left|\sum_{j=1}^{n} (cn)\left(F_{r0j} - \mu_{0r}\right)\right| \ge \varepsilon w_n/8\right)\right\}$$

$$= \sum_{r=1}^{k} (B_1 + B_2) \quad \text{(say)} .$$

But

$$B_1 = EP\left(\sup_{n\le n'\le n(1+c)} \left| \sum_{s=n+1}^{n'} n^{-1} \sum_{j=1}^{n} \left(I\left(X_{0j} \ge X_{rs}\right) - F_{r0j}\right)\right| \ge n\varepsilon w_n/8 \mid X_{0j} , j = 1,\ldots,n\right) .$$

Let the conditional variance of $n^{-1} \sum_{j=1}^{n} I\left(X_{0j} \ge X_{rs}\right)$ given the X_{0j}'s be $\sigma^2\left(X_{0j}\right)$. Now since the $n^{-1} \sum_{j=1}^{n} \left(I\left(X_{0j} \ge X_{rs}\right) - F_{r0j}\right)$ are conditionally independent, we get on applying Kolmogorov's inequality

$$B_1 \le 32\, cE\left[\sigma^2\left(X_{0j}\right)\right] \Big/ \varepsilon^2 \gamma^2 < 32\, c/\varepsilon^2 \gamma^2 .$$

By applying Chebycheff's inequality we may show that

$$B_2 < 32\, c^2/\gamma^2 \varepsilon^2 .$$

Therefore suitable choice of c assures

$$P\left\{\sup_{n\le n'\le n(1+c)} A_{2n'} \ge \varepsilon w_n/4\right\} < \eta/8 .$$

The terms $A_{3n'}$ and $A_{4n'}$ can be handled similarly. (In the case of $A_{4n'}$ we used the identity $\sum_{j=n+1}^{n'} \sum_{s=n+1}^{n'} \equiv \sum_{j=n+1}^{n'} \sum_{s=n+1}^{j} + \sum_{s=n+1}^{n'} \sum_{j=n+1}^{s-1}$). Thus (5.3) is true and consequently so is (1.1).

6. An Upper Bound for $E^{\frac{1}{2}}(N)$. From the stopping rule (4.2) it follows that

$$(k+1)^{-1} \sum_{i=0}^{k} \sum_{s=n_0+1}^{N-1} \sum_{j=s+1}^{N-1} I\left(-2d \le X_{ij} - X_{is} \le 2d\right)$$

$$\le 3^{-\frac{1}{2}} m_{k-1,b} (N-1)^{\frac{1}{2}} (N-2)-(N-1) .$$

Hence

$$(k+1)^{-1} \sum_{i=0}^{k} \sum_{s=n_0+1}^{N} \sum_{j=s+1}^{N} I\left(-2d \le X_{ij} - X_{is} \le 2d\right)$$

$$\le 3^{-\frac{1}{2}} m_{k-1,b} (N-1)^{3/2}$$

Let

$$N_\ell = \min(\ell, N)$$

where ℓ is a finite constant greater than n_0 . Then from the lemma stated and proved at the end of this section we get

$$(2G(d) - 1)E\left(N_\ell - n_0 - 1\right)^2 \le (4/3)^{\frac{1}{2}} m_{k-1,b} E\left(N_\ell - 1\right)^{3/2} .$$

Therefore since $EN^{3/2} \le ((EN^2)(EN))^{\frac{1}{2}}$ and $E^2N \le EN^2$ it follows that

$$E^{\frac{1}{2}}\left(N_\ell - 1\right) - 2E^{-\frac{1}{2}}\left(N_\ell - 1\right)n_0 \le (4/3)^{\frac{1}{2}} m_{k-1,b} [2G(d) - 1]^{-1} = s^{0\frac{1}{2}} .$$

Hence by the monotone convergence theorem we get

$$E^{\frac{1}{2}}(N-1) \le s^{0\frac{1}{2}} + 2n_0^{\frac{1}{2}} + 1$$

from which (1.2) follows.

Lemma 1. Let $\{X_n\}$ be a sequence of iid random variables with cdf $F(X - \gamma)$. Let G denote the cdf of $\frac{1}{2}\left(X_i - X_j\right)$ $i \neq j$. Then for any stopping rule for which $E(N^2) < \infty$

$$E\left(\Sigma_{i=1}^{N} \Sigma_{j=i+1}^{N} I\left(X_i - X_j \leq 2d\right)\right) = \frac{1}{2} EN(N-1)G(d) .$$

Proof. For $i \geq 2$, define random variables u_i, u_i^* and y_i as follows

$$u_i = \Sigma_{j=1}^{i-1} I\left(X_i - X_j \leq 2d\right)$$

$$u_i^* = u_i - \Sigma_{j=1}^{i-1} F\left(2d + X_j\right)$$

$$y_i = \Sigma_{j=2}^{i} u_j^* .$$

Then $\{y_N , N \geq 2\}$ is a martingale. Consequently

$$0 = E(y_2) = E(y_N) = E\left(\Sigma_2^N u_i\right) - E\left[\Sigma_{i=1}^{N} (N-i)F(2d + X_i)\right] .$$

Hence

$$E\left(\Sigma_2^N u_i\right) = E\left[\Sigma_{i=1}^{N} (N-i)F\left(2d + X_i\right)\right]$$

$$= \Sigma_{n=1}^{\infty} E\left[\Sigma_{i=1}^{n} (n-i)F\left(2d + X_i\right) | N = n\right]P(N = n)$$

$$= \Sigma_{i=1}^{\infty} \Sigma_{n=i}^{\infty} (i-1)E\left[F\left(2d + X_i\right) | N = n\right]P(N = n)$$

$$= \Sigma_{i=1}^{\infty} (i-1)E\left[F\left(2d + X_i\right) | N \geq i\right]P(N \geq i)$$

$$= \Sigma_{i=1}^{\infty} (i-1)\ P(N \geq i)\ G(d)$$

$$= \frac{1}{2}\ E\ N(N-1)\ G(d) .$$

The result now follows since

$$\Sigma_{i=1}^{N} \Sigma_{j=i+1}^{N} I\left(X_i - X_j \leq 2d\right) = \Sigma_2^N u_i .$$

7. Comments. (i) We could following Geertsema [4] have based our stopping rule on $(k+1)^{-1} \sum_{i=0}^{k} \left\{ z_i^{(p-b_n)} - z_i^{(b_n)} \right\}$, where the $z_i^{(1)} < z_i^{(2)} \ldots < z_i^{(p)}$ are the $p \equiv \frac{1}{2} n(n+1)$ ordered averages $\frac{1}{2}\left(X_{ij} + X_{is}\right)$ and b_n is of the order of $p/2 - m_{k-1,b}(n-1)(n/12)^{\frac{1}{2}}$. However, such a sequential procedure requires the ranking of the $\frac{1}{2}\left(X_{ij} + X_{is}\right)$'s at every stage, which is very time consuming. Our procedure on the other hand which depends only on counting is much faster.

(ii) The results of this paper can easily be extended to ranking [14] and allied problems - e.g. selecting the $t(\leq k)$ best populations, selecting a subset (of populations) containing all populations better than a standard population.

(iii) The procedures presented in this paper were tried on some simulated data. The cases $d = 1, 1.5, 2, 3$ were studied using standard normal data and the cases $d = 2, 3, 4$ were examined using logistic data. In all cases the simulated values of α and β were found to be the same as or smaller than the corresponding asymptotic values. The comparative merits of the different procedures could not be judged owing to the small number of simulations (about 100 in each case). A detailed Monte Carlo study of the procedures will be communicated in a future paper.

References

[1] Anscombe, F. J. (1952). Large-sample theory of sequential estimation. Proc. Cambridge Philos. Soc. 48, 600-607.

[2] Bechhofer, R. E. (1954). A single-sample multiple decision procedure for ranking means of normal populations with known variances. Ann. Math. Statist. 25, 16-39.

[3] Chow, Y. S. and Robbins, H. (1965). On the asymptotic theory of fixed-width sequential confidence intervals for the mean. Ann. Math. Statist. 36, 457-462.

[4] Geertsema, J. C. (1970). Sequential confidence intervals based on rank tests. Ann. Math. Statist. 41, 1016-1026.

[5] Gupta, S. S. (1963). Probability integrals of multivariate normal and multivariate t. Ann. Math. Statist. 34, 792-828.

[6] Hoeffeding, W. (1948). A class of statistics with asymptotically normal distribution. Ann. Math. Statist. 19, 293-325.

[7] Hoeffeding, W. (1963). Probability inequalities for sums of bounded random variables. J. Amer. Statist. Assoc., 58, 13-30.

[8] Karlin, S. and Truax, D. (1960). Slippage Problems. Ann. Math. Statist. 31, 296-324.

[9] Lehmann, E. L. (1963). Robust Estimation in Analysis of Variance. Ann. Math. Statist. 34, 957-966.

[10] Paulson, E. (1952). An optimal solution to the k-sample slippage problem for the normal distribution. Ann. Math. Statist. 23, 610-616.

[11] Paulson, E. (1962). A sequential procedure for comparing several experimental categories with a standard or control. Ann. Math. Statist. 33, 438-443.

[12] Roberts, C. D. (1963). An asymptotically optimal sequential design for comparing several experiemtal categories with a control. Ann. Math. Statist. 34, 1486-1493.

[13] Roberts, C. D. (1964). An asymptotically optimal fixed sample size procedure for comparing several experimental categories with a control. Ann. Math. Statist. 35, 1571-1575.

[14] Srivastava, M. S. (1966). Some asymptotically efficient sequential procedures for ranking and slippage problems. J. Roy. Statist. Soc. B 28, 370-380.

[15] Srivastava, M. S. (1972). The performance of a sequential procedure for a slippage problem. J. Roy. Statist. Soc. B 34. To appear.

A. K. Sen
Department of Statistics
University of Illinois
at Chicago Circle
Chicago, Illinois
U.S.A.

M. S. Srivastava
Department of Mathematics
University of Toronto
Toronto, Ontario
Canada

SOME ASPECTS OF NONPARAMETRIC PROCEDURES IN MULTIVARIATE STATISTICAL ANALYSIS

Pranab Kumar Sen[1]

1. Introduction. This paper is primarily expository in nature, reviewing the developments on some robust nonparametric procedures in certain specific areas of multivariate statistical analysis. Here we will be mainly concerned with the analysis of the so-called growth curve or longitudinal data, which are quite common in clinical or bio-medical experiments.

In the parametric set-up, statistical procedures for the analysis of linear models arising in problems relating to growth curves have been developed by Rao (1958, 1967), Potthoff and Roy (1964), Grizzle and Allen (1969), among others. In this process, it has been visualized that if the growth curves are characterized by certain linear (in the parameters) models, then under suitable reduction of data, appropriate multivariate analysis of variance (MANOVA) procedures can be applied on the set of estimates of these parameters obtained from different subjects. Recently, Ghosh, Grizzle and Sen (1973) have shown that under similar reduction of data, nonparametric procedures for MANOVA can be applied as well, and often, these can be justified to be more appropriate on the grounds of their robustness and validity for a broader class of distributions.

In the current paper, we shall mainly review this aspect of nonparametric procedures in growth curve models, and supplement the result of Ghosh, Grizzle and Sen (1973) by additional remarks on the enhanced scope for their procedures.

The basic model is considered in section 2. Section 3 is an exposition of how nonparametric methods can be applied in these models. The last section includes certain additional remarks.

2. Statistical models. Consider an index set $I = \{\underset{\sim}{i}=(i_1,\ldots,i_m): 1 \le i_j \le k_j (\le 1)$, for $j=1,\ldots,m\}$, and let $k=k_1 \ldots k_m$. Thus, I contains a set of k distinct points. Also, consider a set T of $q(\ge 1)$ distinct time points i.e.,

$$(2.1) \qquad T = \{(t_1,\ldots,t_q): t_1 < \ldots < t_q\}$$

For each $i \varepsilon I$ and $t_\ell \varepsilon T$, we have a set of $n(\underset{\sim}{i})$ stochastic p-vectors

[1]Work supported by the National Institutes of Health, Grant GM-12868.

(2.3) $\underset{\sim}{X}_s(\underset{\sim}{i}) = (\underset{\sim}{X}_s(\underset{\sim}{i},t_1),\ldots,\underset{\sim}{X}_s(\underset{\sim}{i},t_q)), \quad s=1,\ldots,n(\underset{\sim}{i})$.

The pq-dimensional distribution function (df) of $\underset{\sim}{X}_s(\underset{\sim}{i})$ is denoted by

(2.4) $G(\underset{\sim}{x};\underset{\sim}{i}) \ , \ \underset{\sim}{x}\varepsilon R^{pq} \ , \ \underset{\sim}{i}\varepsilon I$,

where R^t is the t-dimensional Euclidean space; thus it is assumed that $\underset{\sim}{X}_s(\underset{\sim}{i})$, $s=1,\ldots,n(\underset{\sim}{i})$ are independently and identically distributed random matrices with a common df $G(\underset{\sim}{x};\underset{\sim}{i})$, which may depend on $\underset{\sim}{i}\varepsilon I$. Our statistical analysis is concerned with the set of df's

(2.5) $\mathcal{G} = \{G(\ ;\underset{\sim}{i}): \ \underset{\sim}{i}\varepsilon I\}$

In the parametric case, it is usually assumed that for each $\underset{\sim}{i}\varepsilon I$, $G(\underset{\sim}{x};\underset{\sim}{i})$ is a pq-variate normal df with expectation matrix

(2.6) $\underset{\sim}{M}(\underset{\sim}{i}) = (\underset{\sim}{\mu}(\underset{\sim}{i};t_1),\ldots,\underset{\sim}{\mu}(\underset{\sim}{i};t_q))$,

and a dispersion matrix (of order $pq\times pq$)

(2.7) $\underset{\sim}{\Sigma}$, which does not depend on $\underset{\sim}{i}\varepsilon I$.

In terms of this specification, the df's are all characterized explicitly in terms of $(\underset{\sim}{M}(\underset{\sim}{i}), \underset{\sim}{\Sigma})$, $\underset{\sim}{i}\varepsilon I$, so that, basically, we concentrate our attention to the statistical analysis of the parameters $\{\underset{\sim}{M}(\underset{\sim}{i});\ \underset{\sim}{i}\varepsilon I\}$ and $\underset{\sim}{\Sigma}$. In the great majority of cases, we usually treat $\underset{\sim}{\Sigma}$ as a nuisance (unknown) parameter, and desire to draw conclusions for $\mathcal{U} = \{\underset{\sim}{M}(\underset{\sim}{i});\ \underset{\sim}{i}\varepsilon I\}$.

In the characteristic fashion of growth curve models, the vectors $\underset{\sim}{\mu}(\underset{\sim}{i};t_\ell)$, $1\leq\ell\leq q$, $\underset{\sim}{i}\varepsilon I$, can be experssed in terms of certain parameters $\theta(\underset{\sim}{i})$ and the time variable t_ℓ . Specifically, we write for the jth element $\mu_j(\underset{\sim}{i};t_\ell)$ of $\underset{\sim}{\mu}(\underset{\sim}{i};t_\ell)$ as

(2.8) $\mu_j(\underset{\sim}{i};t_\ell) = \psi_j(\underset{\sim}{\theta}_j(\underset{\sim}{i}),t_\ell) \ , \ 1\leq j\leq p \ , \ 1\leq\ell\leq q \ , \ \underset{\sim}{i}\varepsilon I$,

where the $\underset{\sim}{\theta}_j(\underset{\sim}{i})$, some $1\times r$ vectors, $(r\leq q)$, of unknown parameters (constants of the equation ψ_j) and the t_ℓ are known time points. We denote by

(2.9) $\underset{\sim}{\theta}_{p\times r}(\underset{\sim}{i}) = (\theta_1'(\underset{\sim}{i}),\ldots,\theta_p'(\underset{\sim}{i}))' \ , \ \underset{\sim}{i}\varepsilon I$, and let

(2.10) $\Theta = \{\underset{\sim}{\theta}(\underset{\sim}{i}) : \underset{\sim}{i}\varepsilon I\}$.

Thus, we have a reparametrization from u to Θ . In (2.8), we have implicitly assumed that $\underset{\sim}{\mu}(\underset{\sim}{i};t_\ell)$ has the same functional form for all $\underset{\sim}{i}\varepsilon I$, but the parameters $\underset{\sim}{\theta}(\underset{\sim}{i})$ can of course vary from one $\underset{\sim}{i}$ to another.

If we look back at (2.4), then in the parametric case, under the assumed multinormality, we have

(2.11) $G(\underset{\sim}{x};\underset{\sim}{i}) = N_{pq}(\underset{\sim}{M}(\underset{\sim}{i}), \underset{\sim}{\Sigma})$,

so that for all $\underset{\sim}{i}\varepsilon I$,

(2.11) $\underset{\sim}{X}_s(i)-\underset{\sim}{M}(i) \sim N_{pq}(\underset{\sim}{0},\underset{\sim}{\Sigma})$, $s=1,\ldots,n(\underset{\sim}{i})$.

In the nonparametric set-up, we want to dispose of the assumption of multinormality of $G(\underset{\sim}{x};\underset{\sim}{i})$, so by analogy to (2.11), we shall assume that

(2.12) $G(\underset{\sim}{x};\underset{\sim}{i}) = G(\underset{\sim}{x}+\underset{\sim}{M}(\underset{\sim}{i}))$, $\forall$ $\underset{\sim}{i}\varepsilon I$,

i.e., the df of $\underset{\sim}{X}_s(i)-\underset{\sim}{M}(\underset{\sim}{i})$ is $G(\underset{\sim}{x})$, independently of $\underset{\sim}{i}\varepsilon I$. The assumptions (2.8) and (2.12) are the basic ones on which the developments of section 3 rest.

3. Reduction of nonparametric procedures to well-known MANOVA procedures. As in the later developments in the parametric case, we shall first use the following data reduction technique, which amounts to a transformation from $X_s(\underset{\sim}{i})$ to a random matrix $\underset{\sim}{Y}_s(\underset{\sim}{i})$, $s=1,\ldots,n(\underset{\sim}{i})$, $\underset{\sim}{i}\varepsilon I$, where the $Y_s(\underset{\sim}{i})$ are of the order $p\times r$ (compared to the dimension $p\times q$, $q\geq r$, of the $\underset{\sim}{X}_s(\underset{\sim}{i})$). Considering the $p\times q$ matrix $\underset{\sim}{X}_s(\underset{\sim}{i})$, we fit the model (2.8), and by a suitable method of estimation, derive the estimate of $\underset{\sim}{\theta}(\underset{\sim}{i})$. We denote this estimate by

(3.1) $\hat{\theta}_s(\underset{\sim}{i}) = Y_s(\underset{\sim}{i})$, $s=1,\ldots,n(\underset{\sim}{i})$, $\underset{\sim}{i}\varepsilon I$.

We may remark that whereas we allow the flexibility of having a suitable method of estimation in (3.1), once a method is chosen, it is used for all $s=1,\ldots,n(\underset{\sim}{i})$ and $\underset{\sim}{i}\varepsilon I$. These estimates may be the same as in Grizzle and Allen (1969), or the normal theory maximum likelihood estimators, or some nonparametric estimators, such as in Sen and Puri (1969) or Jurečková (1971), among others. Since, the same method of estimation is used, it follows that for each $\underset{\sim}{i} \varepsilon I$, $Y_s(\underset{\sim}{i})$, $s=1,\ldots,n(\underset{\sim}{i})$ are independent and identically distributed random matrices, and we denote the pr-variate df of $Y_s(\underset{\sim}{i})$ by

$$F(\underset{\sim}{x};\underset{\sim}{i}) = P\{\underset{\sim}{Y}_s(\underset{\sim}{i}) \leq \underset{\sim}{x}\}, \quad \underset{\sim}{i} \varepsilon I, \tag{3.2}$$

where for two matrices A and B of the same order, $A \leq B$ means that $a_{ij} \leq b_{ij}$ $\forall$ i,j. Thus, from the set $\mathcal{G}$ of k pq-variate df's, we have a reduction to the set

$$\mathcal{F} = \{F(\ ;\underset{\sim}{i}): \ \underset{\sim}{i} \varepsilon I\} \text{ of pr-variate df's}. \tag{3.3}$$

We shall now consider various hypotheses of special interest and show that in each case, we can apply some well-known nonparametric MANOVA test, a detailed account of which is available with Puri and Sen (1971), and others.

(i) <u>Monoatomic I</u> i.e., k=1, m=1. In this case, we have n independent random matrices $\underset{\sim}{X}_s$, $s=1,\ldots,n$, and dropping $\underset{\sim}{i}$ within the parenthesis in the $Y_s(\underset{\sim}{i})$ and $\underset{\sim}{\theta}(\underset{\sim}{i})$, we conclude that $\underset{\sim}{Y}_1,\ldots,\underset{\sim}{Y}_n$ are n independent and identically distributed stochastic $p\times r$ matrices. Suppose now, we want to test for a specified growth curve model, so that the null hypothesis may be framed as

$$H_o^{(1)}: \ \underset{\sim}{\theta} = \underset{\sim}{\theta}^o = ((\theta_{j\ell}^o))_{j=1,\ldots,p}, \ \ell=1,\ldots,r, \tag{3.4}$$

where the $\theta_{j\ell}^o$ are known quantities.

Let us then write

$$\underset{\sim}{Y}_s^o = \underset{\sim}{Y}_s - \underset{\sim}{\theta}^o, \quad s=1,\ldots,n, \tag{3.5}$$

so that we desire to test the null hypothesis that the $\underset{\sim}{Y}_s^o$ are all distributed around a null location matrix. The problem then reduces to the classical multivariate one-sample location problem in a nonparametric set-up, treated in detail in Chapter four of Puri and Sen (1971). The procedure is based on coordinate-wise ranking of the n observations in accordance to their absolute values and then considering signed-rank statistics, such as the Wilcoxon's one. Under suitable sign-invariance structure, a class of conditionally distribution-free tests for $H_o^{(1)}$ exists, and these test statistics are asymptotically distributed according to a chi-square distribution with pr degrees of freedom (D.F.) when $H_o^{(1)}$ holds. Asymptotic non-central chi-square distributions under suitable sequences of Pitman-type translation alternative, are also studied there, and the allied efficiency results are presented. Thus, the basic data reduction helps one to reduce the problem of testing for a specified growth curve model to that of an one-sample location problem, treated in Puri and Sen (1971, Ch. 4).

(ii) Comparison of several growth curves. Here I consists of $k(\geq 2)$ points, designated, for simplicity, by $1,\ldots,k$. For notational simplicity, we write $\mu(\underset{\sim}{i};t_\ell)$ as $\mu_i(t_\ell)$, $1\leq i\leq k$, $1\leq \ell \leq q$, and as in section 2, we pass on to the reparametrized

$$\underset{\sim}{\theta}(\underset{\sim}{i}) = \underset{\sim}{\theta}_i \ , \ 1\leq i\leq k \ . \tag{3.6}$$

The problem is to test for the homogeneity of $\underset{\sim}{\theta}_1,\ldots,\underset{\sim}{\theta}_k$, i.e., identity of the k growth curves. As a note of explanation, we may add that the k points of I may represent the k groups of subjects or treatments (one criterion MANOVA) or they may be the combination of two or more factors, each at more than one level (factorial layout). We want to test the the homogeneity of the $\underset{\sim}{\theta}_i$, i.e.,

$$H_o^{(2)}: \ \underset{\sim}{\theta}_1 = \ldots = \underset{\sim}{\theta}_k = \underset{\sim}{\theta} \text{ (unknown)}. \tag{3.7}$$

As in the earlier problem, we work with the k sets of random matrices

$$\{\underset{\sim}{Y}_s(i),\ s=1,\ldots,n(i)\}\ ,\ i=1,\ldots,k \tag{3.8}$$

and reduce our problem to that of testing the homogeneity of the df's of $\underset{\sim}{Y}_s(i)$, $i=1,\ldots,k$. This is the multi-sample multivariate nonparametric problem, studied in detail in Chapter 5 of Puri and Sen (1971). Here also, we have a class of conditionally (permutationally) distribution-free rank tests; for large values of $n_1,\ldots,n_k$, the test statistics have all approximately a chi-square distribution with k-1 D.F. when (3.7) holds. Under appropriate sequence of Pitman-type translation alternatives, the non-central chi-square distributions and the allied efficiency results are also discussed there. For a numerical illustration, we refer to Ghosh,

Grizzle and Sen (1973).

In the case of factorial layouts where there are two or more factors, we may be interested in testing component hypotheses relating to the $\underset{\sim}{\theta}_i$. To illustrate, let $i=(i_1,i_2)$, $1\le i_j\le k_j$ (≥ 2) , j=1,2. Then, we may rewrite $\underset{\sim}{\theta}_i$ as

$$(3.9) \qquad \underset{\sim}{\theta}(i_1,i_2) = \underset{\sim}{\theta}_0+\underset{\sim}{\theta}_1(i_1,\cdot) + \underset{\sim}{\theta}_2(\cdot,i_2) + \underset{\sim}{\theta}_3(i_1,i_2) ,$$

where the four components are respectively the mean effect, the main effects of the two factors and their interactions. In such a set-up, we may be interested in the component hypotheses

$$(3.10) \qquad H_{01}^{(2)}: \quad \underset{\sim}{\theta}_1(i_1,\cdot) = \underset{\sim}{0} \text{ for all } 1\le i_1\le k_1 ;$$

$$(3.11) \qquad H_{02}^{(2)}: \quad \underset{\sim}{\theta}_2(\cdot,i_2) = \underset{\sim}{0} \text{ for all } 1\le i_2\le k_2 ;$$

$$(3.12) \qquad H_{03}^{(2)}: \quad \underset{\sim}{\theta}_3(i_1,i_2) = \underset{\sim}{0} \text{ for all } 1\le i_1\le k_1 ,\ 1\le i_2\le k_2 .$$

Tests for these hypotheses can be done in two ways. First, as in Jurečková (1971), we may obtain rank order estimates of $\underset{\sim}{\theta}_1(i_1,\cdot)$, $\underset{\sim}{\theta}_2(\cdot,i_2)$ and $\underset{\sim}{\theta}_3(i_1,i_2)$ from the $Y_s(\underset{\sim}{i})$, s=1,...,$n(\underset{\sim}{i})$, $\underset{\sim}{i}\varepsilon I$. These estimates are (jointly) asymptotically normally distributed with a dispersion matrix which can be estimated from the samples. So a large sample normal theory testing procedure can be employed; for details, we may refer to Chapter six of Puri and Sen (1971). Secondly, using these rank order estimates, we may align the $\underset{\sim}{Y}_s(\underset{\sim}{i})$, and base a rank order test on these aligned matrices. For example, to test $H_{03}^{(2)}$ in (3.12), first, under $H_{03}^{(2)}$, we obtain rank order estimates of $\underset{\sim}{\theta}_1(i_1,\cdot)$ and $\underset{\sim}{\theta}_2(\cdot,i_2)$, and denote these by $\tilde{\theta}_1(i_1,\cdot)$ and $\tilde{\theta}_2(\cdot,i_2)$, respectively. Consider then the aligned matrices

$$(3.13) \qquad \hat{\underset{\sim}{Y}}_s(i_1,i_2) = \underset{\sim}{Y}_s(i_1,i_2)-\hat{\underset{\sim}{\theta}}_1(i_1,\cdot)-\hat{\underset{\sim}{\theta}}_2(\cdot,i_2),\ s=1,\dots,n(\underset{\sim}{i}),\ \underset{\sim}{i}\varepsilon I.$$

On these aligned matrices, we employ the same type of rank order statistics as used on the set (3.8) for testing (3.7). Because of the asymptotic linearity of the rank statistics in shifts [cf. Sen (1969b), Jurečková (1969), and others], it can be shown that the resulting test statistic has, under (3.12), closely a χ^2 distribution with $pr(k_1-1)(k_2-1)$ D.F. For a numerical illustration of such a test, we may refer to Ghosh, Grizzle and Sen (1973).

(iii) Comparison of growth curves in blocked experiments. In many situations, the $n(\underset{\sim}{i})$ subjects (for each $\underset{\sim}{i}\varepsilon I$) serve as a block, so that the p-variates of each $\underset{\sim}{X}_s(\underset{\sim}{i},\ell)$ represent the $p(\geq 2)$ treatments to be compared. In such a case, we want to test the null hypothesis that the p rows of each $\mu(\underset{\sim}{i},\ell)$, $1\leq\ell\leq q$, $i\varepsilon I$ are the same, so that if we reduce our model in terms of the $\underset{\sim}{\theta}(\underset{\sim}{i})$, we may frame our null hypothesis as

$$(3.14) \qquad H_0^{(3)}: \ \underset{\sim}{c}'\underset{\sim}{\theta}(\underset{\sim}{i}) = 0 \ \forall \ \underset{\sim}{i}\varepsilon I \ \text{ and } \ \underset{\sim}{c}' \perp (1,\ldots,1) \ .$$

We consider the following alignment procedure, essentially discussed in Sen (1968, 1969a, 1971). Let $\underset{\sim}{J}_{p\times p}$ be the matrix all whose elements are equal to 1 . Then, let

$$(3.15) \qquad \hat{\underset{\sim}{Y}}_s(\underset{\sim}{i}) = \underset{\sim}{Y}_s(i)-p^{-1}\underset{\sim}{J}\,\underset{\sim}{Y}_s(\underset{\sim}{i}) \ , \ s=1,\ldots,n(\underset{\sim}{i}), \ \underset{\sim}{i}\varepsilon I \ .$$

Note that under $H_0^{(3)}$,

$$(3.16) \qquad \underset{\sim}{\theta}(\underset{\sim}{i}) - p^{-1}\underset{\sim}{J}\underset{\sim}{\theta}(\underset{\sim}{i}) = \underset{\sim}{0} \ \text{ for all } \ \underset{\sim}{i}\varepsilon I \ ,$$

so that we are interested in testing the interchangeability of the p-rows of $\hat{\underset{\sim}{Y}}_s(\underset{\sim}{i})$, $s=1,\ldots,n(\underset{\sim}{i})$ for each $\underset{\sim}{i}\varepsilon I$, against shift alternatives. Aligned rank order tests for this problem, studied by Sen (1968, 1969, 1971), are considered in detail in Chapter 7 of Puri and Sen (1971).

In the same manner, by means of the basic transformation from μ to Θ , related problems of growth curve analysis can be transferred to problems of MANOVA, and solutions for the latter can be adopted for the former.

4. Some additional remarks. There remains some arbitrariness in the transformation from $\underset{\sim}{M}(\underset{\sim}{i})$ to $\underset{\sim}{\theta}(\underset{\sim}{i})$, $\underset{\sim}{i}\varepsilon I$. For example, given the q time points $t_1 < \ldots < t_\ell$, $q\geq 2$, we can always fit a polynomial of degree $q-1$ to the elements within each row of $\underset{\sim}{M}(\underset{\sim}{i})$. As is usual the case, q is moderately large, so that one is naturally interested in fitting a lower degree polynomial, say, a linear, quadratic or a cubic one. If the actual growth curve can be represented by a rth degree polynomial for some $r<q-1$, then the transformation from $\underset{\sim}{M}(\underset{\sim}{i})$ to $\underset{\sim}{\theta}(\underset{\sim}{i})$ is dimension reducing, so that the statistical procedures should be more informative. On the other hand, if the actual degree r is not known, the problem

arises in selecting an appropriate one if other than $q-1$. An incorrect selection may introduce inaccuracies and inefficiencies in the different procedures. A relative comparison of the different procedures in this light seems to be desirable. F. Woolson, in his Ph.D. thesis, under the guidance of the present author, is engaged in the study of this aspect.

Quade (1967), Puri and Sen (1969) and Sen and Puri (1970) have considered analysis of covariance tests in nonparametric set-ups. It seems that such tests can be utilized with advantage in the analysis of growth curves too, For example, if instead of the $q-1^{th}$ degree polynomial, we may partition

$$\underset{p\times q}{\underset{\sim}{\theta}(\underset{\sim}{i})} = (\underset{p\times r}{\underset{\sim}{\theta}(\underset{\sim}{i},1)} , \underset{p\times q-r}{\underset{\sim}{\theta}(i,2)}) \tag{4.1}$$

where under the assumption that the true model involves a polynomial of degree $r-1$, $\underset{\sim}{\theta}(\underset{\sim}{i},2)=\underset{\sim}{0}$ for all $\underset{\sim}{i}\varepsilon I$. If we partition similarly $\underset{\sim}{Y}_s(\underset{\sim}{i})$ as $(\underset{\sim}{Y}_s(\underset{\sim}{i},1),\underset{\sim}{Y}_2(\underset{\sim}{i},2))$, then the stochastic matrices $\underset{\sim}{Y}_s(\underset{\sim}{i},2)$, $s=1,\ldots,n(\underset{\sim}{i})$, $\underset{\sim}{i}\varepsilon I$, are all estimates of $\underset{\sim}{0}$ and, by (2.12), are independent and identically distributed. On the other hand, $\underset{\sim}{Y}_s(\underset{\sim}{i},1)$, $s=1,\ldots,n(\underset{\sim}{i})$, $\underset{\sim}{i}\varepsilon I$, relate to the parameters of interest, i.e., are the primary stochastic matrices. Thus, using (2.12) and the conditional df's of the $\underset{\sim}{Y}_s(\underset{\sim}{i},1)$ given the $\underset{\sim}{Y}_s(\underset{\sim}{i},2)$, $s=1,\ldots,n(\underset{\sim}{i})$, $\underset{\sim}{i}\varepsilon I$, one can test the null hypothesis

$$H_0^*: \ \underset{\sim}{\theta}(\underset{\sim}{i},1) = \underset{\sim}{\theta}(\cdot,1) \ \forall \ \underset{\sim}{i}\varepsilon I \ , \ \underset{\sim}{\theta}(\cdot,1) \text{ unknown} \ , \tag{4.2}$$

based on a class of rank order statistics, studied in detail in Quade (1967), Puri and Sen (1969) and Sen and Puri (1970). These tests are conditionally (permutationally) distribution-free and the statistics have sensibly a χ^2 distribution for large $n(\underset{\sim}{i})$, $\underset{\sim}{i}\varepsilon I$, when H_0^* holds. From the efficiency results of Puri and Sen (1969), we may argue that these procedures will be asymptotically at least as good as the alternative procedures based on the estimates of $\underset{\sim}{\theta}(\underset{\sim}{i},1)$, ignoring totally the concomitant variates.

In the context of MANOVA, Gabriel and Sen (1968), and in the context of m-dependent (multivariate) stochastic processes, Krishnaiah and Sen (1971) have studied simultaneous test and confidence interval procedures based on a general class of rank order statistics and derived estimates. By virtue of the basic reduction of the growth curve model, in terms of $\underset{\sim}{\theta}(\underset{\sim}{i})$ and the estimates

$\underset{\sim}{Y}_s(\underset{\sim}{i})$, s=1,...,n($\underset{\sim}{i}$) , $\underset{\sim}{i}\varepsilon I$, discussed in section 2, these procedures are also applicable to the growth curve models under consideration.

References

[1] Gabriel, K. R., and Sen, P. K. (1968). Simultaneous test procedures for one-way ANOVA and MANOVA based on rank scores. Sankhyā, Ser. A. 30, 303.

[2] Ghosh, M., Grizzle, J. E., and Sen (1973). Nonparametric methods in longitudinal studies. Jour. Amer. Statist. Assoc. 68, in press.

[3] Grizzle, J. E., and Allen, D. M. (1969). Analysis of growth and close response curves. Biometrics, 25, 307.

[4] Jurečková, J. (1969). Asymptotic linearity of a rank statistic in regression parameter. Ann. Math. Statist. 40, 1889.

[5] Jurečková, J. (1971). Nonparametric estimate of regression coefficients. Ann. Math. Statist. 42, 1328.

[6] Krishnaiah, P. R. and Sen, P. K. (1971). Some asymptotic simultaneous tests for multivariate moving average processes. Sankhyā, Ser. A. 33, 81.

[7] Potthoff, R. F. and Roy, S. N. (1964). A generalized multivariate analysis of variance model useful especially for growth curve problems. Biometrika, 51, 313.

[8] Puri, M. L. and Sen, P. K. (1969). Analysis of covariance based on general rank scores. Ann. Math. Statist. 40,610.

[9] Puri, M. L. and Sen, P. K. (1971). Nonparametric Methods in Multivariate Analysis. John Wiley, New York.

[10] Quade, D. (1967). Rank analysis of covariance. Jour. Amer. Statist. Assoc. 62, 1187.

[11] Rao, C. R. (1958). Comparison of growth curves. Biometrics, 14, 1.

[12] Rao, C. R. (1967). Least square theory using as estimated dispersion matrix and its applications to measurement of signals. Proc. 5th Berkeley Symp. Math. Statist. Prob. 1, 355.

[13] Sen, P. K. (1968). On a class of aligned rank order tests in two-way layouts. Ann. Math. Statist. 39, 1115.

[14] Sen, P. K. (1969a). Nonparametric tests for multivariate interchangeability. Part II. The problem of MANOVA in two-way layouts. Sankhyā, Ser. A. 31, 145.

[15] Sen, P. K. (1969b). On a class of rank order tests for the parallelism of several regression lines. Ann. Math. Statist. 40, 1668.

[16] Sen, P. K. (1971). On a class of aligned rank order tests for multiresponse experiments in some incomplete block designs. Ann. Math. Statist. 42, 1104.

[17] Sen, P. K. and Puri, M. L. (1969). Robust nonparametric estimation in some multivariate linear models. Multivariate Analysis-II (ed. P. R. Krishnaiah), Academic Press, New York, p. 33.

[18] Sen, P. K. and Puri, M. L. (1970). Asymptotic theory of likelihood ratio and rank order tests in some multivariate linear models. Ann. Math. Statist. 41, 87.

Pranab Kumar Sen
Department of Biostatistics
University of North Carolina
Chapel Hill, North Carolina
U.S.A.

WHEN DOES LEAST SQUARES GIVE THE
BEST LINEAR UNBIASED ESTIMATE?

George P. H. Styan

1. Introduction

This paper deals with estimating regression coefficients in the usual linear model. Let $\underset{\sim}{y}$ be an n-component vector with expectation

(1.1) $$E(\underset{\sim}{y}) = \underset{\sim}{X}\underset{\sim}{\gamma} \quad ,$$

where $\underset{\sim}{X}$ is an $n \times q$ matrix of numbers and $\underset{\sim}{\gamma}$ is a q-component vector of parameters. The covariance matrix of $\underset{\sim}{y}$ is

(1.2) $$V(\underset{\sim}{y}) = E(\underset{\sim}{y}-\underset{\sim}{X}\underset{\sim}{\gamma})(\underset{\sim}{y}-\underset{\sim}{X}\underset{\sim}{\gamma})' = \underset{\sim}{\Sigma} \ .$$

[All vectors are column vectors unless primed; transposition of a vector or matrix is denoted by a prime.] The problem is to estimate $\underset{\sim}{\gamma}$ on the basis of one observation on $\underset{\sim}{y}$ when $\underset{\sim}{X}$ is known.

When $\underset{\sim}{\Sigma}$ is known, or is known to within a constant multiple, and is positive definite, the Markov or Best Linear Unbiased Estimate (*BLUE*) is given by

(1.3) $$\hat{\underset{\sim}{\gamma}} = (\underset{\sim}{X}'\underset{\sim}{\Sigma}^{-1}\underset{\sim}{X})^{-1}\underset{\sim}{X}'\underset{\sim}{\Sigma}^{-1}\underset{\sim}{y} \ ,$$

when the rank of $\underset{\sim}{X}$ is $q<n$. The least squares estimate is then given by

(1.4) $$\underset{\sim}{\gamma}^* = (\underset{\sim}{X}'\underset{\sim}{X})^{-1}\underset{\sim}{X}'\underset{\sim}{y} \ .$$

The covariance matrix of the Markov estimate is

(1.5) $$V(\hat{\underset{\sim}{\gamma}}) = (\underset{\sim}{X}'\underset{\sim}{\Sigma}^{-1}\underset{\sim}{X})^{-1} \ ,$$

while the covariance matrix of the least squares estimate is

(1.6) $$V(\underset{\sim}{\gamma}^*) = (\underset{\sim}{X}'\underset{\sim}{X})^{-1}\underset{\sim}{X}'\underset{\sim}{\Sigma}\underset{\sim}{X}(\underset{\sim}{X}'\underset{\sim}{X})^{-1} \ .$$

Both estimates are linear (in $\underset{\sim}{y}$) and unbiased (for $\underset{\sim}{\gamma}$). The Gauss-Markov theorem leads to $V(\underset{\sim}{\gamma}^*)-V(\hat{\underset{\sim}{\gamma}})$ positive semidefinite, which in fact equals $V(\underset{\sim}{\gamma}^* - \hat{\underset{\sim}{\gamma}})$. Thus the two estimates are identical if and only if their covariance matrices are equal.

The purpose of this paper is to give a new proof of a theorem of Anderson (1971, 1972) concerning the equality of $\hat{\gamma}$ and γ^* .

THEOREM 1. <u>The least squares estimate</u> γ^* <u>in</u> (1.4) <u>equals the Markov estimate</u> $\hat{\gamma}$ <u>in</u> (1.3) <u>if and only if</u>

$$\sum_{i=1}^{h} \operatorname{rank}(X'P_i) = \operatorname{rank}(X) , \tag{1.7}$$

<u>where the covariance matrix</u> Σ <u>in</u> (1.2) <u>has</u> h <u>distinct characteristic roots</u> $\lambda_1, \lambda_2, \ldots, \lambda_h$ <u>with multiplicities</u> $m_1, m_2, \ldots, m_h$ <u>and corresponding orthonormalized characteristic vector sets</u> $P_1, P_2, \ldots, P_h$.

The $n\times n$ matrix $P = (P_1, P_2, \ldots, P_h)$ is an orthogonal matrix with P_i , $n\times m_i$, $\sum_{i=1}^{h} m_i = n$. The matrix $X'P_i$ in (1.7) is $q\times m_i$ and has the same rank as the $q\times q$ matrix $X'P_iP_i'X = Q_i$. The equation (1.7) may therefore be interpreted as rank additivity of the Q_i , a kind of Cochran's Theorem.

We also prove that (1.7) remains a necessary and sufficient condition for equality of $X\gamma^*$ and $X\hat{\gamma}$ when X has less than full column rank.

We write A^- for any $n\times m$ matrix satisfying $AA^-A = A$, where A is $m\times n$. We call A^- a <u>weak inverse</u> of A . We will use the property that $A(A'A)^-A'A = A$ for any weak inverse $(A'A)^-$ of $A'A$. These ideas are explored in detail by Rao and Mitra (1971).

When X has less than full column rank, we define

$$\hat{\gamma} = (X'\Sigma^{-1}X)^-X'\Sigma^{-1}y \tag{1.8}$$

and

$$\gamma^* = (X'X)^-X'y , \tag{1.9}$$

so that $X\hat{\gamma}$ and $X\gamma^*$ are uniquely determined, and moreover, $V(X\gamma^*) - V(X\hat{\gamma}) = V(X\gamma^* - X\hat{\gamma})$ is positive semidefinite and $X\hat{\gamma} = X\gamma^*$ if and only if $V(X\hat{\gamma}) = V(X\gamma^*)$.

THEOREM 2. <u>The estimate</u> $X\gamma^*$ <u>defined from</u> (1.9) <u>equals the</u>

estimate $\hat{X\gamma}$ *defined from* (1.8) *if and only if* (1.7) *holds*.

2. Intermediate Matrix Results.

Our proofs of Theorems 1 and 2 use the following two lemmas.

LEMMA 1. *Let* $A_1, A_2, \ldots, A_k$ *be* $n \times n$ *matrices such that*

(2.1) $$\operatorname{tr}(A_i A_j) \geq 0; \quad i,j = 1,2,\ldots,k$$

and

(2.2) $$\operatorname{tr}(A_i A_j) = 0 \iff A_i A_j = 0; \quad i,j = 1,2,\ldots,k .$$

Let $c_1, c_2, \ldots, c_k$ *be distinct positive numbers. Then*

(2.3) $$\left(\sum_{i=1}^{k} c_i A_i\right)\left(\sum_{j=1}^{k} c_j^{-1} A_j\right) = \left(\sum_{h=1}^{k} A_h\right)^2$$

if and only if

(2.4) $$A_i A_j = 0; \quad i \neq j; \quad i,j = 1,2,\ldots,k .$$

Proof. Clearly (2.4) implies (2.3). To go the other way, suppose (2.3) holds. Then

(2.5) $$\sum_{i \neq j} c_i c_j^{-1} A_i A_j = \sum_{i \neq j} A_i A_j .$$

Taking traces of both sides and using $\operatorname{tr}(A_i A_j) = \operatorname{tr}(A_j A_i)$, we obtain

(2.6) $$\sum_{i<j} (c_i c_j^{-1} + c_i^{-1} c_j - 2)\operatorname{tr}(A_i A_j)$$

$$= \sum_{i<j} (c_i - c_j)^2 c_i^{-1} c_j^{-1} \operatorname{tr}(A_i A_j) = 0 ,$$

and (2.4) follows using (2.1) and (2.2). (qed)

We note that the matrices $A_1, A_2, \ldots, A_k$ in Lemma 1 need not be symmetric. When they are symmetric and positive semidefinite they certainly satisfy (2.1) and (2.2). The second lemma follows from a result proved in Styan (1970).

LEMMA 2. *Let* $A_1, A_2, \ldots, A_k$ *be* $n \times n$ *matrices such that*

(2.7) $\text{rank}(A_{\sim i}) = \text{rank}(A_{\sim i}^2); \quad i,j = 1,2,\ldots,k$,

and the sum is idempotent,

(2.8) $\sum_{i=1}^{k} A_{\sim i} = (\sum_{i=1}^{k} A_{\sim i})^2$.

Then

(2.9) $\sum_{i=1}^{k} \text{rank}(A_{\sim i}) = \text{rank}(\sum_{i=1}^{k} A_{\sim i})$,

if and only if

(2.10) $A_{\sim i}A_{\sim j} = 0_{\sim}; \quad i \neq j; \quad i,j = 1,2,\ldots,k$.

Proof. Write $A_{\sim i} = B_{\sim i}C_{\sim i}'$, i=1,2,...,k , where $B_{\sim i}$ and $C_{\sim i}$ are of full column rank equal to rank $(A_{\sim i})$. Then $\Sigma A_{\sim i} = \Sigma B_{\sim i}C_{\sim i}' = (B_{\sim 1}, B_{\sim 2}, \ldots, B_{\sim k})(C_{\sim 1}, C_{\sim 2}, \ldots, C_{\sim k})' = B_{\sim}C_{\sim}'$, say. If (2.9) holds then $B_{\sim}$ and $C_{\sim}$ have full column rank and $B_{\sim}C_{\sim}' = B_{\sim}C_{\sim}'B_{\sim}C_{\sim}'$ implies $C_{\sim}'B_{\sim} = I_{\sim}$ and (2.10) follows. To go the other way suppose (2.10) holds. Then multiplying (2.8) by $A_{\sim i}$ yields $A_{\sim i}^2 = A_{\sim i}^3$. Thus $B_{\sim i}C_{\sim i}'B_{\sim i}C_{\sim i}' = B_{\sim i}C_{\sim i}'B_{\sim i}C_{\sim i}'B_{\sim i}C_{\sim i}'$ and $C_{\sim i}'B_{\sim i}$ is idempotent. From (2.7), rank $(B_{\sim i}C_{\sim i}') = \text{rank}(B_{\sim i}C_{\sim i}'B_{\sim i}C_{\sim i}') \leq \text{rank}(C_{\sim i}'B_{\sim i})$. Since $C_{\sim i}'B_{\sim i}$ is of order rank$(B_{\sim i}C_{\sim i}')$, it follows that $C_{\sim i}'B_{\sim i}$ is nonsingular, and being idempotent must equal $I_{\sim}$. Hence $A_{\sim i} = A_{\sim i}^2$ and (2.9) follows since then rank$(A_{\sim i}) = \text{tr}(A_{\sim i})$. (qed)

We note that the matrices $A_{\sim 1}, A_{\sim 2}, \ldots, A_{\sim k}$ in Lemma 2 need not be symmetric. When they are symmetric, however, they always satisfy (2.7).

3. Proofs of Throrems 1 and 2.

We use Lemmas 1 and 2 to prove Theorems 1 and 2.

Proof of Theorem 1. The expressions (1.5) and (1.6) are equal if and only if

(3.1) $X_{\sim}'\Sigma_{\sim}X_{\sim}(X_{\sim}'X_{\sim})^{-1}X_{\sim}'\Sigma_{\sim}^{-1}X_{\sim}(X_{\sim}'X_{\sim})^{-1} = I_{\sim q}$.

If $\Sigma_{\sim}$ has characteristic roots and vectors as given by the statement of Theorem 1, then

(3.2) $\Sigma_{\sim} = \sum_{i=1}^{h} \lambda_i P_{\sim i}P_{\sim i}' , \quad \Sigma_{\sim}^{-1} = \sum_{i=1}^{h} \lambda_i^{-1}P_{\sim i}P_{\sim i}'$.

Substituting (3.2) into (3.1) yields

$$(3.3)\qquad [\sum_{i=1}^{h} \lambda_i X'P_iP_i'X(X'X)^{-1}][\sum_{j=1}^{h} \lambda_j^{-1}X'P_jP_j'X(X'X)^{-1}] = I_q \ .$$

If $A_i = X'P_iP_i'X(X'X)^{-1}$, then $\sum_{i=1}^{h} A_i = I_q$ as $\sum_{i=1}^{h} P_iP_i' = PP' = I_n$. Hence (3.3) is of the form (2.3). Since A_i is the product of two positive semidefinite matrices, (2.1), (2.2) and (2.7) hold. We may, therefore, apply Lemmas 1 and 2 and (1.7) follows, proving Theorem 1.

<u>Proof of Theorem 2</u>. The covariance matrices of $X\gamma^*$ and $X\hat{\gamma}$,

$$(3.4)\qquad V(X\gamma^*) = X(X'X)^{-}X'\Sigma X(X'X)^{-}X' \ ,$$

$$(3.5)\qquad V(X\hat{\gamma}) = X(X'\Sigma^{-1}X)^{-}X'$$

are equal if and only if

$$(3.6)\qquad X'\Sigma X(X'X)^{-}X'\Sigma^{-1}X(X'X)^{-} = X'X(X'X)^{-} \ .$$

To see this, note that if (3.4) and (3.5) are premultiplied by X' and postmultiplied by $\Sigma^{-1}X(X'X)^{-}$ the left and right sides of (3.6) are obtained. Conversely, premultiplying both sides of (3.6) by $X(X'X)^{-}$ and postmultiplying by $X'X(X'\Sigma^{-1}X)^{-}X'$, yield (3.4) and (3.5) respectively. Repeated use has been made of $A(A'A)^{-}A'A = A$, for $A = X$ and for $A = \Sigma^{-\frac{1}{2}}X$, where $\Sigma^{-\frac{1}{2}}$ is the positive definite square root of Σ^{-1}.

Substituting (3.2) into (3.6) yields

$$(3.7)\qquad [\sum_{i=1}^{h} \lambda_i X'P_iP_i'(X'X)^{-}][\sum_{j=1}^{h} \lambda_j^{-1}X'P_jP_j'X(X'X)^{-}] = X'X(X'X)^{-} \ .$$

If $A_i = X'P_iP_i'X(X'X)^{-}$, then $\sum_{i=1}^{h} A_i = X'X(X'X)^{-}$, as $\sum_{i=1}^{h} P_iP_i' = I_n$. Thus $\sum_{i=1}^{h} A_i$ is idempotent; from $tr(A_iA_j) = tr(P_i'X(X'X)^{-}X'P_jP_j'X(X'X)^{-}X'P_i)$ we see that these A_i also satisfy (2.1) and (2.2). They satisfy (2.7) since

$r(A_i^2) \leq r(A_i) = r(X'X(X'X)^- X'P_iP_i'X(X'X)^-) \leq r(X(X'X)^- X'P_iP_i'X(X'X)^-)$
$= r[(X'X)^- X'P_iP_i'X(X'X)^- X'P_iP_i'X(X'X)^-] \leq r(A_i^2)$, where $r(\cdot)$ denotes rank. We have used the facts that for any matrix A, $A(A'A)^- A'$ is symmetric and idempotent and $r(A'A) = r(A) \geq r(AB)$, for any conformable B . We may, therefore, apply Lemmas 1 and 2 and Theorem 2 is proved. (qed)

******

The author is indebted to T. W. Anderson for very helpful discussions.

References

1. Anderson, T. W. (1971). The Statistical Analysis of Time Series. Wiley, New York.

2. Anderson, T. W. (1972). Efficient estimation of regression coefficients in time series. Proceedings of the Sixth Berkeley Symposium on Mathematical Statistics and Probability Volume 1, pp. 471-482. Pub. University of California Press, Berkeley and Los Angeles.

3. Rao, C. Radhakrishna, and Mitra, Sujit Kumar (1971). Generalized Inverse of Matrices and its Applications. Wiley, New York.

4. Styan, George P. H. (1970). Notes on the distribution of quadratic forms in singular normal variables. Biometrika, 57, 567

George P. H. Styan
Department of Mathematics
McGill University
Montreal, Canada

A COMBINATORIAL LOOK AT MULTIVARIATE MOMENT FUNCTIONS

Derrick S. Tracy

1. INTRODUCTION

In the history of statistics, many attempts have been made to find methods to write moments of moments systematically. The aim is to describe sampling moments of sample moments in terms of population moments, sample size and population size. Early authors (Tchouproff, Church, Fisher) considered sampling from infinite populations only. Later workers (Sukhatme, Wishart, Kendall, Tukey) included a consideration of finite populations as well. Some of them transformed moments to k-statistics to make the problem more tractable.

Through the maze of this kind of work, a paper appeared by Carver [2] in the first volume of The Annals of Mathematical Statistics (as the editorial), which attempted at writing the first eight sampling moments of the sample mean, considering a finite population, using an interesting set of polynomials related to the partitions of an integer. Later, Dwyer [5], in his dissertation written under Carver's guidance, explored the whole field quite systematically, giving many new results, putting known results in the proper perspective, and opening up a new range of possibilities. Curiously, and unfortunately, these papers did not attract the attention they deserved, and to this day, we find many formulae being reported by various workers, which can be obtained systematically and much more generally from the considerations set forth in these papers. At a Symposium on Symmetric Functions in Statistics (Windsor, 1971), Professor Dwyer [6] redirected interest in some of these results and showed how generalizations are possible. This paper is an attempt to consider multivariate generalizations and extensions of the results for moments of univariate moments. Naturally, a combinatorial approach comes in very handy.

2. NOTATION AND DEFINITIONS

We use the following scheme of notations, largely following Cramér [3] .

Research supported (in part) by N.R.C. Grant A-3111.

	Population	Sample
Size	N	n
Ordinary moment	α	a
Central moment	μ	m
Cumulant	κ	k .

Carver [2] uses ρ_λ for $\frac{n^{(\lambda)}}{N^{(\lambda)}}$, which later authors, e.g. Sukhatme [12], denote by e_λ . Carver [2] proposes polynomials in e_λ which are related to the partitions of an integer.

Definition 1. Let $P = p_1^{\pi_1} \ldots p_s^{\pi_s}$, $p_1 > p_2 > \ldots > p_s$, denote a partition of integer $p = \Sigma p_i \pi_1$. The weight of P is p and its order is the number of parts $\lambda = \Sigma \pi_i$. The combinatorial coefficient [7] of P is

$$C(P) = \frac{p!}{(p_1!)^{\pi_1} \ldots (p_s!)^{\pi_s} \cdot \pi_1! \ldots \pi_s!}$$

Definition 2. The Carver polynomial

$$C_p = \sum_P (-1)^{\lambda-1} (\lambda-1)! C(P) e_\lambda \quad .$$

Definition 3. $C_{pq} = C_p \times C_q$,where x denotes addition of subscripts. Similarly for $C_{pqr}, \ldots$.

Example: $C_{21} = C_2 \times C_1 = (e_1 - e_2) \times e_1 = e_2 - e_3$.

Dwyer [6] proposes functions $D, G,$ and certain operations on them, which can be generalized.

Definition 4.

$$D_p = \frac{C_p}{n^p} , \; D_{pq} = \frac{C_{pq}}{n^{p+q}} , \; \ldots \quad .$$

Definition 5.

$$G_k = \sum_{r=1}^{k-1} (-1^{r-1} \binom{k}{r-1} D_r + (-1)^{k-1} (k-1) D_k \quad .$$

$$G_{pq} = D_{pq} , \; G_{pqr} = D_{pqr} , \; \ldots \quad .$$

Definition 6. $G_p \times G_q$ denotes expansion of $G_p G_q$ with addition of subscripts.

$G_p \circ G_q$ denotes expansion of $G_p G_q$ with suffixing of subscripts.

Example: $G_2 \times G_{11} = (D_1 - D_2) \times D_{11} = D_{21} - D_{31}$,

$G_2 \circ G_{11} = (D_1 - D_2) \circ D_{11} = D_{111} - D_{211}$.

Definition 7.

$$\text{Power sum } (r) = \sum_1^N x_i^r = N\alpha_r$$

$$(\bar{r}) = \sum_1^N (x_i - \alpha_1)^r = N\mu_r$$

$$(rs) = \sum_1^N x_i^r y_i^s = N\alpha_{rs}$$

$$(\overline{rs}) = \sum_1^N (x_i - \alpha_{10})^r (y_i - \alpha_{01})^s = N\mu_{rs}$$

....

Definition 8. E_N denotes expectation over a finite population of size N .

$$M\binom{r\ u}{s\ v} = E_N(m_{rs}m_{uv}) = \alpha_{11}(m_{rs}, m_{uv}) , \dots .$$

3. MOMENTS OF UNIVARIATE MOMENTS

Dwyer [5,78-79] gives tables through weight 6 from which moments of univariate moments can be read. In the above notation, his tables for weights 2,3 are

Table 3.1

	2	11	2	11
2	D_1		1	
11	D_2	D_{11}	-1	1

Table 3.2

	3	21	111	3	21	111
3	D_1			1		
21	D_2	D_{11}		-3	1	
111	D_3	$3D_{21}$	D_{111}	2	-1	1

The D entry i a cell indicates, through its subscripts, the number of parts in the partition in the left margin which need to be added to yield the partition appearing at the top of the column. Thus D_{21} in Table 3.2 indicates that parts of 111 should be added two together, and one by itself, to yield partition 21. Since the parts can be chosen in $(2\overset{3}{1}) = 3$ ways, the entry is $3D_{21}$.

The entries in the right part of the tables spell out the moment formulae on multiplying the columns in the left part. For example,

$$E_N(m_2) = \alpha_1(m_2) = (D_1 - D_2)(2) - D_{11}(1)(1) \tag{3.1}$$

$$E_N(a_1^2) = \alpha_2(a_1) = D_2(2) + D_{11}(1)(1) \tag{3.2}$$

$$E_N(m_3) = \alpha_1(m_3) = (D_1 - 3D_2 + 2D_3)(3) - 3(D_{11} - 2D_{21})(2)(1) + 2D_{111}(1)^3 \tag{3.3}$$

$$E_N(m_2a_1) = \alpha_{11}(m_2,a_1) = (D_2 - D_3)(3) + (D_{11} - 3D_{21})(2)(1) - D_{111}(1)^3 \tag{3.4}$$

$$E_N(a_1^3) = \alpha_3(a_1) = D_3(3) + 3D_{21}(2)(1) + D_{111}(1)^3 \ . \tag{3.5}$$

These can be converted into expressions involving n, N, α_r . For results in terms of $\mu_r^{\prime}s$, one simply replaces (r)'s by $(\bar{r})$'s , puts $(\bar{1}) = 0$, and uses $(\bar{r}) = N\mu_r$, obtaining reduced formulae.

The formulae may be written more compactly in terms of G's, e.g.

$$E_N(m_2) = G_2(2) - G_{11}(1)^2 \tag{3.6}$$

$$E_N(m_3) = G_3(3) - 3(G_2 \circ G_1 - G_{21})(2)(1) + 2G_{111}(1)^3 \ . \tag{3.7}$$

We are currently studying possibilities of such generalizations in terms of G's .

There are a few features, worth noting, about the tables:

a) When considering sampling without replacement, D's are given in terms of C's as in Definitions 2 and 3. For sampling with replacement (or from infinite population), the tables are still good with $C_{p_1 \cdots p_r} = \frac{n^{(r)}}{N^r}$.

b) Coefficients in the right part of the tables may be obtained by expressing central moments in terms of ordinary moments, the general law being

$$m_p = \sum_{j=o}^{p} (-1)^j a_{p-j} a_1^j .$$

This takes care of partitions with one non-unit part, e.g., p, p1, p11. For others, e.g., pq, one simply multiplies coefficients in m_p, m_q .

c) The sum of the coefficients in each column is 0 . This provides a useful check.

d) Moments of ordinary sample moments are obtained by letting all but the diagonal entries in the right part of the table equal 0 . Thus,

$$E_N(a_2) = D_1(2)$$

$$E_N(a_3) = D_1(3)$$

$$E_N(a_2a_1) = D_2(3) + D_{11}(2)(1) .$$

e) Moments of sample comulants may be obtained by replacing the right part of the tables by coefficients $(-1)^{\lambda-1} (\lambda-1)!\, C(P)$, being the coefficients of $a_{p_1}^{\pi_1} \ldots a_{p_s}^{\pi_s}$ in the expression of the pth sample cumulant

$$k_p = \Sigma(-1)^{\lambda-1} (\lambda-1)!\, C(P)\, a_{p_1}^{\pi_1} \ldots a_{p_s}^{\pi_s} .$$

f) The left part of the tables summarizes all information needed as to how certain partitions of a number coalesce into other partitions of the same number.

The right part of the tables depends upon the particular sample symmetric function being considered. This symmetric function needs to be expressed as a linear function of the partitions (or equivalently, as a polynomial in ordinary moments). Thus we can consider all "moment functions", which can be so expressed.

4. MOMENTS OF MULTIVARIATE MOMENTS

In order to consider moments of multivariate moments, firstly tables for multipartite numbers are made. We present such tables for 11, 111, 1111. Partitions now are written in rows. For example, partitions of multipartite number 111 are

111	110	101	011	100	
	001	010	100	010	(4.1)
				001	

One can use the notation of Livers [10], or of ordered partitions [1], to write these as

$$111 \qquad 112 \qquad 121 \qquad 211 \qquad 123 \quad . \tag{4.2}$$

This second notation is economical for higher weights.

For multivariate sample moments, the coefficients now are quite small, mostly 1's and -1's, so we choose to put the two parts of the tables together. Coefficients are obtained (as for univariate moments) by expressing multivariate central moments in terms of ordinary moments, e.g.,

$$m_{111} = a_{111} - a_{110}a_{001} - a_{101}a_{010} - a_{011}a_{100} + 2a_{100}a_{010}a_{001} \tag{4.3}$$

which is appropriately read in Table 4.2. One can of course see that by combining subscripts,

$$m_3 = a_3 - 3a_2a_1 + 2a_1^3 \quad , \tag{4.4}$$

and, in fact, (4.4) can yield (4.3) using techniques like Kendall's operator [9] or Kaplan's tensor notation [8]. Yet, we look upon the formula (4.3) as the more fundamental one, which can lead to (4.4) or to

$$m_{21} = a_{21} - a_{20}a_{01} - 2a_{11}a_{10} + 2a_{10}^2a_{01} \quad , \tag{4.5}$$

by combining subscripts in the desired manner. Formulae for all such cases can be obtained by properly combining rows of a partition in these tables. If all rows are combined, one obtains univariate tables of Section 3 .

Table 4.1

	11	10 01
11	D_1 1	
10 01	D_2 -1	D_{11} 1

Table 4.2

	111	110 001	101 010	011 100	100 010 001
111	D_1 1				
110 001	D_2 -1	D_{11} 1			
101 010	D_2 -1		D_{11} 1		
011 100	D_2 -1			D_{11} 1	
100 010 001	D_3 2	D_{21} -1	D_{21} -1	D_{21} -1	D_{111} 1

The tables can be read to produce various assortments of formulae. We obtain, for example

$$E_N(m_{11}) = (D_1 - D_2)(11) - D_{11}(10)(01) \tag{4.6}$$

$$E_N(a_{10}a_{01}) = D_2(11) + D_{11}(10)(01) \tag{4.7}$$

$$E_N(m_{110}a_{001}) = (D_2 - D_3)(111) + (D_{11} - D_{21})(110)(001)$$
$$- D_{21}(101)(010) - D_{21}(011)(100) - D_{111}(100)(010)(001). \tag{4.8}$$

Table 4.3

	1111	1110 0001	1101 0010	1011 0100	0111 1000	1100 0011	1010 0101	1001 0110	1100 0010 0001	1010 0100 0001	0110 1000 0001	1001 0100 0010	0101 1000 0010	0011 1000 0100	1000 0100 0010 0001
1111	D_1 1														
1110 0001	D_2 -1	D_{11} 1													
1101 0010	D_2 -1		D_{11} 1												
1011 0100	D_2 -1			D_{11} 1											
0111 1000	D_2 -1				D_{11} 1										
1100 0011	D_2 0					D_{11} 1									
1010 0101	D_2 0						D_{11} 1								
1001 0110	D_2 0							D_{11} 1							
1100 0010 0001	D_3 1	D_{21} -1	D_{21} -1			D_{21} -1			D_{111} 1						
1010 0100 0001	D_3 1	D_{21} -1		D_{21} -1			D_{21} -1			D_{111} 1					
0110 1000 0001	D_3 1	D_{21} -1			D_{21} -1			D_{21} -1			D_{111} 1				
1001 0100 0010	D_3 1		D_{21} -1	D_{21} -1				D_{21} -1				D_{111} 1			
0101 1000 0010	D_3 1		D_{21} -1		D_{21} -1		D_{21} -1						D_{111} 1		
0011 1000 0100	D_3 1			D_{21} -1	D_{21} -1	D_{21} -1								D_{1111} 1	
1000 0100 0010 0001	D_4 -3	D_{31} 2	D_{31} 2	D_{31} 2	D_{31} 2	D_{22} 1	D_{22} 1	D_{22} 1	D_{211} -1	D_{211} -1	D_{211} -1	D_{211} -1	D_{211} -1	D_{211} -1	D_{1111} 1

The formulae above generalize the formulae for their univariate counterparts (3.1), (3.2), (3.4) respectively. Formula (4.8) thus sheds additional light on $D_{11} - 3D_{21}$ as the coefficient of (2)(1) in $E_N(m_2 a_1)$. It is interesting to associate the D-expressions (which are coefficients involving n,N) to the respective partitions in (4.1). Certainly, by combining subscripts in (4.8), one can get $E_N(m_{20}a_{01})$, $E_N(m_{11}a_{10})$, $E_N(m_{11}a_{01})$ or $E_N(m_2 a_1)$.

In order to obtain central moments of these multivariate sample moments, one replaces (rs...) by $(\bar{r}\bar{s}\ldots) = N\mu_{rs\ldots}$ and ignores partitions leading to $(\bar{1}0\ldots0)$, $(0\bar{1}0\ldots0)$, ... Thus for convariance of sample means, although one could use (4.7) and $E_n(a_{10}) = D_1(10)$, $E_n(a_{01}) = D_1(01)$ to obtain

$$\mu_{11}(a_{10}, a_{01}) = D_2(11) + D_{11}(10)(01) - D_1^2(10)(01) \quad ,$$

it is more direct to use

$$\mu_{11}(a_{10}, a_{01}) = D_2(\bar{1}\bar{1}) \quad .$$

One can check that in the case of sampling without replacement, this yields $\frac{N-n}{N-1} \cdot \frac{\mu_{11}}{n}$, and for sampling with replacement, $\frac{\mu_{11}}{n}$.

The multivariate tablesexhibit features similar to the ones remarked in (a) through (f) of Section 3. In particular, moments of other moment functions can be obtained by associating appropriate numerical coefficients with the same D entries.

As an example, we consider the moments of sample cumulants. Up to weight 3, cumulants and moments are identical. For weight 4, the numerical coefficients in the first column only of Table 4.3 under 1111 change to 1, seven -1's, six 2's and -6 . Then, with population means 0 ,

$$E_N(k_{1111}) = (D_1 - 7D_2 + 12D_3 - 6D_4)(1111) - (D_{11} - 4D_{21} + 6D_{22})$$
$$[(1100)(0011) + (1010)(0101) + (1001)(0110)] \quad . \qquad (4.9)$$

By combining subscripts, (4.9) yields E_N for k_{211} , k_{22}, k_{31} and k_4 .

5. FORMULAE USING G FUNCTIONS

The main point though is to be able to write multivariate moments formulae through inspection of the partitions. There seem to be possibilities here through the use of G functions. For formula (4.8), for example, one can write

$$E_n(m_{110}a_{001}) = G_2 \times G_2(111) + G_2 \circ G_1(110)(001) - G_{11} \circ G_1$$
$$[(101)(010) + (011)(100)] - G_1 \circ G_1 \circ G_1(100)(010)(001). \qquad (5.1)$$

Thus, in this case, the G functions associated with the five partitions in (4.1) are $G_2 \times G_1$, $G_2 \circ G_1$, $G_{11} \circ G_1$, $G_{11} \circ G_1$, $G_1 \circ G_1 \circ G_1$. If in the general case, rules can be given to write G functions associated with all partitions, any moment formulae can be written observing such rules. We are currently working on this problem.

When population means are 0, formula (5.1) yields

$$E_N(m_{110}a_{001}) = G_2 \times G_1 \quad (111) \qquad (5.2)$$

which, in terms of e_λ, and extending the notation of Mikhail [11], becomes

$$M\begin{pmatrix}1&0\\1&0\\0&1\end{pmatrix} = \left[(n-1)e_1 - (n-3)e_2 - 2e_3\right]\frac{N\mu_{111}}{n^3}. \qquad (5.3)$$

By combining rows, one obtains the bivariate formulae $M\binom{21}{00}$, $M\binom{20}{01}$, $M\binom{11}{10}$, similar to (5.3) but with μ_{30}, μ_{21}, μ_{21} respectively replacing μ_{111}. Mikhail obtains these using Kendall's operator [9] on $M(21)$ as reported by David, Kendall and Barton [4] from Sukhatme [12].

All bivariate and higher-variate formulae can be written quite easily by this approach. As an example, we write a four-variate formula, considering the population means 0. Using Table 4.3 and converting into G functions, we get

$$E_N(m_{1100}m_{0011}) = G_2 \times G_2(1111) + G_2 \circ G_2(1100)(0011)$$
$$+ G_{11} \times G_{11}\,[(1010)(0101) + (1001)(0110)]. \qquad (5.4)$$

This is $M\begin{pmatrix}10\\10\\01\\01\end{pmatrix}$, and by combining rows, yields the trivariate formula

$$M\begin{pmatrix}20\\01\\01\end{pmatrix} = G_2 \times G_2(211) + G_2 \circ G_2(200)(011)$$
$$+ 2G_{11} \times G_{11}(110)(101),$$

the bivariate formulae [11]

$$M\begin{pmatrix}20\\02\end{pmatrix} = G_2 \times G_2(22) + G_2 \circ G_2(20)(02) + 2G_{11} \times G_{11}$$
$$(11)(11)$$

$$M\begin{pmatrix}11\\11\end{pmatrix} = G_2 \times G_2(22) + G_2 \circ G_2(11)(11) + G_{11} \times G_{11}$$
$$[(20)(02) + (11)(11)]$$

$$M\binom{21}{01} = G_2 \times G_2(31) + (G_2 \circ G_2 + 2G_{11} \times G_{11})(20)(11)$$

$$M\binom{22}{00} = G_2 \times G_2(40) + (G_2 \circ G_2 + 2G_{11} \times G_{11})\ (20)(20),$$

and the univariate formula [4]

$$M(22) = G_2 \times G_2(4) + (G_2 \circ G_2 + 2G_{11} \times G_{11})(2)(2) \quad .$$

6. CONCLUSION AND ACKNOWLEDGEMENT

The aim further is to study partitions together with their G functions, so as to find rules to be able to write the G functions by inspection in a given case. Then plugging in the coefficients for the particular sample multivariate symmetric functions desired, formulae for their sampling moments should be readily available. Dr. Mikhail and I are collaborating on this problem, and many results above are a result of useful discussions with him. The ideas are throughout inspired by Professor Dwyer. I acknowledge my indebtedness to both.

REFERENCES

[1] Carney, Edward J. (1968). Relationship of generalized polykays to unrestricted sums for balanced complete finite populations. Ann. Math. Statist., 39, 643-656.

[2] Carver, H.C. (1930). Fundamentals of the theory of sampling: Editorial. Ann. Math. Statist., 1, 101-121.

[3] Cramér, Harald (1946). Mathematical Methods of Statistics. Princeton University Press.

[4] David, F.N., Kendall, M.G. and Barton, D.E. (1966). Symmertric Function and Allied Tables. Cambridge University Press.

[5] Dwyer, Paul S. (1938). Combined expansions of products of symmetric power sums and of sums of symmetric power products with application to sampling. Ann.Math.Statist., 9, 1-47, 97-132.

[6] Dwyer, Paul S. (1971). Moment functions of sample moment functions. Proceedings of the Dwyer Symposium on Symmetric Functions in Statistics, University of Windsor.

[7] Dwyer, P.S. and Tracy, D.S. (1964). A combinatorial method for products of two polykays with some general formulae. Ann. Math. Statist., 35, 1174-1185.

[8] Kaplan, E.L. (1952). Tensor notation and the sampling cumulants of k-statistics. Biometrika, 39, 319-323.

[9] Kendall, M.G. (1940). The derivation of multivariate sampling formulae from univariate formulae by symbolic operation. Ann. Eugen., 10, 392-402.

[10] Livers, J.J. (1945). Use of partitions in multivariate moment sampling theory. Doctoral dissertation, University of Michigan.

[11] Mikhail, N.N. (1969). Bivariate moments of sampling moments when the population is finite and of size N. Sankhya A, 31, 337-342.

[12] Sukhatme, P.V. (1943). Moments and product moments of moment-statistics for samples of the finite and infinite populations. Sankhya, 6, 363-382.

Derrick S. Tracy
Department of Mathematics
University of Windsor
Windsor, Onatrio
Canada